PENGUIN BOOKS

THE BEGINNING OF INFINITY

Born in Haifa, Israel, David Deutsch was educated at Cambridge and Oxford universities. He is a Fellow of the Royal Society and a professor of physics at the University of Oxford, where he is a member of the Centre for Quantum Computation. His papers on quantum computation laid the foundations for that field, and he is an authority on the theory of parallel universes. His honors include the Institute of Physics' Paul Dirac Prize and Medal. The author of *The Fabric of Reality*, he lives in England.

Praise For *The Beginning of Infinity*

"Making the case for science's open-endedness, Mr. Deutsch mounts a compelling challenge to scientific reductionism. . . . His book is animated by an ambition much greater than defending a particular theory. Mr. Deutsch wants us to share his radically optimistic vision of humanity's future, one in which progress continues forever." —John Horgan, *The Wall Street Journal*

"Mr. Deutsch's previous tome, *The Fabric of Reality*, took a broad-ranging sweep. . . . *The Beginning of Infinity* is equally bold, addressing subjects from artificial intelligence to the evolution of culture and of creativity; its conclusions are just as profound. . . . Deutsch is firmly of the opinion that science is infinite, as is the human thirst for knowledge." —*The Economist*

"Deutsch tells a coherent and compelling story about topics as diverse as why flowers (and faces) are beautiful, the mathematical impossibility of fairly apportioning seats in the U.S. House of Representatives, why it's harder to be a conformist than you think, and which cultures can and cannot sustain themselves (hint: the conformists are doomed). Once you have been exposed to his extraordinarily original worldview—or perhaps multiverse-view—it's impossible to see these and many other topics in quite the same way again." —Neal Stephenson, author of *Anathem* and *Cryptonomicon*

"[Deutsch's books] are among the most ambitious works of nonfiction I have read, in that their aim is no less than an explanation of all reality. . . . They are treatises that weave together not just physics and astronomy but biology, mathematics, computer science, political science, psychology, philosophy, aesthetics, and—most important for Deutsch—epistemology, among other fields, in fashioning a profound new view of the world and the universe. They are guaranteed to be either provocative or life changing." —Nathaniel Stein, *The New Yorker*'s Book Bench

"Deutsch has an important message . . . that our destiny is to be explainers of the world around us, and explaining is the key to mastery. . . . He writes clearly and thinks wisely. His book could help the world toward better ways of dealing with its problems." —Freeman Dyson, *The New York Review of Books*

"As fun as reading Philip K. Dick and as illuminating as reading Bertrand Russell. Deutsch teaches us how to understand that what is counterintuitive may be true; he teaches us this so well that it may even become intuitive. This book is truly marvelous." —Rivka Galchen, author of *Atmospheric Disturbances*

"A dazzling book full of huge ideas, presented with matchless clarity; [Deutsch] argues that explanatory power has changed, can change, and will continue to change our world." —*The Guardian* (London)

"This is Deutsch at his most ambitious, seeking to understand the implications of our scientific explanations of the world. . . . I enthusiastically recommend this rich, wide-ranging, and elegantly written exposition of the unique insights of one of our most original intellectuals." —*The Times Higher Education Supplement* (London)

"An important and wildly illuminating new book on the nature and evolution of human knowledge. Fluidly switching between evolutionary biology, quantum physics, mathematics, philosophy, ancient history, and more, Deutsch offers surprisingly—or perhaps, knowing his work, unsurprisingly—plausible answers to everything from why beauty exists to what is infinity . . . the kind of book that stays with you for an entire lifetime, insights from it finding their way, consciously or unconsciously, into every intellectual conversation you will ever have." —Maria Popova, BrainPickings.org

"Science has never had an advocate quite like David Deutsch. . . . His arguments are so clear that to read him is to experience the thrill of the highest level of discourse available on this planet and to understand it. . . . [He] is a modern Voltaire, lacerating fallacious views with crystal-clear thinking." —*The Independent* (London)

"David Deutsch does something astonishing in this book: he explains the nature of explanation itself. In doing so, he illuminates how strange and remarkable it is that the universe can be explained at all. Deutsch tells us that the creation of knowledge changes the very structure of the universe, and this book is as good an example as any of that idea. I almost cannot believe this book exists." —Charles Yu, author of *How to Live Safely in a Science Fictional Universe*

DAVID DEUTSCH

The Beginning of Infinity

Explanations That Transform the World

PENGUIN BOOKS

PENGUIN BOOKS

Published by the Penguin Group

Penguin Group (USA) Inc., 375 Hudson Street, New York, New York 10014, U.S.A.
Penguin Group (Canada), 90 Eglinton Avenue East, Suite 700, Toronto,
Ontario, Canada M4P 2Y3 (a division of Pearson Penguin Canada Inc.)
Penguin Books Ltd, 80 Strand, London WC2R 0RL, England
Penguin Ireland, 25 St. Stephen's Green, Dublin 2, Ireland (a division of Penguin Books Ltd)
Penguin Books Australia Ltd, 250 Camberwell Road, Camberwell,
Victoria 3124, Australia (a division of Pearson Australia Group Pty Ltd)
Penguin Books India Pvt Ltd, 11 Community Centre, Panchsheel Park, New Delhi – 110 017, India
Penguin Group (NZ), 67 Apollo Drive, Rosedale, Auckland 0632,
New Zealand (a division of Pearson New Zealand Ltd)
Penguin Books (South Africa) (Pty) Ltd, 24 Sturdee Avenue,
Rosebank, Johannesburg 2196, South Africa

Penguin Books Ltd, Registered Offices:
80 Strand, London WC2R 0RL, England

First published in Great Britain by Allen Lane, an imprint of Penguin Books Ltd 2011
First published in the United States of America by Viking Penguin,
a member of Penguin Group (USA) Inc. 2011
Published in Penguin Books (UK) 2012
Published in Penguin Books (USA) 2012

7 9 10 8 6

Excerpts from *The World of Parmenides* by Karl Popper, edited by Arne F. Petersen (Routledge, 1998).
Used by permission of the University of Klagenfurt, Karl Popper Library.

Illustration on page 34: Starfield image from the Digitized Sky Survey (© AURA) courtesy of
the Palomar Observatory and Digitized Sky Survey, created by the Space Telescope Science Institute,
operated by AURA, Inc. for NASA. Reproduced with permission of AURA/STScI.

Illustration on page 426: © Bettmann/Corbis

THE LIBRARY OF CONGRESS HAS CATALOGED THE HARDCOVER EDITION AS FOLLOWS:
Deutsch, David.
The beginning of infinity : explanations that transform the world / David Deutsch.
p. cm.
Includes bibliographical references and index.
ISBN 978-0-670-02275-5 (hc.)
ISBN 978-0-14-312135-0 (pbk.)
1. Explanation. 2. Infinite. 3. Science—Philosophy. I. Title.
Q175.32.E97D48 2011
501—dc22
2011004120

Printed in the United States of America

ALWAYS LEARNING   PEARSON

Contents

Acknowledgements

I am grateful to my friends and colleagues Sarah Fitz-Claridge, Alan Forrester, Herbert Freudenheim, David Johnson-Davies, Paul Tappenden and especially Elliot Temple and my copy-editor, Bob Davenport, for reading earlier drafts of this book and suggesting many corrections and improvements, and also to those who have read and helpfully commented on parts of it, namely Omri Ceren, Artur Ekert, Michael Golding, Alan Grafen, Ruti Regan, Simon Saunders and Lulie Tanett.

I also want to thank the illustrators Nick Lockwood, Tommy Robin and Lulie Tanett for translating explanations into images more accurately than I could have hoped for.

Introduction

Progress that is both rapid enough to be noticed and stable enough to continue over many generations has been achieved only once in the history of our species. It began at approximately the time of the scientific revolution, and is still under way. It has included improvements not only in scientific understanding, but also in technology, political institutions, moral values, art, and every aspect of human welfare.

Whenever there has been progress, there have been influential thinkers who denied that it was genuine, that it was desirable, or even that the concept was meaningful. They should have known better. There is indeed an objective difference between a false explanation and a true one, between chronic failure to solve a problem and solving it, and also between wrong and right, ugly and beautiful, suffering and its alleviation – and thus between stagnation and progress in the fullest sense.

In this book I argue that all progress, both theoretical and practical, has resulted from a single human activity: the quest for what I call good explanations. Though this quest is uniquely human, its effectiveness is also a fundamental fact about reality at the most impersonal, cosmic level – namely that it conforms to universal laws of nature that are indeed good explanations. This simple relationship between the cosmic and the human is a hint of a central role of *people* in the cosmic scheme of things.

Must progress come to an end – either in catastrophe or in some sort of completion – or is it unbounded? The answer is the latter. That unboundedness is the 'infinity' referred to in the title of this book. Explaining it, and the conditions under which progress can and cannot happen, entails a journey through virtually every fundamental field of science and philosophy. From each such field we learn that, although progress has no necessary end, it does have a necessary beginning: a cause, or an event with which it starts, or a necessary condition for it to take off and to thrive. Each of these beginnings is 'the beginning of infinity' as viewed from the perspective of that field. Many seem, superficially, to be unconnected. But they are all facets of a single attribute of reality, which I call *the* beginning of infinity.

I

The Reach of Explanations

Behind it all is surely an idea so simple, so beautiful, that when
we grasp it – in a decade, a century, or a millennium – we will
all say to each other, how could it have been otherwise?
John Archibald Wheeler, *Annals of the*
New York Academy of Sciences, 480 (1986)

To unaided human eyes, the universe beyond our solar system looks
like a few thousand glowing dots in the night sky, plus the faint, hazy
streaks of the Milky Way. But if you ask an astronomer what is out
there in reality, you will be told not about dots or streaks, but about
stars: spheres of incandescent gas millions of kilometres in diameter
and light years away from us. You will be told that the sun is a typical
star, and looks different from the others only because we are much
closer to it – though still some 150 million kilometres away. Yet, even
at those unimaginable distances, we are confident that we know what
makes stars shine: you will be told that they are powered by the nuclear
energy released by *transmutation* – the conversion of one chemical
element into another (mainly hydrogen into helium).

Some types of transmutation happen spontaneously on Earth, in the
decay of radioactive elements. This was first demonstrated in 1901, by
the physicists Frederick Soddy and Ernest Rutherford, but the concept
of transmutation was ancient. Alchemists had dreamed for centuries
of transmuting 'base metals', such as iron or lead, into gold. They never
came close to understanding what it would take to achieve that, so
they never did so. But scientists in the twentieth century did. And so
do stars, when they explode as supernovae. Base metals can be

transmuted into gold by stars, and by intelligent beings who understand the processes that power stars, but by nothing else in the universe.

As for the Milky Way, you will be told that, despite its insubstantial appearance, it is the most massive object that we can see with the naked eye: a *galaxy* that includes stars by the hundreds of billions, bound by their mutual gravitation across tens of thousands of light years. We are seeing it from the inside, because we are part of it. You will be told that, although our night sky appears serene and largely changeless, the universe is seething with violent activity. Even a typical star converts millions of tonnes of mass into energy every second, with each *gram* releasing as much energy as an atom bomb. You will be told that within the range of our best telescopes, which can see more galaxies than there are stars in our galaxy, there are several supernova explosions per second, each briefly brighter than all the other stars in its galaxy put together. We do not know where life and intelligence exist, if at all, outside our solar system, so we do not know how many of those explosions are horrendous tragedies. But we do know that a supernova devastates all the planets that may be orbiting it, wiping out all life that may exist there – including any intelligent beings, unless they have technology far superior to ours. Its neutrino radiation alone would kill a human at a range of billions of kilometres, even if that entire distance were filled with lead shielding. Yet we owe our existence to supernovae: they are the source, through transmutation, of most of the elements of which our bodies, and our planet, are composed.

There are phenomena that outshine supernovae. In March 2008 an X-ray telescope in Earth orbit detected an explosion of a type known as a 'gamma-ray burst', 7.5 billion light years away. That is halfway across the known universe. It was probably a single star collapsing to form a black hole – an object whose gravity is so intense that not even light can escape from its interior. The explosion was intrinsically brighter than a million supernovae, and would have been visible with the naked eye from Earth – though only faintly and for only a few seconds, so it is unlikely that anyone here saw it. Supernovae last longer, typically fading on a timescale of months, which allowed astronomers to see a few in our galaxy even before the invention of telescopes.

Another class of cosmic monsters, the intensely luminous objects known as *quasars*, are in a different league. Too distant to be seen with the naked eye, they can outshine a supernova for millions of years at a time. They are powered by massive black holes at the centres of galaxies, into which entire stars are falling – up to several per day for a large quasar – shredded by tidal effects as they spiral in. Intense magnetic fields channel some of the gravitational energy back out in the form of jets of high-energy particles, which illuminate the surrounding gas with the power of a trillion suns.

Conditions are still more extreme in the black hole's interior (within the surface of no return known as the 'event horizon'), where the very fabric of space and time may be being ripped apart. All this is happening in a relentlessly expanding universe that began about fourteen billion years ago with an all-encompassing explosion, the Big Bang, that makes all the other phenomena I have described seem mild and inconsequential by comparison. And that whole universe is just a sliver of an enormously larger entity, the multiverse, which includes vast numbers of such universes.

The physical world is not only much bigger and more violent than it once seemed, it is also immensely richer in detail, diversity and incident. Yet it all proceeds according to elegant laws of physics that we understand in some depth. I do not know which is more awesome: the phenomena themselves or the fact that we know so much about them.

How do we know? One of the most remarkable things about science is the contrast between the enormous reach and power of our best theories and the precarious, local means by which we create them. No human has ever been at the surface of a star, let alone visited the core where the transmutation happens and the energy is produced. Yet we see those cold dots in our sky and *know* that we are looking at the white-hot surfaces of distant nuclear furnaces. Physically, that experience consists of nothing other than our brains responding to electrical impulses from our eyes. And eyes can detect only light that is inside them at the time. The fact that the light was emitted very far away and long ago, and that much more was happening there than just the emission of light – those are not things that we see. We know them only from theory.

Scientific theories are *explanations*: assertions about what is out there and how it behaves. Where do these theories come from? For most of the history of science, it was mistakenly believed that we 'derive' them from the evidence of our senses – a philosophical doctrine known as *empiricism*:

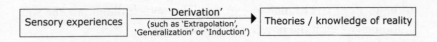

Empiricism

For example, the philosopher John Locke wrote in 1689 that the mind is like 'white paper' on to which sensory experience writes, and that that is where all our knowledge of the physical world comes from. Another empiricist metaphor was that one could *read* knowledge from the 'Book of Nature' by making observations. Either way, the discoverer of knowledge is its passive recipient, not its creator.

But, in reality, scientific theories are not 'derived' from anything. We do not read them in nature, nor does nature write them into us. They are guesses – bold conjectures. Human minds create them by rearranging, combining, altering and adding to existing ideas with the intention of improving upon them. We do not begin with 'white paper' at birth, but with inborn expectations and intentions and an innate ability to improve upon them using thought and experience. Experience is indeed essential to science, but its role is different from that supposed by empiricism. It is not the source from which theories are derived. Its main use is to choose between theories that have already been guessed. That is what 'learning from experience' is.

However, that was not properly understood until the mid twentieth century with the work of the philosopher Karl Popper. So historically it was empiricism that first provided a plausible defence for experimental science as we now know it. Empiricist philosophers criticized and rejected traditional approaches to knowledge such as deference to the authority of holy books and other ancient writings, as well as human authorities such as priests and academics, and belief in traditional lore, rules of thumb and hearsay. Empiricism also contradicted the opposing and surprisingly persistent idea that the senses are little more than sources of error to be ignored. And it was optimistic, being all about obtaining new knowledge, in contrast with the medieval fatalism that had expected everything important to be known already. Thus, despite being quite wrong about where scientific knowledge comes from, empiricism was a great step forward in both the philosophy and the history of science. Nevertheless, the question that sceptics (friendly and unfriendly) raised from the outset always remained: how can knowledge of what has *not* been experienced possibly be 'derived' from what *has*? What sort of thinking could possibly constitute a valid derivation of the one from the other? No one would expect to deduce the *geography* of Mars from a map of Earth, so why should we expect to be able to learn about *physics* on Mars from experiments done on Earth? Evidently, logical deduction alone would not do, because there is a logical gap: no amount of deduction applied to statements describing a set of experiences can reach a conclusion about anything other than those experiences.

The conventional wisdom was that the key is *repetition*: if one repeatedly has similar experiences under similar circumstances, then one is supposed to 'extrapolate' or 'generalize' that pattern and predict that it will continue. For instance, why do we expect the sun to rise tomorrow morning? Because in the past (so the argument goes) we have seen it do so whenever we have looked at the morning sky. From this we supposedly 'derive' the theory that under similar circumstances we shall always have that experience, or that we probably shall. On each occasion when that prediction comes true, and provided that it never fails, the probability that it will always come true is supposed to increase. Thus one supposedly obtains ever more reliable knowledge of the future from the past, and of the general from the particular. That alleged process was called 'inductive inference' or 'induction', and the doctrine that scientific theories are obtained in that way is called *inductivism*. To bridge the logical gap, some inductivists imagine that there is a principle of nature – the 'principle of induction' – that makes inductive inferences likely to be true. 'The future will resemble the past' is one popular version of this, and one could add 'the distant resembles the near,' 'the unseen resembles the seen' and so on.

But no one has ever managed to formulate a 'principle of induction' that is usable in practice for obtaining scientific theories from experiences. Historically, criticism of inductivism has focused on that failure, and on the logical gap that cannot be bridged. But that lets inductivism off far too lightly. For it concedes inductivism's two most serious misconceptions.

First, inductivism purports to explain how science obtains *predictions about experiences*. But most of our theoretical knowledge simply does not take that form. Scientific explanations are about reality, most of which does not consist of anyone's experiences. Astrophysics is not primarily about *us* (what we shall see if we look at the sky), but about what stars are: their composition and what makes them shine, and how they formed, and the universal laws of physics under which that happened. Most of that has never been observed: no one has experienced a billion years, or a light year; no one could have been present at the Big Bang; no one will ever touch a law of physics – except in their minds, through theory. All our predictions of how things will *look* are deduced from such explanations of how things *are*. So inductivism fails even to address how we can know about stars and the universe, as distinct from just dots in the sky.

The second fundamental misconception in inductivism is that scientific theories predict that 'the future will resemble the past', and that 'the unseen resembles the seen' and so on. (Or that it 'probably' will.) But in reality the future is unlike the past, the unseen very different from the seen. Science often predicts – and brings about – phenomena spectacularly different from anything that has been experienced before. For millennia people dreamed about flying, but they experienced only falling. Then they discovered good explanatory theories about flying, and then they flew – in that order. Before 1945, no human being had ever observed a nuclear-fission (atomic-bomb) explosion; there may never have been one in the history of the universe. Yet the first such explosion, and the conditions under which it would occur, had been accurately predicted – but not from the assumption that the future would be like the past. Even sunrise – that favourite example of inductivists – is not always observed every twenty-four hours: when viewed from orbit it may happen every ninety minutes, or not at all. And that was known from theory long before anyone had ever orbited the Earth.

It is no defence of inductivism to point out that in all those cases the future still does 'resemble the past' in the sense that it obeys the same underlying laws of nature. For that is an empty statement: *any* purported law of nature – true or false – about the future and the past is a claim that they 'resemble' each other by both conforming to that law. So that version of the 'principle of induction' could not be used to derive any theory or prediction from experience or anything else.

Even in everyday life we are well aware that the future is unlike the past, and are selective about which aspects of our experience we expect to be repeated. Before the year 2000, I had experienced thousands of times that if a calendar was properly maintained (and used the standard Gregorian system), then it displayed a year number beginning with '19'. Yet at midnight on 31 December 1999 I expected to have the experience of seeing a '20' on every such calendar. I also expected that there would be a gap of 17,000 years before anyone experienced a '19' under those conditions again. Neither I nor anyone else had ever observed such a '20', nor such a gap, but our explanatory theories told us to expect them, and expect them we did.

As the ancient philosopher Heraclitus remarked, 'No man ever steps in the same river twice, for it is not the same river and he is not the same man.' So, when we remember seeing sunrise 'repeatedly' under 'the same' circumstances, we are tacitly relying on explanatory theories to tell us which combinations of variables in our experience we should interpret as being 'repeated' phenomena in the underlying reality, and which are local or irrelevant. For instance, theories about geometry and optics tell us not to expect to see a sunrise on a cloudy day, even if a sunrise is really happening in the unobserved world behind the clouds. Only from those explanatory theories do we know that failing to see the sun on such days does not amount to an experience of its not rising. Similarly, theory tells us that if we see sunrise reflected in a mirror, or in a video or a virtual-reality game, that does not count as seeing it twice. Thus the very idea that an experience has been repeated is not itself a sensory experience, but a theory.

So much for inductivism. And since inductivism is false, empiricism must be as well. For if one cannot derive predictions from experience, one certainly cannot derive explanations. Discovering a new explanation is inherently an act of creativity. To interpret dots in the sky as white-hot, million-kilometre spheres, one must first have thought of the idea of such spheres. And then one must explain why they look small and cold and seem to move in lockstep around us and do not fall down. Such ideas do not create themselves, nor can they be mechanically derived from anything: they have to be guessed – after which they can be criticized and tested. To the extent that experiencing dots 'writes' something into our brains, it does not write explanations but only dots. Nor is nature a book: one could try to 'read' the dots in the sky for a lifetime – many lifetimes – without learning anything about what they really are.

Historically, that is exactly what happened. For millennia, most careful observers of the sky believed that the stars were lights embedded in a hollow, rotating 'celestial sphere' centred on the Earth (or that they were holes in the sphere, through which the light of heaven shone). This *geocentric* – Earth-centred – theory of the universe seemed to have been directly derived from experience, and repeatedly confirmed: anyone who looked up could 'directly observe' the celestial sphere, and the stars maintaining their relative positions on it and being held up just as the theory predicts. Yet in reality, the solar system is *heliocentric* – centred on the sun, not the Earth – and the Earth is not at rest but in complex motion. Although we first noticed a daily rotation by observing stars, it is not a property of the stars at all, but of the Earth, and of the observers who rotate with it. It is a classic example of the deceptiveness of the senses: the Earth looks and feels as though it is at rest beneath our feet, even though it is really rotating. As for the celestial sphere, despite being visible in broad daylight (as the sky), it does not exist at all.

The deceptiveness of the senses was always a problem for empiricism – and thereby, it seemed, for science. The empiricists' best defence was that the senses cannot be deceptive in themselves. What misleads us are only the false interpretations that we place on appearances. That is indeed true – but only because our senses themselves do not say anything. Only our interpretations of them do, and those are very fallible. But the real key to science is that our explanatory theories – which include those interpretations – can be *improved*, through conjecture, criticism and testing.

Empiricism never did achieve its aim of liberating science from authority. It denied the legitimacy of traditional authorities, and that was salutary. But unfortunately it did this by setting up two other false authorities: sensory experience and whatever fictitious process of 'derivation', such as induction, one imagines is used to extract theories from experience.

The misconception that knowledge needs authority to be genuine or reliable dates back to antiquity, and it still prevails. To this day, most courses in the philosophy of knowledge teach that knowledge is some form of *justified, true belief*, where 'justified' means designated as true (or at least 'probable') by reference to some authoritative source or touchstone of knowledge. Thus 'how do we *know* . . . ?' is transformed into 'by what authority do we claim . . . ?' The latter question is a chimera that may well have wasted more philosophers' time and effort than any other idea. It converts the quest for truth into a quest for certainty (a feeling) or for endorsement (a social status). This misconception is called *justificationism*.

The opposing position – namely the recognition that there are no authoritative sources of knowledge, nor any reliable means of justifying ideas as being true or probable – is called *fallibilism*. To believers in the justified-true-belief theory of knowledge, this recognition is the occasion for despair or cynicism, because to them it means that knowledge is unattainable. But to those of us for whom creating knowledge means understanding better what is really there, and how it really behaves and why, fallibilism is part of the very means by which this is achieved. Fallibilists expect even their best and most fundamental explanations to contain misconceptions in addition to truth, and so they are predisposed to try to change them for the better. In contrast, the logic of justificationism is to seek (and typically, to believe that one has found) ways of securing ideas *against* change. Moreover, the logic of fallibilism is that one not only seeks to correct the misconceptions of the past, but hopes in the future to find and change mistaken ideas that no one today questions or finds problematic. So it is fallibilism, not mere rejection of authority, that is essential for the initiation of unlimited knowledge growth – the beginning of infinity.

The quest for authority led empiricists to downplay and even stigmatize *conjecture*, the real source of all our theories. For if the senses were the only source of knowledge, then error (or at least avoidable error) could be caused only by adding to, subtracting from or misinterpreting what that source is saying. Thus empiricists came to believe that, in addition to rejecting ancient authority and tradition, scientists should suppress or ignore any *new* ideas they might have, except those that had been properly 'derived' from experience. As Arthur Conan Doyle's fictional detective Sherlock Holmes put it in the short story 'A Scandal in Bohemia', 'It is a capital mistake to theorize before one has data.'

But that was itself a capital mistake. We never know any data before interpreting it through theories. All observations are, as Popper put it, *theory-laden*,* and hence fallible, as all our theories are. Consider the nerve signals reaching our brains from our sense organs. Far from providing direct or untainted access to reality, even they themselves are never experienced for what they really are – namely crackles of electrical activity. Nor, for the most part, do we experience them as being *where* they really are – inside our brains. Instead, we place them in the reality beyond. We do not just see blue: we see a blue sky up there, far away. We do not just feel pain: we experience a headache, or a stomach ache. The brain attaches those interpretations – 'head', 'stomach' and 'up there' – to events that are in fact within the brain itself. Our sense organs themselves, and all the interpretations that we consciously and unconsciously attach to their outputs, are notoriously fallible – as witness the celestial-sphere theory, as well as every optical illusion and conjuring trick. So we perceive *nothing* as what it really is. It is all theoretical interpretation: conjecture.

Conan Doyle came much closer to the truth when, during 'The Boscombe Valley Mystery', he had Holmes remark that 'circumstantial evidence' (evidence about unwitnessed events) is 'a very tricky thing ... It may seem to point very straight to one thing, but if you shift your own point of view a little, you may find it pointing in an equally uncompromising manner to something entirely different ... There is nothing more deceptive than an obvious fact.' The same holds for scientific discovery. And that again raises the question: how do we know? If all our theories originate locally, as guesswork in our own minds, and can be tested only locally, by experience, how is it that they contain such extensive and accurate knowledge about the reality that we have never experienced?

---

*The term was coined by the philosopher Norwood Russell Hanson.

I am not asking what authority scientific knowledge is derived from, or rests on. I mean, literally, by what process do ever truer and more detailed explanations about the world come to be represented physically in our brains? How do we come to know about the interactions of subatomic particles during transmutation at the centre of a distant star, when even the tiny trickle of light that reaches our instruments from the star was emitted by glowing gas at the star's surface, a million kilometres above where the transmutation is happening? Or about conditions in the fireball during the first few seconds after the Big Bang, which would instantly have destroyed any sentient being or scientific instrument? Or about the future, which we have no way of measuring at all? How is it that we can predict, with some non-negligible degree of confidence, whether a new design of microchip will work, or whether a new drug will cure a particular disease, even though they have never existed before?

For most of human history, we did not know how to do any of this. People were not designing microchips or medications or even the wheel. For thousands of generations, our ancestors looked up at the night sky and wondered what stars are – what they are made of, what makes them shine, what their relationship is with each other and with us – which was exactly the right thing to wonder about. And they were using eyes and brains anatomically indistinguishable from those of modern astronomers. But they discovered nothing about it. Much the same was true in every other field of knowledge. It was not for lack of trying, nor for lack of thinking. People observed the world. They tried to understand it – but almost entirely in vain. Occasionally they recognized simple patterns in the appearances. But when they tried to find out what was really there behind those appearances, they failed almost completely.

I expect that, like today, most people wondered about such things only occasionally – during breaks from addressing their more parochial concerns. But their parochial concerns *also* involved yearning to know – and not only out of pure curiosity. They wished they knew how to safeguard their food supply; how they could rest when tired without risking starvation; how they could be warmer, cooler, safer, in less pain – in every aspect of their lives, they wished they knew how to make progress. But, on the timescale of individual lifetimes, they almost never made any. Discoveries such as fire, clothing, stone tools, bronze, and so on, happened so rarely that from an individual's point of view the world never improved. Sometimes people even realized (with somewhat miraculous prescience) that making progress in practical ways would *depend* on progress in understanding puzzling phenomena in the sky. They even conjectured links between the two, such as myths, which they found compelling enough to dominate their lives – yet which still bore no resemblance to the truth. In short, they wanted to create knowledge, in order to make progress, but they did not know how.

This was the situation from our species' earliest prehistory, through the dawn of civilization, and through its imperceptibly slow increase in sophistication – with many reverses – until a few centuries ago. Then a powerful new mode of discovery and explanation emerged, which later became known as *science*. Its emergence is known as the *scientific revolution*, because it succeeded almost immediately in creating knowledge at a noticeable rate, which has increased ever since.

What had changed? What made science effective at understanding the physical world when all previous ways had failed? What were people now doing, for the first time, that made the difference? This question began to be asked as soon as science began to be successful, and there have been many conflicting answers, some containing truth. But none, in my view, has reached the heart of the matter. To explain my own answer, I have to give a little context first.

The scientific revolution was part of a wider intellectual revolution, the *Enlightenment*, which also brought progress in other fields, especially moral and political philosophy, and in the institutions of society. Unfortunately, the term 'the Enlightenment' is used by historians and philosophers to denote a variety of different trends, some of them violently opposed to each other. What I mean by it will emerge here as we go along. It is one of several aspects of 'the beginning of infinity', and is a theme of this book. But one thing that all conceptions of the Enlightenment agree on is that it was a *rebellion*, and specifically a rebellion against authority in regard to knowledge.

Rejecting authority in regard to knowledge was not just a matter of abstract analysis. It was a necessary condition for progress, because, before the Enlightenment, it was generally believed that everything important that was knowable had already been discovered, and was enshrined in authoritative sources such as ancient writings and traditional assumptions. Some of those sources did contain some genuine knowledge, but it was entrenched in the form of dogmas along with many falsehoods. So the situation was that all the sources from which it was generally believed knowledge came actually knew very little, and were mistaken about most of the things that they claimed to know. And therefore progress depended on learning how to reject their authority. This is why the Royal Society (one of the earliest scientific academies, founded in London in 1660) took as its motto 'Nullius in verba', which means something like 'Take no one's word for it.'

However, rebellion against authority cannot by itself be what made the difference. Authorities have been rejected many times in history, and only rarely has any lasting good come of it. The usual sequel has merely been that new authorities replaced the old. What was needed for the sustained, rapid growth of knowledge was a *tradition of criticism*. Before the Enlightenment, that was a very rare sort of tradition: usually the whole point of a tradition was to keep things the same.

Thus the Enlightenment was a revolution in how people sought knowledge: by trying *not* to rely on authority. That is the context in which empiricism – purporting to rely solely on the senses for knowledge – played such a salutary historical role, despite being fundamentally false and even authoritative in its conception of how science works.

One consequence of this tradition of criticism was the emergence of a methodological rule that a scientific theory must be *testable* (though this was not made explicit at first). That is to say, the theory must make predictions which, if the theory were false, could be contradicted by the outcome of some possible observation. Thus, although scientific theories are not derived from experience, they can be tested by experience – by observation or experiment. For example, before the discovery of radioactivity, chemists had believed (and had verified in countless experiments) that transmutation is impossible. Rutherford and Soddy boldly conjectured that uranium spontaneously transmutes into other elements. Then, by demonstrating the creation of the element radium in a sealed container of uranium, they refuted the prevailing theory and science progressed. They were able to do that because that earlier theory was testable: it was possible to test for the presence of radium. In contrast, the ancient theory that all matter is composed of combinations of the elements earth, air, fire and water was untestable, because it did not include any way of testing for the presence of those components. So it could never be refuted by experiment. Hence it could never be – and never was – improved upon through experiment. The Enlightenment was at root a philosophical change.

The physicist Galileo Galilei was perhaps the first to understand the importance of experimental tests (which he called *cimenti*, meaning 'trials by ordeal') as distinct from other forms of experiment and observation, which can more easily be mistaken for 'reading from the Book of Nature'. Testability is now generally accepted as the defining characteristic of the scientific method. Popper called it the 'criterion of demarcation' between science and non-science.

Nevertheless, testability cannot have been the decisive factor in the scientific revolution either. Contrary to what is often said, testable predictions had always been quite common. Every traditional rule of thumb for making a flint blade or a camp fire is testable. Every would-be prophet who claims that the sun will go out next Tuesday has a testable theory. So does every gambler who has a hunch that 'this is my lucky night – I can feel it'. So what is the vital, progress-enabling ingredient that is present in science, but absent from the testable theories of the prophet and the gambler?

The reason that testability is not enough is that prediction is not, and cannot be, the purpose of science. Consider an audience watching a conjuring trick. The problem facing them has much the same logic as a scientific problem. Although in nature there is no conjurer trying to deceive us intentionally, we can be mystified in both cases for essentially the same reason: appearances are not self-explanatory. If the explanation of a conjuring trick were evident in its appearance, there would be no trick. If the explanations of physical phenomena were evident in their appearance, empiricism would be true and there would be no need for science as we know it.

The problem is not to predict the trick's appearance. I may, for instance, predict that if a conjurer seems to place various balls under various cups, those cups will later appear to be empty; and I may predict that if the conjurer appears to saw someone in half, that person will later appear on stage unharmed. Those are testable predictions. I may experience many conjuring shows and see my predictions vindicated every time. But that does not even address, let alone solve, the problem of how the trick works. Solving it requires an explanation: a statement of the reality that accounts for the appearance.

Some people may enjoy conjuring tricks without ever wanting to know how they work. Similarly, during the twentieth century, most philosophers, and many scientists, took the view that science is incapable of discovering anything about reality. Starting from empiricism, they drew the inevitable conclusion (which would nevertheless have horrified the early empiricists) that science cannot validly do more than predict the outcomes of observations, and that it should never purport to describe the reality that brings those outcomes about. This is known as *instrumentalism*. It denies that what I have been calling 'explanation' can exist at all. It is still very influential. In some fields (such as statistical analysis) the very word 'explanation' has come to mean prediction, so that a mathematical formula is said to 'explain' a set of experimental data. By 'reality' is meant merely the *observed data* that the formula is supposed to approximate. That leaves no term for assertions about reality itself, except perhaps 'useful fiction'.

Instrumentalism is one of many ways of denying *realism*, the common-sense, and true, doctrine that the physical world really exists, and is accessible to rational inquiry. Once one has denied this, the logical implication is that all claims about reality are equivalent to myths, none of them being better than the others in any objective sense. That is *relativism*, the doctrine that statements in a given field cannot be objectively true or false: at most they can be judged so relative to some cultural or other arbitrary standard.

Instrumentalism, even aside from the philosophical enormity of reducing science to a collection of statements about human experiences, does not make sense in its own terms. For there is no such thing as a purely predictive, explanationless theory. One cannot make even the simplest prediction without invoking quite a sophisticated explanatory framework. For example, those predictions about conjuring tricks apply specifically to conjuring tricks. That is explanatory information, and it tells me, among other things, not to 'extrapolate' the predictions to another type of situation, however successful they are at predicting conjuring tricks. So I know not to predict that saws in general are harmless to humans; and I continue to predict that if *I* were to place a ball under a cup, it really would go there and stay there.

The concept of a conjuring trick, and of the distinction between it and other situations, is familiar and unproblematic – so much so that it is easy to forget that it depends on substantive explanatory theories about all sorts of things such as how our senses work, how solid matter and light behave, and also subtle cultural details. Knowledge that is both familiar and uncontroversial is *background knowledge*. A predictive theory whose explanatory content consists only of background knowledge is a *rule of thumb*. Because we usually take background knowledge for granted, rules of thumb may seem to be explanationless predictions, but that is always an illusion.

There is always an explanation, whether we know it or not, for why a rule of thumb works. Denying that some regularity in nature has an explanation is effectively the same as believing in the supernatural – saying, 'That's not conjuring, it's actual magic.' Also, there is always an explanation when a rule of thumb *fails*, for rules of thumb are always parochial: they hold only in a narrow range of familiar circumstances. So, if an unfamiliar feature were introduced into a cups-and-balls trick, the rule of thumb I stated might easily make a false prediction. For instance, I could not tell from the rule of thumb whether it would be possible to perform the trick with lighted candles instead of balls. If I had an explanation of how the trick worked, I could tell.

Explanations are also essential for arriving at a rule of thumb in the first place: I could not have guessed those predictions about conjuring tricks without having a great deal of explanatory information in mind – even before any specific explanation of how the trick works. For instance, it is only in the light of explanations that I could have abstracted the concept of *cups* and *balls* from my experience of the trick, rather than, say, *red* and *blue*, even if it so happened that the cups were red and the balls blue in every instance of the trick that I had witnessed.

The essence of experimental testing is that there are at least two apparently viable theories known about the issue in question, making conflicting predictions that can be distinguished by the experiment. Just as conflicting predictions are the occasion for experiment and observation, so *conflicting ideas* in a broader sense are the occasion for all rational thought and inquiry. For example, if we are simply curious about something, it means that we believe that our existing ideas do not adequately capture or explain it. So, we have some *criterion* that our best existing explanation fails to meet. The criterion and the existing explanation are conflicting ideas. I shall call a situation in which we experience conflicting ideas a *problem*.

The example of a conjuring trick illustrates how observations provide problems for science – dependent, as always, on prior explanatory theories. For a conjuring trick is a trick only if it makes us think that *something happened* that *cannot happen*. Both halves of that proposition depend on our bringing quite a rich set of explanatory theories to the experience. That is why a trick that mystifies an adult may be uninteresting to a young child who has not yet learned to have the expectations on which the trick relies. Even those members of the audience who are incurious about how the trick works can detect that it *is* a trick only because of the explanatory theories that they brought with them into the auditorium. *Solving* a problem means creating an explanation that does not have the conflict.

Similarly, no one would have wondered what stars are if there had not been existing expectations – explanations – that unsupported things fall, and that lights need fuel, which runs out, and so on, which conflicted with interpretations (which are also explanations) of what was seen, such as that the stars shine constantly and do not fall. In this case it was those interpretations that were false: stars are indeed in free fall and do need fuel. But it took a great deal of conjecture, criticism and testing to discover how that can be.

A problem can also arise purely theoretically, without any observations. For instance, there is a problem when a theory makes a prediction that we did not expect. Expectations are theories too. Similarly, it is a problem when the way things *are* (according to our best explanation) is not the way they *should be* – that is, according to our current criterion of how they should be. This covers the whole range of ordinary meanings of the word 'problem', from unpleasant, as when the Apollo 13 mission reported, 'Houston, we've had a problem here,' to pleasant, as when Popper wrote:

I think that there is only one way to science – or to philosophy, for that matter: to meet a problem, to see its beauty and fall in love with it; to get married to it and to live with it happily, till death do ye part – unless you should meet another and even more fascinating problem or unless, indeed, you should obtain a solution. But even if you do obtain a solution, you may then discover, to your delight, the existence of a whole family of enchanting, though perhaps difficult, problem children . . .

Realism and the Aim of Science (1983)

Experimental testing involves many prior explanations in addition to the ones being tested, such as theories of how measuring instruments work. The refutation of a scientific theory has, from the point of view of someone who expected it to be true, the same logic as a conjuring trick – the only difference being that a conjurer does not normally have access to unknown laws of nature to make a trick work.

Since theories can contradict each other, but there are no contradictions in reality, every problem signals that our knowledge must be flawed or inadequate. Our misconception could be about the reality we are observing, or about how our perceptions are related to it, or both. For instance, a conjuring trick presents us with a problem only because we have misconceptions about what 'must' be happening – which implies that the knowledge that we used to interpret what we were seeing is defective. To an expert steeped in conjuring lore, it may be obvious what is happening – even if the expert did not observe the trick at all but merely heard a misleading account of it from a person who was fooled by it. This is another general fact about scientific explanation: if one has a misconception, observations that conflict with one's expectations may (or may not) spur one into making further conjectures, but no amount of observing will *correct* the misconception until after one has thought of a better idea; in contrast, if one has the right idea one can explain the phenomenon even if there are large errors in the data. Again, the very term 'data' ('givens') is misleading. Amending the 'data', or rejecting some as erroneous, is a frequent concomitant of scientific discovery, and the crucial 'data' cannot even be obtained until theory tells us what to look for and how and why.

A new conjuring trick is never totally unrelated to existing tricks. Like a new scientific theory, it is formed by creatively modifying, rearranging and combining the ideas from existing tricks. It requires pre-existing knowledge of how objects work and how audiences work, as well as how existing tricks work. So where did the earliest conjuring tricks come from? They must have been modifications of ideas that were not originally conjuring tricks – for instance, ideas for hiding objects in earnest. Similarly, where did the first scientific ideas come from? Before there was science there were rules of thumb, and explanatory assumptions, and myths. So there was plenty of raw material for criticism, conjecture and experiment to work with. Before that, there were our inborn assumptions and expectations: we are born with ideas, and with the ability to make progress by changing them. And there were patterns of cultural behaviour – about which I shall say more in Chapter 15.

But even *testable, explanatory theories* cannot be the crucial ingredient that made the difference between no-progress and progress. For they, too, have always been common. Consider, for example, the ancient Greek myth for explaining the annual onset of winter. Long ago, Hades, god of the underworld, kidnapped and raped Persephone, goddess of spring. Then Persephone's mother, Demeter, goddess of the earth and agriculture, negotiated a contract for her daughter's release, which specified that Persephone would marry Hades and eat a magic seed that would compel her to visit him once a year thereafter. Whenever Persephone was away fulfilling this obligation, Demeter became sad and would command the world to become cold and bleak so that nothing could grow.

That myth, though comprehensively false, does constitute an explanation of seasons: it is a claim about the reality that brings about our experience of winter. It is also eminently testable: if the cause of winter is Demeter's periodic sadness, then winter must happen everywhere on Earth at the same time. Therefore, if the ancient Greeks had known that a warm growing season occurs in Australia at the very moment when, as they believed, Demeter is at her saddest, they could have inferred that there was something wrong with their explanation of seasons.

Yet, when myths were altered or superseded by other myths over the course of centuries, the new ones were almost never any closer to the truth. Why? Consider the role that the specific elements of the Persephone myth play in the explanation. For example, the gods provide the *power* to affect a large-scale phenomenon (Demeter to command the weather, and Hades and his magic seed to command Persephone and hence to affect Demeter). But why those gods and not others? In Nordic mythology, seasons are caused by the changing fortunes of Freyr, the god of spring, in his eternal war with the forces of cold and darkness. Whenever Freyr is winning, the Earth is warm; when he is losing, it is cold.

That myth accounts for the seasons about as well as the Persephone myth. It is slightly better at explaining the randomness of weather, but worse at explaining the regularity of seasons, because real wars do not ebb and flow so regularly (except insofar as that is due to seasons themselves). In the Persephone myth, the role of the marriage contract and the magic seed is to explain that regularity. But why is it specifically a magic seed and not any other kind of magic? Why is it a conjugal-visits contract and not some other reason for someone to repeat an action annually? For instance, here is a variant explanation that fits the facts just as well: Persephone was not released – she escaped. Each year in spring, when her powers are at their height, she takes revenge on Hades by raiding the underworld and cooling all the caverns with spring air. The hot air thus displaced rises into the human world, causing summer. Demeter celebrates Persephone's revenge and the anniversary of her escape by commanding plants to grow and adorn the Earth. This myth accounts for the same observations as the original, and it is testable (and in fact refuted) by the same observations. Yet what it asserts about reality is markedly different from – in many ways it is the opposite of – the original myth.

Every other detail of the story, apart from its bare prediction that winter happens once a year, is just as easily variable. So, although the myth was created to explain the seasons, it is only superficially adapted to that purpose. When its author was wondering what could possibly make a goddess do something once a year, he did not shout, 'Eureka! It must have been a marriage contract enforced by a magic seed.' He made that choice – and all his substantive choices as author – for cultural and artistic reasons, and not because of the attributes of winter at all. He may also have been trying to explain aspects of human nature metaphorically – but here I am concerned with the myth only in its capacity as an explanation *of seasons*, and in that respect even its author could not have denied that the role of all the details could be played equally well by countless other things.

The Persephone and Freyr myths assert radically incompatible things about what is happening in reality to cause seasons. Yet no one, I guess, has ever adopted either myth as a result of comparing it on its merits with the other, because there is no way of distinguishing between them. If we ignore all the parts of both myths whose role could be easily replaced, we are left with the same core explanation in both cases: *the gods did it*. Although Freyr is a very different god of spring from Persephone, and his battles very different events from her conjugal visits, none of those differing attributes has any function in the myths' respective accounts of why seasons happen. Hence none of them provides any reason for choosing one explanation over the other.

The reason those myths are so easily variable is that their details are barely connected to the details of the phenomena. Nothing in the problem of why winter happens is addressed by postulating specifically a marriage contract or a magic seed, or the gods Persephone, Hades and Demeter – or Freyr. Whenever a wide range of variant theories can account equally well for the phenomenon they are trying to explain, there is no reason to prefer one of them over the others, so advocating a particular one in preference to the others is irrational.

That freedom to make drastic changes in those mythical explanations of seasons is the fundamental flaw in them. It is the reason that myth-making in general is not an effective way to understand the world. And that is so whether the myths are testable or not, for whenever it is easy to vary an explanation without changing its predictions, one could just as easily vary it to make different predictions if they were needed. For example, if the ancient Greeks *had* discovered that the seasons in the northern and southern hemispheres are out of phase, they would have had a choice of countless slight variants of the myth that would be consistent with that observation. One would be that when Demeter is sad she banishes warmth *from her vicinity*, and it has to go elsewhere – into the southern hemisphere. Similarly, slight variants of the Perse-phone explanation could account just as well for seasons that were marked by green rainbows, or seasons that happened once a week, or sporadically, or not at all. Likewise for the superstitious gambler or the end-of-the-world prophet: when their theory is refuted by experience, they do indeed switch to a new one; but, because their underlying explanations are bad, they can easily accommodate the new experience without changing the substance of the explanation. Without a good explanatory theory, they can simply reinterpret the omens, pick a new date, and make essentially the same prediction. In such cases, testing one's theory and abandoning it when it is refuted constitutes no progress towards understanding the world. If an explanation could easily explain anything in the given field, then it actually explains nothing.

In general, when theories are easily variable in the sense I have described, experimental testing is almost useless for correcting their errors. I call such theories *bad explanations*. Being proved wrong by experiment, and changing the theories to other bad explanations, does not get their holders one jot closer to the truth.

Because explanation plays this central role in science, and because testability is of little use in the case of bad explanations, I myself prefer to call myths, superstitions and similar theories *un*scientific even when they make testable predictions. But it does not matter what terminology you use, so long as it does not lead you to conclude that there is something worthwhile about the Persephone myth, or the prophet's apocalyptic theory or the gambler's delusion, just because it is testable. Nor is a person capable of making progress merely by virtue of being willing to drop a theory when it is refuted: one must also be seeking a better explanation of the relevant phenomena. That is the scientific frame of mind.

As the physicist Richard Feynman said, 'Science is what we have learned about how to keep from fooling ourselves.' By adopting easily variable explanations, the gambler and prophet are ensuring that they will be able to continue fooling themselves no matter what happens. Just as thoroughly as if they had adopted untestable theories, they are insulating themselves from facing evidence that they are mistaken about what is really there in the physical world.

The quest for good explanations is, I believe, the basic regulating principle not only of science, but of the Enlightenment generally. It is the feature that distinguishes those approaches to knowledge from all others, and it implies all those other conditions for scientific progress I have discussed: It trivially implies that prediction alone is insufficient.

Somewhat less trivially, it leads to the rejection of authority, because if we adopt a theory on authority, that means that we would also have accepted a range of different theories on authority. And hence it also implies the need for a tradition of criticism. It also implies a methodological rule – a *criterion for reality* – namely that we should conclude that a particular thing is real if and only if it figures in our best explanation of something.

Although the pioneers of the Enlightenment and of the scientific revolution did not put it this way, seeking good explanations was (and remains) the spirit of the age. This is how they began to think. It is what they began to do, systematically for the first time. It is what made that momentous difference to the rate of progress of all kinds.

Long before the Enlightenment, there were individuals who sought good explanations. Indeed, my discussion here suggests that all progress then, as now, was due to such people. But in most ages they lacked contact with a tradition of criticism in which others could carry on their ideas, and so created little that left any trace for us to detect. We do know of sporadic traditions of good-explanation-seeking in narrowly defined fields, such as geometry, and even short-lived traditions of criticism – mini-enlightenments – which were tragically snuffed out, as I shall describe in Chapter 9. But the sea change in the values and patterns of thinking of a whole community of thinkers, which brought about a sustained and accelerating creation of knowledge, happened only once in history, with *the* Enlightenment and its scientific revolution. An entire political, moral, economic and intellectual culture – roughly what is now called 'the West' – grew around the values entailed by the quest for good explanations, such as tolerance of dissent, openness to change, distrust of dogmatism and authority, and the aspiration to progress both by individuals and for the culture as a whole. And the progress made by that multifaceted culture, in turn, promoted those values – though, as I shall explain in Chapter 15, they are nowhere close to being fully implemented.

Now consider the true explanation of seasons. It is that the Earth's axis of rotation is tilted relative to the plane of its orbit around the sun. Hence for half of each year the northern hemisphere is tilted towards the sun while the southern hemisphere is tilted away, and for the other half it is the other way around. Whenever the sun's rays are falling vertically in one hemisphere (thus providing more heat per unit area of the surface) they are falling obliquely in the other (thus providing less).

The true explanation of seasons (not to scale!)

That is a good explanation – hard to vary, because all its details play a functional role. For instance, we know – and can test independently of our experience of seasons – that surfaces tilted away from radiant heat are heated less than when they are facing it, and that a spinning sphere in space points in a constant direction. And we can explain why, in terms of theories of geometry, heat and mechanics. Also, the same tilt appears in our explanation of where the sun appears relative to the horizon at different times of year. In the Persephone myth, in contrast, the coldness of the world is caused by Demeter's sadness – but people do not generally cool their surroundings when they are sad, and we have no way of knowing that Demeter *is* sad, or that she ever cools the world, other than the onset of winter itself. One could not substitute the moon for the sun in the axis-tilt story, because the position of the moon in the sky does not repeat itself once a year, and because the sun's rays heating the Earth are integral to the explanation. Nor could one easily incorporate any stories about how the sun god feels about all this, because if the true explanation of winter is in the geometry of the Earth–sun motion, then how anyone feels about it is irrelevant, and if there were some flaw in that explanation, then no story about how anyone felt would put it right.

The axis-tilt theory also predicts that the seasons will be out of phase in the two hemispheres. So if they had been found to be in phase, the theory would have been refuted, just as, in the event, the Persephone and Freyr myths were refuted by the opposite observation. But the difference is, if the axis-tilt theory had been refuted, its defenders would have had nowhere to go. No easily implemented change could make tilted axes cause the same seasons all over the planet. Fundamentally new ideas would have been needed. That is what makes good explanations essential to science: it is only when a theory is a good explanation – hard to vary – that it even matters whether it is testable. Bad explanations are equally useless whether they are testable or not.

Most accounts of the differences between myth and science make too much of the issue of testability – as if the ancient Greeks' great mistake was that they did not send expeditions to the southern hemisphere to observe the seasons. But in fact they could never have guessed that such an expedition might provide evidence about seasons unless they had already guessed that seasons would be out of phase in the two hemispheres – and if that guess was hard to vary, which it could have been only if it had been part of a good explanation. If their guess was *easy* to vary, they might just as well have saved themselves the boat fare, stayed at home, and tested the easily testable theory that winter can be staved off by yodelling.

So long as they had no better explanation than the Persephone myth, there should have been no need for testing. Had they been seeking good explanations, they would immediately have tried to improve upon the myth, without testing it. That is what we do today. We do not test every testable theory, but only the few that we find are good explanations. Science would be impossible if it were not for the fact that the overwhelming majority of false theories can be rejected out of hand without any experiment, simply for being bad explanations.

Good explanations are often strikingly simple or elegant – as I shall discuss in Chapter 14. Also, a common way in which an explanation can be bad is by containing superfluous features or arbitrariness, and sometimes removing those yields a good explanation. This has given rise to a misconception known as 'Occam's razor' (named after the fourteenth-century philosopher William of Occam, but dating back to antiquity), namely that one should always seek the 'simplest explanation'. One statement of it is 'Do not multiply assumptions beyond necessity.' However, there are plenty of very simple explanations that are nevertheless easily variable (such as 'Demeter did it'). And, while assumptions 'beyond necessity' make a theory bad by definition, there have been many mistaken ideas of what is 'necessary' in a theory. Instrumentalism, for instance, considers explanation itself unnecessary, and so do many other bad philosophies of science, as I shall discuss in Chapter 12.

When a formerly good explanation has been falsified by new observations, it is no longer a good explanation, because the problem has expanded to include those observations. Thus the standard scientific methodology of dropping theories when refuted by experiment is implied by the requirement for good explanations. The best explanations are the ones that are most constrained by existing knowledge – including other good explanations as well as other knowledge of the phenomena to be explained. That is why testable explanations that have passed stringent tests become extremely good explanations, which is in turn why the maxim of testability promotes the growth of knowledge in science.

Conjectures are the products of creative imagination. But the problem with imagination is that it can create fiction much more easily than truth. As I have suggested, historically, virtually all human attempts to explain experience in terms of a wider reality have indeed been fiction, in the form of myths, dogma and mistaken common sense – and the rule of testability is an insufficient check on such mistakes. But the quest for good explanations does the job: inventing falsehoods is easy, and therefore they are easy to vary once found; discovering good explanations is hard, but the harder they are to find, the harder they are to vary once found. The ideal that explanatory science strives for is nicely described by the quotation from Wheeler with which I began this chapter: 'Behind it all is surely an idea so simple, so beautiful, that when we grasp it – in a decade, a century, or a millennium – we will all say to each other, *how could it have been otherwise? [my italics]*.' Now we shall see how this explanation-based conception of science answers the question that I asked above: how do we know so much about *unfamiliar* aspects of reality?

Put yourself in the place of an ancient astronomer thinking about the axis-tilt explanation of seasons. For the sake of simplicity, let us assume that you have also adopted the heliocentric theory. So you might be, say, Aristarchus of Samos, who gave the earliest known arguments for the heliocentric theory in the third century BCE.

Although you know that the Earth is a sphere, you possess no evidence about any location on Earth south of Ethiopia or north of the Shetland Islands. You do not know that there is an Atlantic or a Pacific ocean; to you, the known world consists of Europe, North Africa and parts of Asia, and the coastal waters nearby. Nevertheless, from the axis-tilt theory of seasons, you can make predictions about the weather in the literally unheard-of places beyond your known world. Some of these predictions are mundane and could be mistaken for induction: you predict that due east or west, however far you travel, you will experience seasons at about the same time of year (though the timings of sunrise and sunset will gradually shift with longitude). But you will also make some counter-intuitive predictions: if you travel only a little further north than the Shetlands, you will reach a frozen region where each day and each night last six months; if you travel further south than Ethiopia, you will first reach a place where there are no seasons, and then, still further south, you will reach a place where there are seasons, but they are perfectly out of phase with those everywhere in your known world. You have never travelled more than a few hundred kilometres from your home island in the Mediterranean. You have never experienced any seasons other than Mediterranean ones. You have never read, nor heard tell, of seasons that were out of phase with the ones you have experienced. But you know about them.

What if you'd rather not know? You may not like these predictions. Your friends and colleagues may ridicule them. *You may try to modify the explanation* so that it will not make them, without spoiling its agreement with observations and with other ideas for which you have no good alternatives. You will fail. That is what a good explanation will do for you: it makes it harder for you to fool yourself.

For instance, it may occur to you to modify your theory as follows: 'In the known world, the seasons happen at the times of year predicted by the axis-tilt theory; everywhere else on Earth, they *also* happen at those times of year.' This theory correctly predicts all evidence known to you. And it is just as testable as your real theory. But now, in order to deny what the axis-tilt theory predicts in the faraway places, you have had to deny what it says about reality, everywhere. The modified theory is no longer an explanation of seasons, just a (purported) rule of thumb. So denying that the original explanation describes the true cause of seasons in the places about which you have no evidence has forced you to deny that it describes the true cause even on your home island.

Suppose for the sake of argument that you thought of the axis-tilt theory yourself. It is your conjecture, your own original creation. Yet because it is a good explanation – hard to vary – it is not yours to modify. It has an autonomous meaning and an autonomous domain of applicability. You cannot confine its predictions to a region of your choosing. Whether you like it or not, it makes predictions about places both known to you and unknown to you, predictions that you have thought of and ones that you have not thought of. Tilted planets in similar orbits in other solar systems must have seasonal heating and cooling – planets in the most distant galaxies, and planets that we shall never see because they were destroyed aeons ago, and also planets that have yet to form. The theory reaches out, as it were, from its finite origins inside one brain that has been affected only by scraps of patchy evidence from a small part of one hemisphere of one planet – to infinity. This *reach* of explanations is another meaning of 'the beginning of infinity'. It is the ability of some of them to solve problems beyond those that they were created to solve.

The axis-tilt theory is an example: it was originally proposed to explain the changes in the sun's angle of elevation during each year. Combined with a little knowledge of heat and spinning bodies, it then explained seasons. And, without any further modification, it also explained why seasons are out of phase in the two hemispheres, and why tropical regions do not have them, and why the summer sun shines at midnight in polar regions – three phenomena of which its creators may well have been unaware.

The reach of an explanation is not a 'principle of induction'; it is not something that the creator of the explanation can use to obtain or justify it. It is not part of the creative process at all. We find out about it only after we have the explanation – sometimes long after. So it has nothing to do with 'extrapolation', or 'induction', or with 'deriving' a theory in any other alleged way. It is exactly the other way round: the reason that the explanation of seasons reaches far outside the experience of its creators is precisely that it *does not* have to be extrapolated. By its nature as an explanation, when its creators first thought of it, it already applied in our planet's other hemisphere, and throughout the solar system, and in other solar systems, and at other times.

Thus the reach of an explanation is neither an additional assumption nor a detachable one. It is determined by the content of the explanation itself. The better an explanation is, the more rigidly its reach is determined – because the harder it is to vary an explanation, the harder it is in particular to construct a variant with a different reach, whether larger or smaller, that is still an explanation. We expect the law of gravity to be the same on Mars as on Earth because only one viable explanation of gravity is known – Einstein's general theory of relativity – and that is a universal theory; but we do not expect the *map* of Mars to resemble the map of Earth, because our theories about how Earth looks, despite being excellent explanations, have no reach to the appearance of any other astronomical object. Always, it is explanatory theories that tell us which (usually few) aspects of one situation can be 'extrapolated' to others.

It also makes sense to speak of the reach of non-explanatory forms of knowledge – rules of thumb, and also knowledge that is implicit in the genes for biological adaptations. So, as I said, my rule of thumb about cups-and-balls tricks has reach to a certain class of tricks; but I could not know what that class is without the explanation for why the rule works.

Old ways of thought, which did not seek good explanations, permitted no process such as science for correcting errors and misconceptions. Improvements happened so rarely that most people never experienced one. Ideas were static for long periods. Being bad explanations, even the best of them typically had little reach and were therefore brittle and unreliable beyond, and often within, their traditional applications. When ideas did change, it was seldom for the better, and when it did happen to be for the better, that seldom increased their reach. The emergence of science, and more broadly what I am calling the Enlightenment, was the beginning of the end of such static, parochial systems of ideas. It initiated the present era in human history, unique for its sustained, rapid creation of knowledge with ever-increasing reach. Many have wondered how long this can continue. Is it inherently bounded? Or is this the beginning of infinity – that is to say, do these methods have unlimited potential to create further knowledge? It may seem paradoxical to claim anything so grand (even if only potentially) on behalf of a project that has swept away all the ancient myths that used to assign human beings a special significance in the scheme of things. For if the power of the human faculties of reason and creativity, which have driven the Enlightenment, were indeed unlimited, would humans not have just such a significance?

And yet, as I mentioned at the beginning of this chapter, gold can be created only by stars and by intelligent beings. If you find a nugget of gold anywhere in the universe, you can be sure that in its history there was either a supernova or an intelligent being with an explanation. And if you find an explanation anywhere in the universe, you know that there must have been an intelligent being. A supernova alone would not suffice.

But – so what? Gold is important *to us*, but in the cosmic scheme of things it has little significance. Explanations are important to us: we need them to survive. But is there anything significant, in the cosmic scheme of things, about explanation, that apparently puny physical process that happens inside brains? I shall address that question in Chapter 3, after some reflections about appearance and reality.

TERMINOLOGY

Explanation   Statement about what is there, what it does, and how and why.

Reach   The ability of some explanations to solve problems beyond those that they were created to solve.

Creativity   The capacity to create new explanations.

Empiricism   The misconception that we 'derive' all our knowledge from sensory experience.

Theory-laden   There is no such thing as 'raw' experience. All our experience of the world comes through layers of conscious and unconscious interpretation.

Inductivism   The misconception that scientific theories are obtained by generalizing or extrapolating repeated experiences, and that the more often a theory is confirmed by observation the more likely it becomes.

Induction   The non-existent process of 'obtaining' referred to above.

Principle of induction   The idea that 'the future will resemble the past', combined with the misconception that this asserts anything about the future.

Realism   The idea that the physical world exists in reality, and that knowledge of it can exist too.

Relativism   The misconception that statements cannot be objectively true or false, but can be judged only relative to some cultural or other arbitrary standard.

Instrumentalism   The misconception that science cannot describe reality, only predict outcomes of observations.

Justificationism   The misconception that knowledge can be genuine or reliable only if it is justified by some source or criterion.

Fallibilism   The recognition that there are no authoritative sources of knowledge, nor any reliable means of justifying knowledge as true or probable.

Background knowledge   Familiar and currently uncontroversial knowledge.

Rule of thumb   'Purely predictive theory' (theory whose explanatory content is all background knowledge).

Problem   A problem exists when a conflict between ideas is experienced.

Good/bad explanation   An explanation that is hard/easy to vary while still accounting for what it purports to account for.

The Enlightenment   (The beginning of) a way of pursuing knowledge with a tradition of criticism and seeking good explanations instead of reliance on authority.

Mini-enlightenment   A short-lived tradition of criticism.

Rational   Attempting to solve problems by seeking good explanations; actively pursuing error-correction by creating criticisms of both existing ideas and new proposals.

The West   The political, moral, economic and intellectual culture that has been growing around the Enlightenment values of science, reason and freedom.

MEANINGS OF 'THE BEGINNING OF INFINITY'
ENCOUNTERED IN THIS CHAPTER

- The fact that some explanations have reach.
- The universal reach of some explanations.
- The Enlightenment.
- A tradition of criticism.
- Conjecture: the origin of all knowledge.
- The discovery of how to make progress: science, the scientific revolution, seeking good explanations, and the political principles of the West.
- Fallibilism.

SUMMARY

Appearances are deceptive. Yet we have a great deal of knowledge about the vast and unfamiliar reality that causes them, and of the elegant, universal laws that govern that reality. This knowledge consists of explanations: assertions about what is out there beyond the appearances, and how it behaves. For most of the history of our species, we had almost no success in creating such knowledge. Where does it come from? Empiricism said that we derive it from sensory experience. This is false. The real source of our theories is conjecture, and the real source of our knowledge is conjecture alternating with criticism. We create theories by rearranging, combining, altering and adding to existing ideas with the intention of improving upon them. The role of experiment and observation is to choose between existing theories, not to be the source of new ones. We interpret experiences through explanatory theories, but true explanations are not obvious. Fallibilism entails not looking to authorities but instead acknowledging that we may always be mistaken, and trying to correct errors. We do so by seeking good explanations – explanations that are hard to vary in the sense that changing the details would ruin the explanation. This, not experimental testing, was the decisive factor in the scientific revolution, and also in the unique, rapid, sustained progress in other fields that have participated in the Enlightenment. That was a rebellion against authority which, unlike most such rebellions, tried not to seek authoritative justifications for theories, but instead set up a tradition of criticism. Some of the resulting ideas have enormous reach: they explain more than what they were originally designed to. The reach of an explanation is an intrinsic attribute of it, not an assumption that we make about it as empiricism and inductivism claim.

Now I'll say some more about appearance and reality, explanation and infinity.

2

Closer to Reality

A galaxy is a mind-bogglingly huge thing. For that matter, a star is a mind-bogglingly huge thing. Our own planet is. A human brain is – in terms of both its internal complexity and the reach of human ideas. And there can be thousands of galaxies in a cluster, which can be millions of light years across. The phrase 'thousands of galaxies' trips lightly off the tongue, but it takes a while to make room in one's mind for the reality of it.

I was first stunned by the concept when I was a graduate student. Some fellow students were showing me what they were working on: observing clusters of galaxies – through *microscopes*. That is how astronomers used to use the Palomar Sky Survey, a collection of 1,874 photographic negatives of the sky, on glass plates, which showed the stars and galaxies as dark shapes on a white background.

They mounted one of the plates for me to look at. I focused the eyepiece of the microscope and saw something like this:

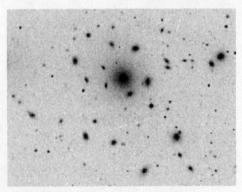

The Coma cluster of galaxies

Those fuzzy things are galaxies, and the sharply defined dots are stars in our own galaxy, thousands of times closer. The students' job was to catalogue the positions of the galaxies by lining them up in cross-hairs and pressing a button. I tried my hand at it – just for fun, since of course I was not qualified to make serious measurements. I soon found that it was not as easy as it had seemed. One reason is that it is not always obvious which are the galaxies and which are merely stars or other foreground objects. Some galaxies are easy to recognize: for instance, stars are never spiral, or noticeably elliptical. But some shapes are so faint that it is hard to tell whether they are sharp. Some galaxies appear small, faint and circular, and some are partly obscured by other objects. Nowadays such measurements are made by computers using sophisticated pattern-matching algorithms. But in those days one just had to examine each object carefully and use clues such as how fuzzy the edges looked – though there are also fuzzy objects, such as supernova remnants, in our galaxy. One used rules of thumb.

How would one test such a rule of thumb? One way is to select a region of the sky at random, and then take a photograph of it at higher resolution, so that the identification of galaxies is easier. Then one compares those identifications with the ones made using the rule of thumb. If they differ, the rule is inaccurate. If they do not differ, then one cannot be sure. One can never be sure, of course.

I was wrong to be impressed by the mere scale of what I was looking at. Some people become depressed at the scale of the universe, because it makes them feel insignificant. Other people are *relieved* to feel insignificant, which is even worse. But, in any case, those are mistakes. Feeling insignificant because the universe is large has exactly the same logic as feeling inadequate for not being a cow. Or a herd of cows. The universe is not there to overwhelm us; it is our home, and our resource. The bigger the better.

But then there is the *philosophical* magnitude of a cluster of galaxies. As I moved the cross-hairs to one nondescript galaxy after another, clicking at what I guessed to be the centre of each, some whimsical thoughts occurred to me. I wondered whether I would be the first and last human being ever to pay conscious attention to a particular galaxy. I was looking at the blurry object for only a few seconds, yet it might be laden with meaning for all I knew. It contains billions of planets. Each planet is a *world*. Each has its own unique history – sunrises and sunsets; storms, seasons; in some cases continents, oceans, earthquakes, rivers. Were any of those worlds inhabited? Were there astronomers there? Unless they were an exceedingly ancient, and advanced, civilization, those people would never have travelled outside their galaxy. So they would never have seen what it looked like from my perspective – though they might know from theory. Were any of them at that moment staring at the Milky Way, asking the same questions about us as I was about them? If so, then they were looking at our galaxy as it was when the most advanced forms of life on Earth were fish.

The computers that nowadays catalogue galaxies may or may not do it better than the graduate students used to. But they certainly do not experience such reflections as a result. I mention this because I often hear scientific research described in rather a bleak way, suggesting that it is mostly mindless toil. The inventor Thomas Edison once said, 'None of my inventions came by accident. I see a worthwhile need to be met and I make trial after trial until it comes. What it boils down to is one per cent inspiration and ninety-nine per cent perspiration.' Some people say the same about theoretical research, where the 'perspiration' phase is supposedly uncreative intellectual work such as doing algebra or translating algorithms into computer programs. But the fact that a computer or a robot can perform a task mindlessly does not imply that it is mindless when scientists do it. After all, computers play chess mindlessly – by exhaustively searching the consequences of all possible moves – but humans achieve a similar-looking functionality in a completely different way, by creative and enjoyable thought. Perhaps those galaxy-cataloguing computer programs were written by those same graduate students, distilling what they had learned into reproducible algorithms. Which means that they must have learned something while performing a task that a computer performs without learning anything. But, more profoundly, I expect that Edison was misinterpreting his own experience. A trial that fails is still fun. A repetitive experiment is not repetitive if one is thinking about the ideas that it is testing and the reality that it is investigating. That galaxy project was intended to discover whether 'dark matter' (see the next chapter) really exists – and it succeeded. If Edison, or those graduate students, or any scientific researcher engaged upon the 'perspiration' phase of discovery, had really been doing it mindlessly, they would be missing most of the fun – which is also what largely powers that 'one per cent inspiration'.

As I reached one particularly ambiguous image I asked my hosts, 'Is that a galaxy or a star?'

'Neither,' was the reply. 'That's just a defect in the photographic emulsion.'

The drastic mental gear change made me laugh. My grandiose speculations about the deep meaning of what I was seeing had turned out to be, in regard to this particular object, about nothing at all: suddenly there were no astronomers in that image, no rivers or earthquakes. They had disappeared in a puff of imagination. I had overestimated the mass of what I was looking at by some fifty powers of ten. What I had taken to be the largest object I had ever seen, and the most distant in space and time, was in reality just a speck barely visible without a microscope, within arm's reach. How easily, and how thoroughly, one can be misled.

But wait. Was I ever looking at a galaxy? All the other blobs were in fact microscopic smudges of silver too. If I misclassified the *cause* of one of them, because it looked too like the others, why was that such a big error?

Because an error in experimental science *is* a mistake about the cause of something. Like an accurate observation, it is a matter of theory. Very little in nature is detectable by unaided human senses. Most of what happens is too fast or too slow, too big or too small, or too remote, or hidden behind opaque barriers, or operates on principles too different from anything that influenced our evolution. But in some cases we can arrange for such phenomena to become perceptible, via scientific instruments.

We experience such instruments as bringing us closer to the reality – just as I felt while looking at that galactic cluster. But in purely physical terms they only ever separate us further from it. I could have looked up at the night sky in the direction of that cluster, and there would have been nothing between it and my eye but a few grams of air – but I would have seen nothing at all. I could have interposed a telescope, and then I might have seen it. In the event, I was interposing a telescope, a camera, a photographic development laboratory, another camera (to make copies of the plates), a truck to bring the plates to my university, and a microscope. I could see the cluster far better with all that equipment in the way.

Astronomers nowadays never look up at the sky (except perhaps in their spare time), and hardly ever look through telescopes. Many telescopes do not even have eyepieces suitable for a human eye. Many do not even detect visible light. Instead, instruments detect invisible signals which are then digitized, recorded, combined with others, and processed and analysed by computers. As a result, images may be produced – perhaps in 'false colours' to indicate radio waves or other radiation, or to display still more indirectly inferred attributes such as temperature or composition. In many cases, no image of the distant object is ever produced, only lists of numbers, or graphs and diagrams, and only the outcome of those processes affects the astronomers' senses.

Every additional layer of physical separation requires further levels of theory to relate the resulting perceptions to reality. When the astronomer Jocelyn Bell discovered pulsars (extremely dense stars that emit regular bursts of radio waves), this is what she was looking at:

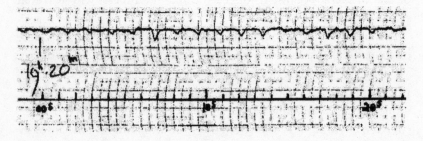

Radio-telescope output from the first known pulsar

Only through a sophisticated chain of theoretical interpretation could she 'see', by looking at that shaky line of ink on paper, a powerful, pulsating object in deep space, and recognize that it was of a hitherto unknown type.

The better we come to understand phenomena remote from our everyday experience, the longer those chains of interpretation become, and every additional link necessitates more theory. A single unexpected or misunderstood phenomenon anywhere in the chain can, and often does, render the resulting sensory experience arbitrarily misleading. Yet, over time, the conclusions that science has drawn have become ever truer to reality. Its quest for good explanations corrects the errors, allows for the biases and misleading perspectives, and fills in the gaps. This is what we can achieve when, as Feynman said, we keep learning more about how not to fool ourselves.

Telescopes contain automatic tracking mechanisms that continuously realign them so as to compensate for the effect of the Earth's motion; in some, computers continuously change the shape of the mirror so as to compensate for the shimmering of the Earth's atmosphere. And so, observed through such a telescope, stars do not appear to twinkle or to move across the sky as they did to generations of observers in the past. Those things are only appearance – parochial error. They have nothing to do with the reality of stars. The primary function of the telescope's optics is to reduce the illusion that the stars are few, faint, twinkling and moving. The same is true of every feature of the telescope, and of all other scientific instruments: each layer of indirectness, through its associated theory, corrects errors, illusions, misleading perspectives and gaps. Perhaps it is the mistaken empiricist ideal of 'pure', theory-free observation that makes it seem odd that truly accurate observation is always so hugely indirect. But the fact is that progress requires the application of ever more knowledge *in advance* of our observations.

So I was indeed looking at galaxies. Observing a galaxy via specks of silver is no different in that regard from observing a garden via images on a retina. In all cases, to say that we have genuinely observed any given thing is to say that we have accurately attributed our evidence (ultimately always evidence inside our own brains) to that thing. Scientific truth consists of such correspondence between theories and physical reality.

Scientists operating giant particle accelerators likewise look at pixels and ink, numbers and graphs, and thereby observe the microscopic reality of subatomic particles like nuclei and quarks. Others operate electron microscopes and fire the beam at cells that are as dead as dodos, having been stained, quick-frozen by liquid nitrogen, and mounted in a vacuum – but they thereby learn what *living* cells are like. It is a marvellous fact that objects can exist which, when we observe them, accurately take on the appearance and other attributes of other objects that are elsewhere and very differently constituted. Our sensory systems are such objects too, for it is only they that are directly affecting our brains when we perceive anything.

Such instruments are rare and fragile configurations of matter. Press one wrong button on the telescope's control panel, or code one wrong instruction into its computer, and the whole immensely complex artefact may well revert to revealing nothing other than itself. The same would be true if, instead of making that scientific instrument, you were to assemble those raw materials into almost any other configuration: stare at them, and you would see nothing other than them.

Explanatory theories tell us how to build and operate instruments in exactly the right way to work this miracle. Like conjuring tricks in reverse, such instruments fool our senses into seeing what is really there. Our minds, through the methodological criterion that I mentioned in Chapter 1, conclude that a particular thing is real if and only if it figures in our best explanation of something. Physically, all that has happened is that human beings, on Earth, have dug up raw materials such as iron ore and sand, and have rearranged them – still on Earth – into complex objects such as radio telescopes, computers and display screens, and now, instead of looking at the sky, they look at those objects. They are focusing their *eyes* on human artefacts that are close enough to touch. But their *minds* are focused on alien entities and processes, light years away.

Sometimes they are still looking at glowing dots just as their ancestors did – but on computer monitors instead of the sky. Sometimes they are looking at numbers or graphs. But in all cases they are inspecting local phenomena: pixels on a screen, ink on paper, and so on. These things are physically very unlike stars: they are much smaller; they are not dominated by nuclear forces and gravity; they are not capable of transmuting elements or creating life; they have not been there for billions of years. But when astronomers look at them, they see stars.

SUMMARY

It may seem strange that scientific instruments bring us closer to reality when in purely physical terms they only ever separate us further from it. But we observe nothing directly anyway. All observation is theory-laden. Likewise, whenever we make an error, it is an error in the explanation of something. That is why appearances can be deceptive, and it is also why we, and our instruments, can correct for that deceptiveness. The growth of knowledge consists of correcting misconceptions in our theories. Edison said that research is one per cent inspiration and ninety-nine per cent perspiration – but that is misleading, because people can apply creativity even to tasks that computers and other machines do uncreatively. So science is not mindless toil for which rare moments of discovery are the compensation: the toil can be creative, and fun, just as the discovery of new explanations is.

Now, can this creativity – and this fun – continue indefinitely?

3

The Spark

Most ancient accounts of the reality beyond our everyday experience were not only false, they had a radically different character from modern ones: they were *anthropocentric*. That is to say, they centred on human beings, and more broadly on *people* – entities with intentions and human-like thoughts – which included powerful, supernatural people such as spirits and gods. So, winter might be attributed to someone's sadness, harvests to someone's generosity, natural disasters to someone's anger, and so on. Such explanations often involved cosmically significant beings caring what humans did, or having intentions about them. This conferred cosmic significance on humans too. Then the geocentric theory placed humans at the physical hub of the universe as well. Those two kinds of anthropocentrism – explanatory and geometrical – made each other more plausible, and, as a result, pre-Enlightenment thinking was more anthropocentric than we can readily imagine nowadays.

A notable exception was the science of geometry itself, especially the system developed by the ancient Greek mathematician Euclid. Its elegant axioms and modes of reasoning about impersonal entities such as points and lines would later be an inspiration to many of the pioneers of the Enlightenment. But until then it had little impact on prevailing world views. For example, most astronomers were also astrologers: despite using sophisticated geometry in their work, they believed that the stars foretold political and personal events on Earth.

Before anything was known about how the world works, trying to explain physical phenomena in terms of purposeful, human-like thought and action may have been a reasonable approach. After all, that is how we explain much of our everyday experience even today: if a jewel is mysteriously missing from a locked safe, we seek human-level explanations such as error or theft (or, under some circumstances, conjuring), not new laws of physics. But that anthropocentric approach has never yielded any good explanations beyond the realm of human affairs. In regard to the physical world at large, it was colossally misconceived. We now know that the patterns of stars and planets in our night sky have no significance for human affairs. We know that we are not at the centre of the universe – it does not even have a geometrical centre. And we know that, although some of the titanic astrophysical phenomena that I have described played a significant role in our past, we have never been significant to them. We call a phenomenon *significant* (or *fundamental*) if parochial theories are inadequate to explain it, or if it appears in the explanation of many other phenomena; so it may seem that human beings and their wishes and actions are extremely insignificant in the universe at large.

Anthropocentric misconceptions have also been overturned in every other fundamental area of science: our knowledge of physics is now expressed entirely in terms of entities that are as impersonal as Euclid's points and lines, such as elementary particles, forces and spacetime – a four-dimensional continuum with three dimensions of space and one of time. Their effects on each other are explained not in terms of feelings and intentions, but through mathematical equations expressing laws of nature. In biology, it was once thought that living things must have been designed by a supernatural person, and that they must contain some special ingredient, a 'vital principle', to make them behave with apparent purposefulness. But biological science discovered new modes of explanation through such impersonal things as chemical reactions, genes and evolution. So we now know that living things, including humans, all consist of the same ingredients as rocks and stars, and obey the same laws, and that they were not designed by anyone. Modern science, far from explaining physical phenomena in terms of the thoughts and intentions of unseen people, considers our own thoughts and intentions to be aggregates of unseen (though not un-seeable) microscopic physical processes in our brains.

So fruitful has this abandonment of anthropocentric theories been, and so important in the broader history of ideas, that *anti*-anthropocentrism has increasingly been elevated to the status of a universal principle, sometimes called the 'Principle of Mediocrity': *there is nothing significant about humans (in the cosmic scheme of things)*. As the physicist Stephen Hawking put it, humans are 'just a chemical scum on the surface of a typical planet that's in orbit round a typical star on the outskirts of a typical galaxy'. The proviso 'in the cosmic scheme of things' is necessary because the chemical scum evidently does have a special significance according to values that it applies to itself, such as moral values. But the Principle says that all such values are themselves anthropocentric: they explain only the behaviour of the scum, which is itself insignificant.

It is easy to mistake quirks of one's own, familiar environment or perspective (such as the rotation of the night sky) for objective features of what one is observing, or to mistake rules of thumb (such as the prediction of daily sunrises) for universal laws. I shall refer to that sort of error as *parochialism*.

Anthropocentric errors are examples of parochialism, but not all parochialism is anthropocentric. For instance, the prediction that the seasons are in phase all over the world is a parochial error but not an anthropocentric one: it does not involve explaining seasons in terms of people.

Another influential idea about the human condition is sometimes given the dramatic name *Spaceship Earth*. Imagine a 'generation ship' – a spaceship on a journey so long that many generations of passengers live out their lives in transit. This has been proposed as a means of colonizing other star systems. In the Spaceship Earth idea, that generation ship is a metaphor for the *biosphere* – the system of all living things on Earth and the regions they inhabit. Its passengers represent all humans on Earth. Outside the spaceship, the universe is implacably hostile, but the interior is a vastly complex life-support system, capable of providing everything that the passengers need to thrive. Like the spaceship, the biosphere recycles all waste and, using its capacious nuclear power plant (the sun), it is completely self-sufficient.

Just as the spaceship's life-support system is designed to sustain its passengers, so the biosphere has the 'appearance of design': it seems highly adapted to sustaining us (claims the metaphor) because *we* were adapted to *it* by evolution. But its capacity is finite: if we overload it, either by our sheer numbers or by adopting lifestyles too different from those that we evolved to live (the ones that it is 'designed' to support), it will break down. And, like the passengers on that spaceship, we get no second chances: if our lifestyle becomes too careless or profligate and we ruin our life-support system, we have nowhere else to go.

The Spaceship Earth metaphor and the Principle of Mediocrity have both gained wide acceptance among scientifically minded people – to the extent of becoming truisms. This is despite the fact that, on the face of it, they argue in somewhat opposite directions: the Principle of Mediocrity stresses how *typical* the Earth and its chemical scum are (in the sense of being unremarkable), while Spaceship Earth stresses how *untypical* they are (in the sense of being uniquely suited to each other). But when the two ideas are interpreted in broad, philosophical ways, as they usually are, they can easily converge. Both see themselves as correcting much the same parochial misconceptions, namely that our experience of life on Earth is representative of the universe, and that the Earth is vast, fixed and permanent. They both stress instead that it is tiny and ephemeral. Both oppose arrogance: the Principle of Mediocrity opposes the pre-Enlightenment arrogance of believing ourselves significant in the world; the Spaceship Earth metaphor opposes the Enlightenment arrogance of aspiring to control the world. Both have a moral element: we *should not* consider ourselves significant, they assert; we *should not* expect the world to submit indefinitely to our depredations.

Thus the two ideas generate a rich conceptual framework that can inform an entire world view. Yet, as I shall explain, they are both false, even in the straightforward factual sense. And in the broader sense they are so misleading that, if you were seeking maxims worth being carved in stone and recited each morning before breakfast, you could do a lot worse than to use their *negations*. That is to say, the truth is that

> People *are* significant in the cosmic scheme of things; and

> The Earth's biosphere is *incapable* of supporting human life.

Consider Hawking's remark again. It is true that we are on a (somewhat) typical planet of a typical star in a typical galaxy. But we are far from typical of the matter in the universe. For one thing, about

80 per cent of that matter is thought to be invisible 'dark matter', which can neither emit nor absorb light. We currently detect it only through its indirect gravitational effects on galaxies. Only the remaining 20 per cent is matter of the type that we parochially call 'ordinary matter'. It is characterized by glowing continuously. We do not usually think of ourselves as glowing, but that is another parochial misconception, due to the limitations of our senses: we emit radiant heat, which is infrared light, and also light in the visible range, too faint for our eyes to detect.

Concentrations of matter as dense as ourselves and our planet and star, though numerous, are not exactly typical either. They are isolated, uncommon phenomena. The universe is mostly vacuum (plus radiation and dark matter). Ordinary matter is familiar to us only because we are made of it, and because of our untypical location near large concentrations of it.

Moreover, we are an uncommon form of ordinary matter. The commonest form is plasma (atoms dissociated into their electrically charged components), which typically emits bright, visible light because it is in stars, which are rather hot. We scums are mainly infra-red emitters because we contain liquids and complex chemicals which can exist only at a much lower range of temperatures.

The universe is pervaded with microwave radiation – the afterglow of the Big Bang. Its temperature is about 2.7 kelvin, which means 2.7 degrees above the coldest possible temperature, absolute zero, or about 270 degrees Celsius colder than the freezing point of water. Only very unusual circumstances can make anything colder than those microwaves. Nothing in the universe is known to be cooler than about *one* kelvin – except in certain physics laboratories on Earth. There, the record low temperature achieved is below one *billionth* of a kelvin. At those extraordinary temperatures, the glow of ordinary matter is effectively extinguished. The resulting 'non-glowing ordinary matter' on our planet is an exceedingly exotic substance in the universe at large. It may well be that the interiors of refrigerators constructed by physicists are by far the coldest and darkest places in the universe. Far from typical.

What is a typical place in the universe like? Let me assume that you are reading this on Earth. In your mind's eye, travel straight upwards a few hundred kilometres. Now you are in the slightly more typical environment of space. But you are still being heated and illuminated by the sun, and half your field of view is still taken up by the solids, liquids and scums of the Earth. A typical location has none of those features. So, travel a few trillion kilometres further in the same direction. You are now so far away that the sun looks like other stars. You are at a much colder, darker and emptier place, with no scum in sight. But it is not yet typical: you are still inside the Milky Way galaxy, and most places in the universe are not in any galaxy. Continue until you are clear outside the galaxy – say, a hundred thousand light years from Earth. At this distance you could not glimpse the Earth even if you used the most powerful telescope that humans have yet built. But the Milky Way still fills much of your sky. To get to a typical place in the universe, you have to imagine yourself at least a thousand times as far out as that, deep in intergalactic space.

What is it like there? Imagine the whole of space notionally divided into cubes the size of our solar system. If you were observing from a typical one of them, the sky would be pitch black. The nearest star would be so far away that if it were to explode as a supernova, and you were staring directly at it when its light reached you, you would not see even a glimmer. That is how big and dark the universe is. And it is cold: it is at that background temperature of 2.7 kelvin, which is cold enough to freeze every known substance except helium. (Helium is believed to remain liquid right down to absolute zero, unless highly pressurized.)

And it is empty: the density of atoms out there is below one per cubic metre. That is a million times sparser than atoms in the space between the stars, and those atoms are themselves sparser than in the best vacuum that human technology has yet achieved. Almost all the atoms in intergalactic space are hydrogen or helium, so there is no chemistry. No life could have evolved there, nor any intelligence. Nothing changes there. Nothing happens. The same is true of the next cube and the next, and if you were to examine a million consecutive cubes in any direction the story would be the same.

Cold, dark and empty. That unimaginably desolate environment is typical of the universe – and is another measure of how *un*typical the Earth and its chemical scum are, in a straightforward physical sense.

The issue of the cosmic significance of this type of scum will shortly take us back out into intergalactic space. But let me first return to Earth, and consider the Spaceship Earth metaphor, in *its* straightforward physical version.

This much is true: if, tomorrow, physical conditions on the Earth's surface were to change even slightly by astrophysical standards, then no humans could live here unprotected, just as they could not survive on a spaceship whose life-support system had broken down. Yet I am writing this in Oxford, England, where winter nights are likewise often cold enough to kill any human unprotected by clothing and other technology. So, while intergalactic space would kill me in a matter of seconds, Oxfordshire in its primeval state might do it in a matter of hours – which can be considered 'life support' only in the most contrived sense. There *is* a life-support system in Oxfordshire today, but it was not provided by the biosphere. It has been built by humans. It consists of clothes, houses, farms, hospitals, an electrical grid, a sewage system and so on. Nearly the whole of the Earth's biosphere in its primeval state was likewise incapable of keeping an unprotected human alive for long. It would be much more accurate to call it a death trap for humans rather than a life-support system. Even the Great Rift Valley in eastern Africa, where our species evolved, was barely more hospitable than primeval Oxfordshire. Unlike the life-support system in that imagined spaceship, the Great Rift Valley lacked a safe water supply, and medical equipment, and comfortable living quarters, and was infested with predators, parasites and disease organisms. It frequently injured, poisoned, drenched, starved and sickened its 'passengers', and most of them died as a result.

It was similarly harsh to all the other organisms that lived there: few individuals live comfortably or die of old age in the supposedly beneficent biosphere. That is no accident: most populations, of most species, are living close to the edge of disaster and death. It has to be that way, because as soon as some small group, somewhere, begins to have a slightly easier life than that, for any reason – for instance, an increased food supply, or the extinction of a competitor or predator – then its numbers increase. As a result, its other resources are depleted by the increased usage; so an increasing proportion of the population now has to colonize more marginal habitats and make do with inferior resources, and so on. This process continues until the disadvantages caused by the increased population have exactly balanced the advantage conferred by the beneficial change. That is to say, the new birth rate is again just barely keeping pace with the rampant disabling and killing of individuals by starvation, exhaustion, predation, overcrowding and all those other natural processes.

That is the situation to which evolution adapts organisms. And that, therefore, is the lifestyle in which the Earth's biosphere 'seems adapted' to sustaining them. The biosphere only ever achieves stability – and only temporarily at that – by continually neglecting, harming, disabling and killing individuals. Hence the metaphor of a spaceship or a life-support system, is quite perverse: when humans design a life-support system, they design it to provide the maximum possible comfort, safety and longevity for its users within the available resources; the biosphere has no such priorities.

Nor is the biosphere a great preserver of *species*. In addition to being notoriously cruel to individuals, evolution involves continual extinctions of entire species. The average rate of extinction since the beginning of life on Earth has been about ten species per year (the number is known only very approximately), becoming much higher during the relatively brief periods that palaeontologists call 'mass extinction events'. The rate at which species have come into existence has on balance only slightly exceeded the extinction rate, and the net effect is that the overwhelming majority of species that have ever existed on Earth (perhaps 99.9 per cent of them) are now extinct. Genetic evidence suggests that our own species narrowly escaped extinction on at least one occasion. Several species closely related to ours did become extinct. Significantly, the 'life-support system' itself wiped them out – by means such as natural disasters, evolutionary changes in other species, and climate change. Those cousins of ours had not invited extinction by changing their lifestyles or overloading the biosphere: on the contrary, it wiped them out because they *were* living the lifestyles that they had evolved to live, and in which, according to the Spaceship Earth metaphor, the biosphere had been 'supporting' them.

Yet that still overstates the degree to which the biosphere is hospitable to humans in particular. The first people to live at the latitude of Oxford (who were actually from a species related to us, possibly the Neanderthals) could do so only because they brought knowledge with them, about such things as tools, weapons, fire and clothing. That knowledge was transmitted from generation to generation not genetically but culturally. Our pre-human ancestors in the Great Rift Valley used such knowledge too, and our own species must have come into existence already dependent on it for survival. As evidence of that, note that I would soon die if I tried to live in the Great Rift Valley in its primeval state: I do not have the requisite knowledge. Since then, there have been human populations who, for instance, knew how to survive in the Amazon jungle but not in the Arctic, and populations for whom it was the other way round. Therefore that knowledge was not part of their genetic inheritance. It was created by human thought, and preserved and transmitted in human culture.

Today, almost the entire capacity of the Earth's 'life-support system for humans' has been provided not *for* us but *by* us, using our ability to create new knowledge. There are people in the Great Rift Valley today who live far more comfortably than early humans did, and in far greater numbers, through knowledge of things like tools, farming and hygiene. The Earth did provide the raw materials for our survival – just as the sun has provided the energy, and supernovae provided the elements, and so on. But a heap of raw materials is not the same thing as a life-support system. It takes knowledge to convert the one into the other, and biological evolution never provided us with enough knowledge to survive, let alone to thrive. In this respect we differ from almost all other species. They do have all the knowledge that they need, genetically encoded in their brains. And that knowledge was indeed provided for them by evolution – and so, in the relevant sense, 'by the biosphere'. So *their* home environments do have the appearance of having been designed as life-support systems for them, albeit only in the desperately limited sense that I have described. But the biosphere no more provides humans with a life-support system than it provides us with radio telescopes.

So the biosphere is incapable of supporting human life. From the outset, it was only human knowledge that made the planet even marginally habitable by humans, and the enormously increased capacity of our life-support system since then (in terms both of numbers and of security and quality of life) has been entirely due to the creation of human knowledge. To the extent that we are on a 'spaceship', we have never been merely its passengers, nor (as is often said) its stewards, nor even its maintenance crew: we are its designers and builders. Before the designs created by humans, it was not a vehicle, but only a heap of dangerous raw materials.

The 'passengers' metaphor is a misconception in another sense too. It implies that there was a time when humans lived unproblematically: when they were provided for, like passengers, without themselves having to solve a stream of problems in order to survive and to thrive. But in fact, even with the benefit of their cultural knowledge, our ancestors continually faced desperate problems, such as where the next meal was coming from, and typically they barely solved these problems or they died. There are very few fossils of old people.

The moral component of the Spaceship Earth metaphor is therefore somewhat paradoxical. It casts humans as ungrateful for gifts which, in reality, they never received. And it casts all other species in morally positive roles in the spaceship's life-support system, with humans as the only negative actors. But humans are part of the biosphere, and the supposedly immoral behaviour is identical to what all other species do when times are good – except that humans alone try to mitigate the effect of that response on their descendants and on other species.

The Principle of Mediocrity is paradoxical too. Since it singles out anthropocentrism for special opprobrium among all forms of parochial misconception, it is itself anthropocentric. Also, it claims that all value judgements are anthropocentric, yet it itself is often expressed in value-laden terminology, such as 'arrogance', 'just scum' and the very word 'mediocrity'. With respect to whose values are those disparagements to be understood? Why is arrogance even relevant as a criticism? Also, even if holding an arrogant opinion is morally wrong, morality is supposed to refer only to the internal organization of chemical scum. So how can it tell us anything about how the world *beyond* the scum is organized, as the Principle of Mediocrity purports to do?

In any case, it was not arrogance that made people adopt anthropocentric explanations. It was merely a parochial error, and quite a reasonable one originally. Nor was it arrogance that prevented people from realizing their mistake for so long: they didn't realize *anything*, because they did not know how to seek better explanations. In a sense their whole problem was that they were not arrogant *enough*: they assumed far too easily that the world was fundamentally incomprehensible to them.

The misconception that there was once an unproblematic era for humans is present in ancient myths of a past Golden Age, and of a Garden of Eden. The theological notions of *grace* (unearned benefit from God) and *Providence* (which is God regarded as the provider of human needs) are also related to this. In order to connect the supposed unproblematic past with their own less-than-pleasant experiences, the authors of such myths had to include some past transition, such as a Fall from Grace when Providence reduced its level of support. In the Spaceship Earth metaphor, the Fall from Grace is usually deemed to be imminent or under way.

The Principle of Mediocrity contains a similar misconception. Consider the following argument, which is due to the evolutionary biologist Richard Dawkins: Human attributes, like those of all other organisms, evolved under natural selection in an ancestral environment. That is why our senses are adapted to detecting things like the colours and smell of fruit, or the sound of a predator: being able to detect such things gave our ancestors a better chance of surviving to have offspring. But, for the same reason, Dawkins points out, evolution did not waste our resources on detecting phenomena that were never relevant to our survival. We cannot, for instance, distinguish between the colours of most stars with the naked eye. Our night vision is poor and monochromatic because not enough of our ancestors died of that limitation to create evolutionary pressure for anything better. So Dawkins argues – and here he is invoking the Principle of Mediocrity – that there is no reason to expect our brains to be any different from our eyes in this regard: they evolved to cope with the narrow class of phenomena that commonly occur in the biosphere, on approximately human scales of size, time, energy and so on. Most phenomena in the universe happen far above or below those scales. Some would kill us instantly; others could never affect anything in the lives of early humans. So, just as our senses cannot *detect* neutrinos or quasars or most other significant phenomena in the cosmic scheme of things, there is no reason to expect our brains to *understand* them. To the extent that they already do understand them, we have been lucky – but a run of luck cannot be expected to continue for long. Hence

Dawkins agrees with an earlier evolutionary biologist, John Haldane, who expected that 'the universe is not only queerer than we suppose, but queerer than we *can* suppose.'

That is a startling – and paradoxical – consequence of the Principle of Mediocrity: it says that all human abilities, including the distinctive ones such as the ability to create new explanations, are necessarily parochial. That implies, in particular, that progress in science cannot exceed a certain limit defined by the biology of the human brain. And we must expect to reach that limit sooner rather than later. Beyond it, the world stops making sense (or seems to). The answer to the question that I asked at the end of Chapter 2 – whether the scientific revolution and the broader Enlightenment could be a beginning of infinity – would then be a resounding no. Science, for all its successes and aspirations, would turn out to be inherently parochial – and, ironically, anthropocentric.

So here the Principle of Mediocrity and Spaceship Earth converge. They share a conception of a tiny, human-friendly bubble embedded in the alien and uncooperative universe. The Spaceship Earth metaphor sees it as a physical bubble, the biosphere. For the Principle of Mediocrity, the bubble is primarily conceptual, marking the limits of the human capacity to understand the world. Those two bubbles are related, as we shall see. In both views, anthropocentrism is true in the interior of the bubble: there the world is unproblematic, uniquely compliant with human wishes and human understanding. Outside it there are only insoluble problems.

Dawkins would prefer it to be otherwise. As he wrote:

> I believe that an orderly universe, one indifferent to human preoccupations, in which everything has an explanation even if we still have a long way to go before we find it, is a more beautiful, more wonderful place than a universe tricked out with capricious ad hoc magic.

> *Unweaving the Rainbow* (1998)

An 'orderly' (explicable) universe is indeed more beautiful (see Chapter 14) – though the assumption that to be orderly it has to be 'indifferent to human preoccupations' is a misconception associated with the Principle of Mediocrity.

Any assumption that the world is *in*explicable can lead only to extremely bad explanations. For an inexplicable world is indistinguishable from one 'tricked out with capricious ad hoc magic': by definition, no hypothesis about the world outside the bubble of explicability can be a better explanation than that Zeus rules there – or practically any myth or fantasy one likes.

Moreover, since the outside of the bubble affects our explanations of the inside (or else we may as well do without it), the inside is not really explicable either. It seems so only if we carefully refrain from asking certain questions. This bears an uncanny resemblance to the intellectual landscape before the Enlightenment, with its distinction between Earth and heaven. It is a paradox inherent in the Principle of Mediocrity: contrary to its motivation, here it is forcing us back to an archaic, anthropocentric, pre-scientific conception of the world.

At root, the Principle of Mediocrity and the Spaceship Earth metaphor overlap in a claim about *reach*: they both claim that the reach of the distinctively human way of being – that is to say, the way of problem-solving, knowledge-creating and adapting the world around us – is bounded. And they argue that its bounds cannot be very far beyond what it has already reached. Trying to go beyond that range must lead to failure and catastrophe respectively.

Both ideas also rely on essentially the same argument, namely that if there were no such limit, there would be no explanation for the continued effectiveness of the adaptations of the human brain beyond the conditions under which they evolved. Why should one adaptation out of the trillions that have ever existed on Earth have unlimited reach, when all others reach only inside the tiny, insignificant, untypical biosphere? Fair enough: all reach has an explanation. But what if there *is* an explanation, and what if it has nothing to do with evolution or the biosphere?

Imagine that a flock of birds from a species that evolved on one island happens to fly to another. Their wings and eyes still work. That is an example of the reach of those adaptations. It has an explanation, the essence of which is that wings and eyes exploit universal laws of physics (of aerodynamics and optics respectively). They exploit those laws only imperfectly; but the atmospheric and lighting conditions on the two islands are sufficiently similar, by the criteria defined by those laws, for the same adaptations to work on both.

Thus the birds may well be able to fly to an island many kilometres away horizontally, but if they were transported only a few kilometres upwards their wings would stop working because the density of the air would be too low. Their implicit knowledge about how to fly fails at high altitude. A little further up, their eyes and other organs would stop working. The design of these too does not have that much reach: all vertebrate eyes are filled with liquid water, but water freezes at stratospheric temperatures and boils in the vacuum of space. Less dramatically, the birds might also die if they merely had no good night vision and they reached an island where the only suitable prey organisms were nocturnal. For the same reason, biological adaptations also have limited reach in regard to changes in their *home* environment – which can and do cause extinctions.

If those birds' adaptations do have enough reach to make the species viable on the new island, they will set up a colony there. In subsequent generations, mutants slightly better adapted to the new island will end up having slightly more offspring on average, so evolution will adapt the population more accurately to contain the knowledge needed to make a living there. The ancestor species of humans colonized new habitats and embarked on new lifestyles in exactly that way. But by the time our species had evolved, our fully human ancestors were achieving much the same thing thousands of times faster, by evolving their cultural knowledge instead. Because they did not yet know how to do science, their knowledge was only a little less parochial than biological knowledge. It consisted of rules of thumb. And so progress, though rapid compared to biological evolution, was sluggish compared to what the Enlightenment has accustomed us to.

Since the Enlightenment, technological progress has depended specifically on the creation of explanatory knowledge. People had dreamed for millennia of flying to the moon, but it was only with the advent of Newton's theories about the behaviour of invisible entities such as forces and momentum that they began to understand what was needed in order to go there.

This increasingly intimate connection between *explaining* the world and *controlling* it is no accident, but is part of the deep structure of the world. Consider the set of all conceivable transformations of physical objects. Some of those (like faster-than-light communication) never happen because they are forbidden by laws of nature; some (like the formation of stars out of primordial hydrogen) happen spontaneously; and some (such as converting air and water into trees, or converting raw materials into a radio telescope) are possible, but happen only when the requisite knowledge is present – for instance, embodied in genes or brains. But those are the only possibilities. That is to say, every putative physical transformation, to be performed in a given time with given resources or under any other conditions, is either

– impossible because it is forbidden by the laws of nature; or
– achievable, given the right knowledge.

That momentous dichotomy exists because if there were transformations that technology could never achieve regardless of what knowledge was brought to bear, then this fact would itself be a testable regularity in nature. But all regularities in nature have explanations, so the explanation of that regularity would itself be a law of nature, or a consequence of one. And so, again, everything that is not forbidden by laws of nature is achievable, given the right knowledge.

This fundamental connection between explanatory knowledge and technology is why the Haldane–Dawkins queerer-than-we-can-suppose argument is mistaken – why the reach of human adaptations does have a different character from that of all the other adaptations in the biosphere. The ability to create and use explanatory knowledge gives *people* a power to transform nature which is ultimately not limited by parochial factors, as all other adaptations are, but only by universal laws. This is the cosmic significance of explanatory knowledge – and hence of people, whom I shall henceforward define as entities that can create explanatory knowledge.

For every other species on Earth, we can determine its reach simply by making a list of all the resources and environmental conditions on which its adaptations depend. In principle one could determine those from a study of its DNA molecules – because that is where all its genetic information is encoded (in the form of sequences of small constituent molecules called 'bases'). As Dawkins has pointed out:

A gene pool is carved and whittled through generations of ancestral natural selection to fit [a particular] environment. In theory a knowledge-able zoologist, presented with the complete transcript of a genome [the set of all the genes of an organism], should be able to reconstruct the environmental circumstances that did the carving. In this sense the DNA is a coded description of ancestral environments.

In Art Wolfe, *The Living Wild*, ed. Michelle A. Gilders (2000)

To be precise, the 'knowledgeable zoologist' would be able to reconstruct only those aspects of the organism's ancestral environment that exerted selection pressure – such as the types of prey that existed there, what behaviours would catch them, what chemicals would digest them and so on. Those are all regularities in the environment. A genome contains coded descriptions of them, and hence implicitly specifies the environments in which the organism can survive. For example, all primates require vitamin C. Without it, they fall ill and die of the disease scurvy, but their genes do not contain the knowledge of how to synthesize it. So, whenever any non-human primate is in an environment that does not supply vitamin C for an extended period, it dies. Any account that overlooks this fact will overestimate the reach of those species. Humans are primates, yet *their* reach has nothing to do with which environments supply vitamin C. Humans can create and apply new knowledge of how to cause it to be synthesized from a wide range of raw materials, by agriculture or in chemical factories. And, just as essentially, humans can discover for themselves that, in most environments, they *need* to do that in order to survive.

Similarly, whether humans could live entirely outside the biosphere – say, on the moon – does not depend on the quirks of human biochemistry. Just as humans currently cause over a tonne of vitamin C to appear in Oxfordshire every week (from their farms and factories), so they could do the same on the moon – and the same goes for breathable air, water, a comfortable temperature and all their other parochial needs. Those needs can all be met, given the right knowledge, by transforming other resources. Even with present-day technology, it would be possible to build a self-sufficient colony on the moon, powered by sunlight, recycling its waste, and obtaining raw materials from the moon itself. Oxygen is plentiful on the moon in the form of metal oxides in moon rock. Many other elements could easily be extracted too. Some elements are rare on the moon, and so in practice these would be supplied from the Earth, but in principle the colony could be entirely independent of the Earth if it sent robot space vehicles to mine asteroids for such elements, or if it manufactured them by transmutation.

I specified *robot* space vehicles because all technological knowledge can eventually be implemented in automated devices. This is another reason that 'one per cent inspiration and ninety-nine per cent perspiration' is a misleading description of how progress happens: *the 'perspiration' phase can be automated* – just as the task of recognizing galaxies on astronomical photographs was. And the more advanced technology becomes, the shorter is the gap between inspiration and automation. The more this happens in the moon colony, the less human effort will be required to live there. Eventually the moon colonists will take air for granted, just as the people now living in Oxfordshire take for granted that water will flow if they turn on a tap. If either of those populations lacked the right knowledge, their environment would soon kill them.

We are accustomed to thinking of the Earth as hospitable and the moon as a bleak, faraway deathtrap. But that is how our ancestors would have regarded Oxfordshire, and, ironically, it is how I, today, would regard the primeval Great Rift Valley. In the unique case of humans, the difference between a hospitable environment and a death-trap depends on what knowledge they have created. Once enough knowledge has been embodied in the lunar colony, the colonists can devote their thoughts and energies to creating even more knowledge, and soon it will cease to be a colony and become simply home. No one will think of the moon as a fringe habitat, distinguished from our 'natural' environment on Earth, any more than we now think of Oxfordshire as being fundamentally different from the Great Rift Valley as a place to live.

Using knowledge to cause automated physical transformations is, in itself, not unique to humans. It is the basic method by which all organisms keep themselves alive: every cell is a chemical factory. The difference between humans and other species is in what kind of knowledge they can use (explanatory instead of rule-of-thumb) and in how they create it (conjecture and criticism of ideas, rather than the variation and selection of genes). It is precisely those two differences that explain why every other organism can function only in a certain range of environments that are hospitable to it, while humans transform *in*hospitable environments like the biosphere into support systems for themselves. And, while every other organism is a factory for converting resources of a fixed type into more such organisms, human bodies (including their brains) are factories for transforming *anything into anything* that the laws of nature allow. They are 'universal constructors'.

This universality in the human condition is part of a broader phenomenon that I shall discuss in Chapter 6. We do not share it with any other species currently on Earth. But, since it is a consequence of the ability to create explanations, we do necessarily share it with any other people that might exist in the universe. The opportunities provided by the laws of nature for transforming resources are universal, and all entities with universal reach necessarily have the same reach.

A few species other than humans are known to be capable of having cultural knowledge. For example, some apes can discover new methods of cracking nuts, and pass that knowledge on to other apes. As I shall discuss in Chapter 16, the existence of such knowledge is suggestive of how ape-like species evolved into people. But it is irrelevant to the arguments of this chapter, because no such organism is capable of creating or using explanatory knowledge. Hence the cultural knowledge of such organisms is of essentially the same type as genetic knowledge, and does indeed have only a small and inherently limited reach. They are not universal constructors, but highly specialized ones. For them, the Haldane–Dawkins argument is valid: the world is stranger than they can conceive.

In some environments in the universe, the most efficient way for humans to thrive might be to alter their own genes. Indeed, we are already doing that in our present environment, to eliminate diseases that have in the past blighted many lives. Some people object to this on the grounds (in effect) that a genetically altered human is no longer human. This is an anthropomorphic mistake. The only uniquely significant thing about humans (whether in the cosmic scheme of things or according to any rational human criterion) is our ability to create new explanations, and we have that in common with all people. You do not become less of a person if you lose a limb in an accident; it is only if you lose your brain that you do. Changing our genes in order to improve our lives and to facilitate further improvements is no different in this regard from augmenting our skin with clothes or our eyes with telescopes.

One might wonder whether the reach of people in general might be greater than the reach of humans. What if, for instance, the reach of technology is indeed unlimited, but only to creatures with two opposable thumbs on each hand; or if the reach of scientific knowledge is unlimited, but only to beings whose brains are twice the size of ours? But our faculty of being universal constructors makes these issues as irrelevant as that of access to vitamins. If progress at some point were to depend on having two thumbs per hand, then the outcome would depend not on the knowledge we inherit in our genes, but on whether we could discover how to build robots, or gloves, with two thumbs per hand, or alter ourselves to have a second thumb. If it depends on having more memory capacity, or speed, than a human brain, then the outcome would depend on whether we could build computers to do the job. Again, such things are already commonplace in technology.

The astrophysicist Martin Rees has speculated that somewhere in the universe 'there could be life and intelligence out there in forms we can't conceive. Just as a chimpanzee can't understand quantum theory, it could be there are aspects of reality that are beyond the capacity of our brains.' But that cannot be so. For if the 'capacity' in question is mere computational speed and amount of memory, then we can understand the aspects in question with the help of computers – just as we have understood the world for centuries with the help of pencil and paper. As Einstein remarked, 'My pencil and I are more clever than I.' In terms of computational repertoire, our computers – and brains – are already universal (see Chapter 6). But if the claim is that we may be *qualitatively* unable to understand what some other forms of intelligence can – if our disability cannot be remedied by mere automation – then this is just another claim that the world is not explicable. Indeed, it is tantamount to an appeal to the supernatural, with all the arbitrariness that is inherent in such appeals, for if we wanted to incorporate into our world view an imaginary realm explicable only to superhumans, we need never have bothered to abandon the myths of Persephone and her fellow deities.

So human reach is essentially the same as the reach of explanatory knowledge itself. An environment is within human reach if it is possible to create an open-ended stream of explanatory knowledge there. That means that if knowledge of a suitable kind were instantiated in such an environment in suitable physical objects, it would cause itself to survive and would then continue to increase indefinitely. Can there really be such an environment? This is essentially the question that I asked at the end of the last chapter – *can this creativity continue indefinitely?* – and it is the question to which the Spaceship Earth metaphor assumes a negative answer.

The issue comes down to this: if such an environment can exist, what are the minimal physical features that it must have? Access to *matter* is one. For example, the trick of extracting oxygen from moon rocks depends on having compounds of oxygen available. With more advanced technology, one could manufacture oxygen by transmutation; but, no matter how advanced one's technology is, one still needs raw materials of some sort. And, although mass can be recycled, creating an open-ended stream of knowledge depends on having an ongoing supply of it, both to make up for inevitable inefficiencies and to make the additional memory capacity to store new knowledge as it is created.

Also, many of the necessary transformations require *energy*: something must power conjectures and scientific experiments and all those manufacturing processes; and, again, the laws of physics forbid the creation of energy from nothing. So access to an energy supply is also a necessity. To some extent, energy and mass can be transformed into each other. For instance, transmuting hydrogen into any other element releases energy through nuclear fusion. Energy can also be converted into mass by various subatomic processes (but I cannot imagine naturally occurring circumstances in which those would be the best way of obtaining matter).

In addition to matter and energy, there is one other essential requirement, namely *evidence*: the information needed to test scientific theories. The Earth's surface is rich in evidence. We happened to get round to testing Newton's laws in the seventeenth century, and Einstein's in the twentieth, but the evidence with which we did that – light from the sky – had been deluging the surface of the Earth for billions of years before that, and will continue to do so for billions more. Even today we have barely begun to examine that evidence: on any clear night, the chances are that your roof will be struck by evidence falling from the sky which, if you only knew what to look for and how, would win you a Nobel prize. In chemistry, every stable element that exists anywhere is also present on or just below the Earth's surface. In biology, copious evidence of the nature of life is ubiquitous in the biosphere – and within arm's reach, in our own DNA. As far as we know, all the fundamental constants of nature can be measured here, and every fundamental law can be tested here. Everything needed for the open-ended creation of knowledge is here in abundance, in the Earth's biosphere.

And the same is true of the moon. It has essentially the same resources of mass, energy and evidence as the Earth has. Parochial details differ, but the fact that humans living on the moon will have to make their own air is no more significant than the fact that laboratories on Earth have to make their own vacuum. Both tasks can be automated so as to require arbitrarily little human effort or attention. Likewise, because humans are universal constructors, *every* problem of finding or transforming resources can be no more than a transient factor limiting the creation of knowledge in a given environment. And therefore matter, energy and evidence are the only requirements that an environment needs to have in order to be a venue for open-ended knowledge creation.

Though any particular problem is a transient factor, the condition of having to solve problems in order to survive and continue to create knowledge is permanent. I have mentioned that there has never been an unproblematic time for humans; that applies as much to the future as to the past. Today on Earth, in the short run, there are still countless problems to be solved to eliminate even starvation and other forms of extreme human suffering that date back to prehistory. On a timescale of decades, we shall be faced with choices to make substantial modifications to the biosphere, or to keep it the same, or anything in between. Whichever option we choose, it will be a project of planet-wide control, requiring the creation of a great deal of scientific and technological knowledge as well as knowledge about how to make such decisions rationally (see Chapter 13). In the even longer run, it is not only our comfort and aesthetic sensibilities, and the suffering of individuals, that are problematic, but, as always, the survival of our species. For instance, at present during any given century there is about one chance in a thousand that the Earth will be struck by a comet or asteroid large enough to kill at least a substantial proportion of all human beings. That means that a typical child born in the United States today is more likely to die as a result of an astronomical event than a plane crash. Both are very low-probability events, but, unless we create a great deal more scientific and technological knowledge than we have now, we shall have no defence against those and other forms of natural disaster that must, eventually, strike. Arguably there are more immediate existential threats too – see Chapter 9.

Setting up self-sufficient colonies on the moon and elsewhere in the solar system – and eventually in other solar systems – will be a good hedge against the extinction of our species or the destruction of civilization, and is a highly desirable goal for that reason among others. As Hawking has said:

> I don't think the human race will survive the next thousand years, unless we spread into space. There are too many accidents that can befall life on a single planet. But I'm an optimist. We will reach out to the stars.
>
> *Daily Telegraph*, 16 October 2001

But even that will be far from an unproblematic state. And most people are not satisfied merely to be confident in the survival of the *species*: they want to survive personally. Also, like our earliest human ancestors, they want to be free from physical danger and suffering. In future, as various causes of suffering and death such as disease and ageing are successively addressed and eliminated, and human life spans increase, people will care about ever longer-term risks.

In fact people will always want still more than that: they will want to make progress. For, in addition to threats, there will always be problems in the benign sense of the word: errors, gaps, inconsistencies and inadequacies in our knowledge that we wish to solve – including, not least, moral knowledge: knowledge about *what to want*, what to strive for. The human mind seeks explanations; and now that we know how to find them, we are not going to stop voluntarily. Here is another misconception in the Garden of Eden myth: that the supposed unproblematic state would be a *good* state to be in. Some theologians have denied this, and I agree with them: an unproblematic state is a state without creative thought. Its other name is death.

All those kinds of problem (survival-related, progress-related, moral, and sheer-curiosity-driven problems) are connected. We can, for instance, expect that our ability to cope with existential threats will continue to depend on knowledge that was originally created for its own sake. And we can expect disagreements about goals and values always to exist, because, among other reasons, moral explanations depend partly on facts about the physical world. For instance, the moral stances in the Principle of Mediocrity and the Spaceship Earth idea depend on the physical world not being explicable in the sense that I have argued it must be.

Nor will we ever run out of problems. The deeper an explanation is, the more new problems it creates. That must be so, if only because there can be no such thing as an ultimate explanation: just as 'the gods did it' is always a bad explanation, so any other purported foundation of all explanations must be bad too. It must be easily variable because it cannot answer the question: why that foundation and not another? Nothing can be explained only in terms of itself. That holds for philosophy just as it does for science, and in particular it holds for *moral* philosophy: no utopia is possible, but only because our values and our objectives can continue to improve indefinitely.

Thus fallibilism alone rather understates the error-prone nature of knowledge-creation. Knowledge-creation is not only *subject* to error: errors are common, and significant, and always will be, and correcting them will always reveal further and better problems. And so the maxim that I suggested should be carved in stone, namely 'The Earth's biosphere is *incapable* of supporting human life' is actually a special case of a much more general truth, namely that, for people, *problems are inevitable*. So let us carve *that* in stone:

PROBLEMS
ARE
INEVITABLE

It is inevitable that we face problems, but no particular problem is inevitable. We survive, and thrive, by solving each problem as it comes up. And, since the human ability to transform nature is limited only by the laws of physics, none of the endless stream of problems will ever constitute an impassable barrier. So a complementary and equally important truth about people and the physical world is that *problems are soluble*. By 'soluble' I mean that the right knowledge would solve them. It is not, of course, that we can possess knowledge just by wishing for it; but it is in principle accessible to us. So let us carve that in stone too:

PROBLEMS
ARE
SOLUBLE

That *progress* is both possible and desirable is perhaps the quintessential idea of the Enlightenment. It motivates all traditions of criticism, as well as the principle of seeking good explanations. But it can be interpreted in two almost opposite ways, both of which, confusingly, are known as 'perfectibility'. One is that humans, or human societies, are capable of attaining a state of supposed perfection – such as the Buddhist or Hindu 'nirvana', or various political utopias. The other is that every attainable state can be indefinitely improved. Fallibilism rules out that first position in favour of the second. Neither the human condition in particular nor our explanatory knowledge in general will ever be perfect, nor even approximately perfect. We shall always be at the *beginning* of infinity.

These two interpretations of human progress and perfectibility have historically inspired two broad branches of the Enlightenment which, though they share attributes such as their rejection of authority, are so different in important respects in that it is most unfortunate that they share the same name. The utopian 'Enlightenment' is sometimes called the Continental (European) Enlightenment to distinguish it from the more fallibilist British Enlightenment, which began a little earlier and took a very different course. (See, for instance, the historian Roy Porter's book *Enlightenment*.) In my terminology, the Continental Enlightenment understood that problems are soluble but not that they are inevitable, while the British Enlightenment understood both equally. Note that this is a classification of ideas, not of nations or even individual thinkers: not all Enlightenment thinkers belong wholly to one branch or the other; nor were all thinkers of the respective Enlightenments born in the eponymous part of the world. The mathematician and philosopher Nicholas de Condorcet, for instance, was French yet belonged more to what I am calling the 'British' Enlightenment, while Karl Popper, the twentieth century's foremost proponent of the British Enlightenment, was born in Austria.

The Continental Enlightenment was impatient for the perfected state – which led to intellectual dogmatism, political violence and new forms of tyranny. The French Revolution of 1789 and the Reign of Terror that followed it are the archetypal examples. The British Enlightenment, which was evolutionary and cognizant of human fallibility, was impatient for institutions that did not stifle gradual, continuing change. It was also enthusiastic for small improvements, unbounded in the future. (See, for instance, the historian Jenny Uglow's book *Lunar Men*.) This is, I believe, the movement that was successful in its pursuit of progress, so in this book when I refer to 'the' Enlightenment I mean the 'British' one.

To investigate the ultimate reach of humans (or of people, or of progress), we should not be considering places like the Earth and the moon, which are unusually rich in resources. Let us go back to that typical place. While the Earth is inundated with matter, energy and evidence, out there in intergalactic space all three are at their lowest possible supply. There is no rich supply of minerals, no vast nuclear reactor overhead delivering free energy, no lights in the sky or diverse local events to provide evidence of the laws of nature. It is empty, cold and dark.

Or is it? Actually, that is yet another parochial misconception. Intergalactic space is indeed very empty by human standards. But each of those solar-system-sized cubes still contains over a billion tonnes of matter – mostly in the form of ionized hydrogen. A billion tonnes is more than enough mass to build, say, a space station and a colony of scientists creating an open-ended stream of knowledge – *if* anyone were present who knew how to do that.

No human today knows how. For instance, one would first have to transmute some of the hydrogen into other elements. Collecting it from such a diffuse source would be far beyond us at present. And, although some types of transmutation are already routine in the nuclear industry, we do not know how to transmute hydrogen into other elements on an industrial scale. Even a simple nuclear-fusion reactor is currently beyond our technology. But physicists are confident that it is not forbidden by any laws of physics, in which case, as always, it can only be a matter of knowing how.

No doubt a billion-tonne space station is not large enough to thrive in the very long run. The inhabitants will want to enlarge it. But that presents no problem of principle. As soon as they started to trawl their cube for hydrogen, more would drift in from the surrounding space, supplying the cube with millions of tonnes of hydrogen per year. (There is also believed to be an even greater mass of 'dark matter' in the cube, but we do not know how to do anything useful with it, so let us ignore it in this thought experiment.)

As for the cold, and the lack of available energy – as I said, the transmutation of hydrogen releases the energy of nuclear fusion. That would be a sizeable power supply, orders of magnitude more than the combined power consumption of everyone on Earth today. So the cube is not as lacking in resources as a parochial first glance would suggest.

How would the space station get its vital supply of evidence? Using the elements created by transmutation, one could construct scientific laboratories, as in the projected moon base. On Earth, when chemistry was in its infancy, making discoveries often depended on travelling all over the planet to find materials to experiment on. But transmutation makes that irrelevant; and chemical laboratories on the space station would be able to synthesize arbitrary compounds of arbitrary elements. The same is true of elementary particle physics: in that field, almost anything will do as a source of evidence, because every atom is potentially a cornucopia of particles just waiting to display themselves if one hits the atom hard enough (using a particle accelerator) and observes with the right instruments. In biology, DNA and all other biochemical molecules could be synthesized and experimented on. And, although biology field trips would be difficult (because the closest natural eco-system would be millions of light years away), arbitrary life forms could be created and studied in artificial ecosystems, or in virtual-reality simulations of them. As for astronomy – the sky there is pitch black to the human eye, but to an observer with a telescope, even one of present-day design, it would be packed with galaxies. A somewhat bigger telescope could see stars in those galaxies in sufficient detail to test most of our present-day theories of astrophysics and cosmology.

Even aside from those billion tonnes of matter, the cube is not empty. It is full of faint light, and the amount of evidence in that light is staggering: enough to construct a map of every star, planet and moon in all the nearest galaxies to a resolution of about ten kilometres. To extract that evidence in full, the telescope would need to use something like a mirror of the same width as the cube itself, which would require at least as much matter as building a planet. But even that would not be beyond the bounds of possibility, given the level of technology we are considering. To gather that much matter, those intergalactic scientists would merely have to trawl out to a distance of a few thousand cube-widths – still a piffling distance by intergalactic stand-ards. But even with a mere million-tonne telescope they could do a lot of astronomy. The fact that planets with tilted axes have annual seasons would be plain to see. They could detect life if it was present on any of the planets, via the composition of its atmosphere. With more subtle measurements they could test theories about the nature and history of life – or intelligence – on the planet. At any instant, a typical cube contains evidence, at that level of detail, about more than a trillion stars and their planets, simultaneously.

And that is only one instant. Additional evidence of all those kinds is pouring into the cube all the time, so astronomers there could track changes in the sky just as we do. And visible light is only one band of the electromagnetic spectrum. The cube is receiving evidence in every other band too – gamma rays, X-rays, all the way down to the micro-wave background radiation and radio waves, as well as a few cosmic-ray particles. In short, nearly all the channels by which we on Earth currently receive evidence about any of the fundamental sciences are available in intergalactic space too.

And they carry much the same content: not only is the universe full of evidence, it is full of the same evidence everywhere. All people in the universe, once they have understood enough to free themselves from parochial obstacles, face essentially the same opportunities. This is an underlying unity in the physical world more significant than all the dissimilarities I have described between our environment and a typical one: the fundamental laws of nature are so uniform, and evidence about them so ubiquitous, and the connections between understanding and control so intimate, that, whether we are on our parochial home planet or a hundred million light years away in the intergalactic plasma, we can do the same science and make the same progress.

So a typical location in the universe is amenable to the open-ended creation of knowledge. And therefore so are almost all other kinds of environment, since they have more matter, more energy and easier access to evidence than intergalactic space. The thought experiment considered almost the worst possible case. Perhaps the laws of physics do not allow knowledge-creation inside, say, the jet of a quasar. Or perhaps they do. But either way, in the universe at large, knowledge-friendliness is the rule, not the exception. That is to say, the rule is person-friendliness *to people who have the relevant knowledge*. Death is the rule for those who do not. These are the same rules that prevailed in the Great Rift Valley from whence we came, and have prevailed ever since.

Oddly enough, that quixotic space station in our thought experiment is none other than the 'generation ship' in the Spaceship Earth metaphor – except that we have removed the unrealistic assumption that the inhabitants never improve it. Hence presumably they have long since solved the problem of how to avoid dying, and so 'generations' are no longer essential to the way their ship works. In any case, with hindsight, a generation ship was a poor choice for dramatizing the claim that the human condition is fragile and dependent on support from an unaltered biosphere, for that claim is contradicted by the very possibility of such a spaceship. If it is possible to live indefinitely in a spaceship in space, then it would be much more possible to use the same technology to live on the surface of the Earth – and to make continuing progress which would make it ever easier. It would make little practical difference whether the biosphere had been ruined or not. Whether or not it could support any other species, it could certainly accommodate people – including humans – if they had the right knowledge.

Now I can turn to the significance of knowledge – and therefore of people – in the cosmic scheme of things.

Many things are more *obviously* significant than people. Space and time are significant because they appear in almost all explanations of other physical phenomena. Similarly, electrons and atoms are significant. Humans seem to have no place in that exalted company. Our history and politics, our science, art and philosophy, our aspirations and moral values – all these are tiny side effects of a supernova explosion a few billion years ago, which could be extinguished tomorrow by another such explosion. Supernovae, too, are moderately significant in the cosmic scheme of things. But it seems that one can explain everything about supernovae, and almost everything else, without ever mentioning people or knowledge at all.

However, that is merely another parochial error, due to our current, untypical, vantage point in an Enlightenment that is mere centuries old. In the longer run, humans may colonize other solar systems and, by increasing their knowledge, control ever more powerful physical processes. If people ever choose to live near a star that is capable of exploding, they may well wish to prevent such an explosion – probably by removing some of the material from the star. Such a project would use many orders of magnitude more energy than humans currently control, and more advanced technology as well. But it is a fundamentally simple task, not requiring any steps that are even close to limits imposed by the laws of physics. So, with the right knowledge, it could be achieved. Indeed, for all we know, engineers elsewhere in the universe are already achieving it routinely. And consequently it is not true that the attributes of supernovae in general are independent of the presence or absence of people, or of what those people know and intend.

More generally, if we want to predict what a star will do, we first have to guess whether there are any people near it, and, if so, what knowledge they may have and what they may want to achieve. Outside our parochial perspective, astrophysics is incomplete without a theory of people, just as it is incomplete without a theory of gravity or nuclear reactions. Note that this conclusion does not depend on the assumption that humans, or anyone, *will* colonize the galaxy and take control of any supernovae: the assumption that they will not is equally a theory about the future behaviour of knowledge. Knowledge is a significant phenomenon in the universe, because to make almost any prediction about astrophysics one must *take a position* about what types of knowledge will or will not be present near the phenomena in question. So all explanations of what is out there in the physical world mention knowledge and people, if only implicitly.

But knowledge is more significant even than that. Consider any physical object – for instance, a solar system, or a microscopic chip of silicon – and then consider all the transformations that it is physically possible for it to undergo. For instance, the silicon chip might be melted and solidify in a different shape, or be transformed into a chip with different functionality. The solar system might be devastated when its star becomes a supernova, or life might evolve on one of its planets, or it might be transformed, using transmutation and other futuristic technologies, into microprocessors. In all cases, the class of transformations that could happen spontaneously – in the absence of knowledge – is negligibly small compared with the class that could be effected artificially by intelligent beings who wanted those transformations to happen. So the explanations of *almost all physically possible phenomena* are about how knowledge would be applied to bring these phenomena about. If you want to explain how an object might possibly reach a temperature of ten degrees or a million, you can refer to spontaneous processes and can avoid mentioning people explicitly (even though *most* processes at those temperatures can be brought about only by people). But if you want to explain how an object might possibly cool down to a millionth of a degree above absolute zero, you cannot avoid explaining in detail what people would do.

And that is still only the least of it. In your mind's eye, continue your journey from that point in intergalactic space to another, at least ten times as far away. Our destination this time is inside one of the jets of a quasar. What would it be like in one of those jets? Language is barely capable of expressing it: it would be rather like facing a supernova explosion at point-blank range, but for millions of years at a time. The survival time for a human body would be measured in picoseconds. As I said, it is unclear whether the laws of physics permit any knowledge to grow there, let alone a life-support system for humans. It is about as different from our ancestral environment as it could possibly be. The laws of physics that explain it bear no resemblance to any rules of thumb that were ever in our ancestors' genes or in their culture. Yet human brains today know in considerable detail what is happening there.

Somehow that jet happens in such a way that billions of years later, on the other side of the universe, a chemical scum can know and predict what the jet will do, and can understand why. That means that one physical system – say, an astrophysicist's brain – contains an accurate working model of the other, the jet. Not just a superficial image (though it contains that as well), but an explanatory theory that embodies the same mathematical relationships and causal structure. That is scientific knowledge. Furthermore, the faithfulness with which the one structure resembles the other is steadily increasing. That constitutes the creation of knowledge. Here we have physical objects very unlike each other, and whose behaviour is dominated by different laws of physics, embodying the same mathematical and causal structures – and doing so ever more accurately over time. Of all the physical processes that can occur in nature, only the creation of knowledge exhibits that underlying unity.

In Arecibo, Puerto Rico, there is a giant radio telescope, one of whose many uses is in the Search For Extraterrestrial Intelligence (SETI). In an office in a building near the telescope there is a small domestic refrigerator. Inside that refrigerator is a bottle of champagne, sealed by a cork. Consider that cork.

It is going to be removed from the bottle if and when SETI succeeds in its mission to detect radio signals transmitted by an extraterrestrial intelligence. Hence, if you were to keep a careful watch on the cork, and one day saw it popping from the bottle, you could infer that an extraterrestrial intelligence exists. The configuration of the cork is what experimentalists call a 'proxy': a physical variable which can be measured as a way of measuring another variable. (All scientific measurements involve chains of proxies.) Thus we can also regard the entire Arecibo observatory, including its staff and that bottle and its cork, as a scientific instrument to detect distant people.

The behaviour of that humble cork is therefore extraordinarily difficult to explain or predict. To predict it, you have to know whether there really are people sending radio signals from various solar systems. To explain it, you have to explain how you know about those people and their attributes. Nothing less than that specific knowledge, which depends among other things on subtle properties of the chemistry on the planets of distant stars, can explain or predict with any accuracy whether, and when, that cork will pop.

The SETI instrument is also remarkably finely tuned to its purpose. Completely insensitive to the presence of several tonnes of people a few metres away, and even to the tens of millions of tonnes of people on the same planet, it detects only people on planets orbiting other stars, and only if they are radio engineers. No other type of phenomenon on Earth, or in the universe, is sensitive to what people are doing at locations hundreds of light years away, let alone with that enormous degree of discrimination.

This is made possible in part by the corresponding fact that few types of matter are as prominent, at those distances, as that type of scum. Specifically, the only phenomena that our best current instruments can detect at stellar distances are (1) extraordinarily luminous ones such as stars (or, to be precise, only their surfaces); (2) a few objects that obscure our view of those luminous objects; and (3) the effects of certain types of knowledge. We can detect devices such as lasers and radio transmitters that have been designed for the purpose of communication; and we can detect components of planetary atmospheres that could not be present in the absence of life. Thus those types of knowledge are among the most prominent phenomena in the universe.

Note also that the SETI instrument is exquisitely adapted to detecting something that has never yet been detected. Biological evolution could never produce such an adaptation. Only scientific knowledge can. This illustrates why non-explanatory knowledge cannot be universal. Like all science, the SETI project can conjecture the existence of something, calculate what some of its observable attributes would be, and then construct an instrument to detect it. Non-explanatory systems cannot cross the conceptual gap that an explanatory conjecture crosses, to engage with unexperienced evidence or non-existent phenomena. Nor is that true only of fundamental science: *if* such-and-such a load were put on the proposed bridge it would collapse, says the engineer, and such statements can be true and immensely valuable even if the bridge is never even built, let alone subjected to such a load.

Similar champagne bottles are stored in other laboratories. The popping of each such cork signals a discovery about something significant in the cosmic scheme of things. Thus the study of the behaviour of champagne corks and other proxies for what people do is logically equivalent to the study of *everything* significant. It follows that humans, people and knowledge are not only objectively significant: they are by far the most significant phenomena in nature – the only ones whose behaviour cannot be understood without understanding everything of fundamental importance.

Finally, consider the enormous difference between how an environment will behave spontaneously (that is to say, in the absence of knowledge) and how it behaves once a tiny sliver of knowledge, of just the right kind, has reached it. We would normally regard a lunar colony, even after it has become self-sufficient, as having originated on Earth. But what, exactly, will have originated on Earth? In the long run, all its atoms have originated on the moon (or the asteroids). All the energy that it uses has originated in the sun. Only some proportion of its *knowledge* came from Earth, and, in the hypothetical case of a perfectly isolated colony, that would be a rapidly dwindling proportion. What has happened, physically, is that the *moon* has been changed – initially only minimally – by matter that came from the Earth. And what made the difference was not the matter, but the knowledge that it encoded. In response to that knowledge, the substance of the moon reorganized itself in a new, increasingly extensive and complex way, and started to create an indefinitely long stream of ever-improving explanations. A beginning of infinity.

Similarly, in the intergalactic thought experiment, we imagined 'priming' a typical cube, and as a result intergalactic space itself began to produce a stream of ever-improving explanations. Notice how different, physically, the transformed cube is from a typical one. A typical cube has about the same mass as any of the millions of nearby cubes, and that mass barely changes over many millions of years. The transformed cube is more massive than its neighbours, and its mass is increasing continuously as the inhabitants systematically capture matter and use it to embody knowledge. The mass of a typical cube is spread thinly throughout its whole volume; most of the mass of the transformed cube is concentrated at its centre. A typical cube contains mostly hydrogen; the transformed cube contains every element. A typical cube is not producing any energy; the transformed cube is converting mass to energy at a substantial rate. A typical cube is full of evidence, but most of it is just passing through, and none of it ever causes any changes. The transformed cube contains even more evidence, most of it having been created locally, and is detecting it with ever-improving instruments and changing rapidly as a result. A typical cube is not emitting any energy; the transformed cube may well be broadcasting explanations into space. But perhaps the biggest physical difference is that, like all knowledge-creating systems, the transformed cube corrects errors. You would notice this if you tried to modify or harvest the matter in it: it would resist!

It appears, nevertheless, that most environments are not yet creating any knowledge. We know of none that is, except on or near the Earth, and what we see happening elsewhere is radically different from what would happen if knowledge-creation were to become widespread. But the universe is still young. An environment that is not currently creating anything may do so in the future. What will be typical in the distant future could be very different from what is typical now.

Like an explosive awaiting a spark, unimaginably numerous environments in the universe are waiting out there, for aeons on end, doing nothing at all or blindly generating evidence and storing it up or pouring it out into space. Almost any of them would, if the right knowledge ever reached it, instantly and irrevocably burst into a radically different type of physical activity: intense knowledge-creation, displaying all the various kinds of complexity, universality and reach that are inherent in the laws of nature, and transforming that environment from what is typical today into what could become typical in the future. If we want to, we could be that spark.

TERMINOLOGY

Person   An entity that can create explanatory knowledge.
Anthropocentric   Centred on humans, or on persons.
Fundamental or significant phenomenon: One that plays a necessary

role in the explanation of many phenomena, or whose distinctive features require distinctive explanation in terms of fundamental theories.

Principle of Mediocrity   'There is nothing significant about humans.'

Parochialism   Mistaking appearance for reality, or local regularities for universal laws.

Spaceship Earth   'The biosphere is a life-support system for humans.'

Constructor   A device capable of causing other objects to undergo transformations without undergoing any net change itself.

Universal constructor   A constructor that can cause any raw materials to undergo any physically possible transformation, given the right information.

MEANINGS OF 'THE BEGINNING OF INFINITY' ENCOUNTERED IN THIS CHAPTER

- The fact that everything that is not forbidden by laws of nature is achievable, given the right knowledge. 'Problems are soluble.'
- The 'perspiration' phase can always be automated.
- The knowledge-friendliness of the physical world.
- People are universal constructors.
- The beginning of the open-ended creation of explanations.
- The environments that could create an open-ended stream of knowledge, if suitably primed – i.e. almost all environments.
- The fact that new explanations create new problems.

SUMMARY

Both the Principle of Mediocrity and the Spaceship Earth idea are, contrary to their motivations, irreparably parochial and mistaken. From the least parochial perspectives available to us, *people* are the most significant entities in the cosmic scheme of things. They are not 'supported' by their environments, but support themselves by creating knowledge. Once they have suitable knowledge (essentially, the knowledge of the Enlightenment), they are capable of sparking unlimited further progress.

Apart from the thoughts of people, the only process known to be capable of creating knowledge is biological evolution. The knowledge it creates (other than via people) is inherently bounded and parochial. Yet it also has close similarities with human knowledge. The similarities and the differences are the subject of the next chapter.

4

Creation

The knowledge in human brains and the knowledge in biological adaptations are both created by *evolution* in the broad sense: the variation of existing information, alternating with selection. In the case of human knowledge, the variation is by conjecture, and the selection is by criticism and experiment. In the biosphere, the variation consists of mutations (random changes) in genes, and natural selection favours the variants that most improve the ability of their organisms to reproduce, thus causing those variant genes to spread through the population.

That a gene is *adapted* to a given function means that few, if any, small changes would improve its ability to perform that function. Some changes might make no practical difference to that ability, but most of those that did would make it worse. In other words good adaptations, like good explanations, are distinguished by being hard to vary while still fulfilling their functions.

Human brains and DNA molecules each have many functions, but among other things they are general-purpose information-storage media: they are in principle capable of storing any kind of information. Moreover, the two types of information that they respectively evolved to store have a property of cosmic significance in common: *once they are physically embodied in a suitable environment, they tend to cause themselves to remain so.* Such information – which I call *knowledge* – is very unlikely to come into existence other than through the error-correcting processes of evolution or thought.

There are also important differences between those two kinds of knowledge. One is that biological knowledge is non-explanatory, and therefore has limited reach; explanatory human knowledge can have broad or even unlimited reach. Another difference is that mutations are random, while conjectures can be constructed intentionally for a purpose. Nevertheless, the two kinds of knowledge share enough of their underlying logic for the theory of evolution to be highly relevant to human knowledge. In particular, some historic misconceptions about biological evolution have counterparts in misconceptions about human knowledge. So in this chapter I shall describe some of those misconceptions in addition to the actual explanation of biological adaptations, namely modern Darwinian evolutionary theory, sometimes known as 'neo-Darwinism'.

Creationism

Creationism is the idea that some supernatural being or beings designed and created all biological adaptations. In other words, 'the gods did it.' As I explained in Chapter 1, theories of that form are bad explanations. Unless supplemented by hard-to-vary specifics, they do not even address the problem – just as 'the laws of physics did it' will never win you a Nobel prize, and 'the conjurer did it' does not solve the mystery of the conjuring trick.

Before a conjuring trick is ever performed, its explanation must be known to the person who invented it. The origin of that knowledge is the origin of the trick. Similarly, the problem of explaining the biosphere is that of explaining how the knowledge embodied in its adaptations could possibly have been created. In particular, a putative designer of any organism must also have created the knowledge of how that organism works. Creationism thus faces an inherent dilemma: is the designer a purely supernatural being – one who was 'just there', complete with all that knowledge – or not? A being who was 'just there' would serve no explanatory purpose (in regard to the biosphere), since then one could more economically say that the biosphere itself 'just happened', complete with that same knowledge, embodied in organisms. On the other hand, to whatever extent a creationist theory provides explanations about how supernatural beings designed and created the biosphere, they are no longer supernatural beings but merely unseen ones. They might, for instance, be an extraterrestrial civilization. But then the theory is not really creationism – unless it proposes that the extraterrestrial designers themselves had supernatural designers.

Moreover, the designer of any adaptation must by definition have had the *intention* that the adaptation be as it is. But that is hard to reconcile with the designer envisaged in virtually all creationist theories, namely a deity or deities worthy of worship; for the reality is that many biological adaptations have distinctly suboptimal features. For instance, the eyes of vertebrates have their 'wiring' and blood supply *in front* of the retina, where they absorb and scatter incoming light and so degrade the image. There is also a blind spot where the optic nerve passes through the retina on its way to the brain. The eyes of some invertebrates, such as squids, have the same basic design but without those design flaws. The effect of the flaws on the efficiency of the eye is small; but the point is that they are wholly contrary to the eye's functional purpose, and so conflict with the idea that that purpose was intended by a divine designer. As Charles Darwin put it in *The Origin of Species*, 'On the view of each organism with all its separate parts having been specially created, how utterly inexplicable is it that organs bearing the plain stamp of inutility . . . should so frequently occur.'

There are even examples of *non*-functional design. For instance, most animals have a gene for synthesizing vitamin C, but in primates, including humans, though that gene is recognizably present, it is faulty: it does not do anything. This is very difficult to account for except as a vestigial feature that primates have inherited from non-primate ancestors. One could retreat to the position that all these apparently poor design features do have some undiscovered purpose. But that is a bad explanation: it could be used to claim that *any* poorly designed or undesigned entity was perfectly designed.

Another assumed characteristic of the designer according to most religions is benevolence. But, as I mentioned in Chapter 3, the biosphere is much less pleasant for its inhabitants than anything that a benevolent, or even halfway decent, human designer would design. In theological contexts this is known as 'the problem of suffering' or 'the problem of evil', and is frequently used as an argument against the existence of God. But in that role it is easily brushed off. Typical defences are that perhaps morality is different for a supernatural being; or perhaps we are too limited intellectually to be able to understand how moral the biosphere really is. However, here I am concerned not with whether

God exists, only with how to explain biological adaptations, and in that regard those defences of creationism have the same fatal flaw as the Haldane–Dawkins argument (Chapter 3): a world that is 'queerer than we *can* suppose' is indistinguishable from a world 'tricked out with magic'. So all such explanations are bad.

The central flaw of creationism – that its account of how the knowledge in adaptations could possibly be created is either missing, supernatural or illogical – is also the central flaw of pre-Enlightenment, authoritative conceptions of *human* knowledge. In some versions it is literally the same theory, with certain types of knowledge (such as cosmology or moral knowledge and other rules of behaviour) being spoken to early humans by supernatural beings. In others, parochial features of society (such as the existence of monarchs in government, or indeed the existence of God in the universe) are protected by taboos or taken so uncritically for granted that they are not even recognized as ideas. And I shall discuss the *evolution* of such ideas and institutions in Chapter 15.

The prospect of the unlimited creation of knowledge in the future conflicts with creationism by undercutting its motivation. For eventually, with the assistance of what we would consider stupendously powerful computers, any child will be capable of designing and implementing a better, more complex, more beautiful, and also far more moral biosphere than the Earth's, within a video game – perhaps by placing it in such a state by fiat, or perhaps by inventing fictional laws of physics that are more conducive to enlightenments than the actual laws. At that point, a supposed designer of *our* biosphere will seem not only morally deficient, but intellectually unremarkable. And the latter attribute is not so easy to brush aside. Religions will no longer want to claim the design of the biosphere as one of the achievements of their deities, just as today they no longer bother to claim thunder.

Spontaneous generation

Spontaneous generation is the formation of organisms not as offspring of other organisms, but entirely from non-living precursors – for example, the generation of mice from a pile of rags in a dark corner. The theory that small animals are being spontaneously generated like that all the time (in addition to reproducing in the normal way) was part of unquestioned conventional wisdom for millennia, and was taken seriously until well into the nineteenth century. Its defenders gradually retreated to ever smaller animals as knowledge of zoology grew, until eventually the debate was confined to what are now called micro-organisms – things like fungi and bacteria that grow on nutrient media. For those, it proved remarkably difficult to refute spontaneous generation experimentally. For instance, experiments could not be done in airtight containers in case air was necessary for spontaneous generation. But it was finally refuted by some ingenious experiments conducted by the biologist Louis Pasteur in 1859 – the same year in which Darwin published his theory of evolution.

But experiment should never have been needed to convince scientists that spontaneous generation is a bad theory. A conjuring trick cannot have been performed by real magic – by the magician simply commanding events to happen – but must have been brought about by knowledge that was somehow created beforehand. Similarly, biologists need only have asked: how does the knowledge to construct a mouse get to those rags, and how is it then applied to transform the rags into a mouse?

One attempted explanation of spontaneous generation, which was advocated by the theologian St Augustine of Hippo (354–430), was that all life comes from 'seeds', some of which are carried by living organisms and others of which are distributed all over the Earth. Both kinds of seed were created during the original creation of the world. Both could, under the right conditions, develop into new individuals of the appropriate species. Augustine ingeniously suggested that this might explain why Noah's Ark did not have to carry impossibly large numbers of animals: most species could regenerate after the Flood without Noah's help. However, under that theory organisms are *not* being formed purely from non-living raw materials. That distributed kind of seed would be a life form, just as a real seed is: it would contain all the knowledge in its organism's adaptations. So Augustine's theory – as he himself stressed – is really just a form of creationism, not spontaneous generation. Some religions regard the universe as an ongoing act of supernatural creation. In such a world, all spontaneous generation would fall under the heading of creationism.

But, if we insist on good explanations, we must rule out creationism, as I have explained. So, in regard to spontaneous generation, that leaves only the possibility that the laws of physics might simply mandate it. For instance, mice might simply *form* under suitable circumstances, like crystals, rainbows, tornadoes and quasars do.

That seems absurd today, because the actual molecular mechanisms of life are now known. But is there anything wrong with that theory itself, as an explanation? Phenomena such as rainbows have a distinctive appearance that is endlessly repeated without any information having been transmitted from one instance to the next. Crystals even behave in ways that are reminiscent of living things: when placed in a suitable solution, a crystal attracts more molecules of the right kind and arranges them in such a way as to make more of the same crystal. Since crystals and mice both obey the same laws of physics, why is spontaneous generation a good explanation of the former and not of the latter? The answer, ironically, comes from an argument that was originally intended to justify creationism:

The argument from design

The 'argument from design' has been used for millennia as one of the classic 'proofs' of the existence of God, as follows. Some aspects of the world appear to have been designed, but they were not designed by humans; since 'design requires a designer', there must therefore be a God. As I said, that is a bad explanation because it does not address how the knowledge of how to create such designs could possibly have been created. ('Who designed the designer?', and so on.) But the argument from design can be used in valid ways too, and indeed its earliest known use, by the ancient Athenian philosopher Socrates, was valid. This issue was: *given* that the gods have created the world, do they care what happens in it? Socrates' pupil Aristodemus had argued that they do not. Another pupil, the historian Xenophon, recalled Socrates' reply:

SOCRATES: Because our eyes are delicate, they have been shuttered with eyelids that open when we have occasion to use them ... And our foreheads have been fringed with eyebrows to prevent damage from the sweat of the head ... And the mouth set close to the eyes and nostrils as a portal of ingress for all our supplies, whereas, since matter passing out of the body is unpleasant, the outlets are directed hindwards, as far away from the senses as possible. I ask you, when you see all these things constructed with such show of foresight, can you doubt whether they are products of chance or design?

ARISTODEMUS: Certainly not! Viewed in this light they seem very much like the contrivances of some wise craftsman, full of love for all things living.

SOCRATES: And what of the implanting of the instinct to procreate; and in the mother, the instinct to rear her young; and in the young, the intense desire to live and the fear of death?

ARISTODEMUS: These provisions too seem like the contrivances of someone who has determined that there shall be living creatures.

Socrates was right to point out that the *appearance of design* in living things is something that needs to be explained. It cannot be the 'product of chance'. And that is specifically because it signals the presence of knowledge. How was that knowledge created?

However, Socrates never stated what constitutes an appearance of design, and why. Do crystals and rainbows have it? Does the sun, or summer? How are they different from biological adaptations such as eyebrows?

The issue of what exactly needs to be explained in an 'appearance of design' was first addressed by the clergyman William Paley, the finest exponent of the argument from design. In 1802, before Darwin was born, he published the following thought experiment in his book *Natural Theology*. He imagined walking across a heath and finding a stone, or alternatively a watch. In either case, he imagined wondering how the object came to exist. And he explained why the watch would require a wholly different kind of explanation from that of the stone. For all he knew, he said, the stone might have lain there for ever. Today we know more about the history of the Earth, so we should refer instead to supernovae, transmutation and the Earth's cooling crust. But that would make no difference to Paley's argument. His point was: that sort of account can explain how the stone came to exist, or the raw materials for the watch, but it could never explain the watch itself.

A watch *could not* have been lying there for ever, nor could it have formed during the solidification of the Earth. Unlike the stone, or a rainbow or a crystal, it could not have assembled itself by spontaneous generation from its raw materials, nor could it *be* a raw material. But why not, exactly, asked Paley: 'Why should not this answer serve for the watch as well as for the stone; why is it not as admissible in the second case as in the first?' And he knew why. Because the watch not only *serves* a purpose, it is *adapted* to that purpose:

> For this reason, and for no other, viz., that, when we come to inspect the watch, we perceive (what we could not discover in the stone) that its several parts are framed and put together for a purpose, e.g., that they are so formed and adjusted as to produce motion, and that motion so regulated as to point out the hour of the day.

One cannot explain why the watch is as it is without referring to its purpose of keeping accurate time. Like the telescopes that I discussed in Chapter 2, it is a rare configuration of matter. It is not a coincidence that it can keep time accurately, nor that its components are well suited to that task, nor that they were put together in that way rather than another. Hence *people* must have designed that watch. Paley was of course implying that all this is even more true of a living organism – say, a mouse. *Its* 'several parts' are all constructed (and appear to be designed) for a purpose. For instance, the lenses in its eyes have a purpose similar to that of a telescope, of focusing light to form an image on its retina, which in turn has the purpose of recognizing food, danger and so on.

Actually, Paley did not know the overall purpose of the mouse (though we do now – see 'Neo-Darwinism' below). But even a single eye would suffice to make Paley's triumphant point – namely that the evidence of apparent design for a purpose is not only that the parts all serve that purpose, but that if they were slightly altered they would serve it less well, or not at all. A good design is *hard to vary*:

> If the different parts had been differently shaped from what they are, of a different size from what they are, or placed after any other manner, or in any other order, than that in which they are placed, either no motion at all would have been carried on in the machine, or none which would have answered the use that is now served by it.

Merely being useful for a purpose, without being hard to vary while still serving that purpose, is not a sign of adaptation or design. For instance, one can also use the sun to keep time, but all its features would serve that purpose equally well if slightly (or even massively) altered. Just as we transform many of the Earth's non-adapted raw materials to meet our purposes, so we also find uses for the sun that it was never designed or adapted to provide. The knowledge, in that case, is entirely in us – and in our sundials – not in the sun. But it *is* embodied in the watch, and in the mouse.

So, how did all that knowledge come to be embodied in those things? As I said, Paley could conceive of only one explanation. That was his first mistake:

> The inference we think is inevitable, that the watch must have had a maker . . . There cannot be design without a designer; contrivance without a contriver; order without choice; arrangement without anything capable of arranging; subserviency and relation to a purpose without that which could intend a purpose; means suitable to an end . . . without the end ever having been contemplated or the means accommodated to it. Arrangement, disposition of parts, subserviency of means to an end, relation of instruments to a use imply the presence of intelligence and mind.

We now know that there *can* be 'design without a designer': knowledge without a person who created it. Some types of knowledge can be created by evolution. I shall come to that shortly. But it is no criticism of Paley that he was unaware of a discovery that had yet to be made – one of the greatest discoveries in the history of science.

However, although Paley was spot on in his understanding of the *problem*, he somehow did not realize that his proposed solution, creationism, does not solve it, and is even ruled out by his own argument. For the ultimate designer for whose existence Paley was arguing would also be a purposeful and complex entity – certainly no less so than a watch or a living organism. Hence, as many critics have since noticed, if we substitute 'ultimate designer' for 'watch' in Paley's text above, we force Paley to 'the [inevitable] inference . . . that the ultimate designer must have had a maker'. Since that is a contradiction, the argument from design as perfected by Paley rules out the existence of an ultimate designer.

Note that this is not a disproof of the existence of God, any more than the original argument was a proof. But it does show that, in any good explanation of the origin of biological adaptations, God cannot play the role assigned by creationism. Though this is the opposite of what Paley believed he had achieved, none of us can choose what our ideas imply. His argument has universal reach for anything that has, by his criterion, the appearance of design. As an elucidation of the special status of living things, and in setting a benchmark that explanations of knowledge-laden entities must meet if they are to make sense, it is essential to understanding the world.

Lamarckism

Before Darwin's theory of evolution, people had already been wondering whether the biosphere and its adaptations might have come into existence gradually. Darwin's grandfather Erasmus Darwin (1731–1802), a stalwart of the Enlightenment, was among them. They called that process 'evolution', but the meaning of the word then was different from its primary one today. *All* processes of gradual improvement, regardless of their mechanism, were known as 'evolution'. (That terminology survives to this day in casual usage and as a technical term in, of all places, theoretical physics, where 'evolution' means any sort of continuous change that one is explaining through laws of physics.) Charles Darwin distinguished the process that he discovered by calling it 'evolution by natural selection' – though a better name would have been 'evolution by variation and selection'.

As Paley might well have recognized if he had lived to hear of it, 'evolution by natural selection' is a much more substantive mode of explanation than mere 'evolution'. For the latter does not solve his problem, while the former does. *Any* theory about improvement raises the question: how is the knowledge of how to make that improvement created? Was it already present at the outset? The theory that it was is creationism. Did it 'just happen'? The theory that it did is spontaneous generation.

During the early years of the nineteenth century, the naturalist Jean-Baptiste Lamarck proposed an answer that is now known as *Lamarckism*. Its key idea is that improvements acquired by an organism during its lifetime can be inherited by its offspring. Lamarck was thinking mainly of improvements in the organism's organs, limbs and so on – such as, for instance, the enlargement and strengthening of muscles that an individual uses heavily, and the weakening of those that it seldom uses. This 'use-and-disuse' explanation had also been arrived at independently by Erasmus Darwin. A classic Lamarckian explanation is that giraffes, when eating leaves from trees whose lower-lying leaves were already eaten, stretched their necks to get at the higher ones. This supposedly lengthened their necks slightly, and then their offspring inherited the trait of having slightly longer necks. Thus, over many generations, long-necked giraffes evolved from ancestors with unremarkable necks. In addition, Lamarck proposed that improvements were driven by a tendency, built into the laws of nature, towards ever greater complexity.

The latter is a fudge, for not just any complexity could account for the evolution of adaptations: it has to be *knowledge*. And so that part of the theory is just invoking spontaneous generation – unexplained knowledge. Lamarck might not have minded that, because, like many thinkers of his day, he took the existence of spontaneous generation for granted. He even incorporated it explicitly into his theory of evolution: he guessed that, as successive generations of organisms are forced by his law of nature to take ever more complex forms, we still see simple creatures because a continuous supply of them is formed spontaneously.

Some have considered this a pretty vision. But it bears hardly any resemblance to the facts. Its most glaring mismatch is that, in reality, evolutionary adaptations are of a wholly different character from the changes that take place in an individual during its lifetime. The former involve the creation of new knowledge; the latter happen only when there is already an adaptation for making that change. For instance, the tendency of muscles to become stronger or weaker with use and disuse is controlled by a sophisticated (knowledge-laden) set of genes. The animal's distant ancestors did not have those genes. Lamarckism cannot possibly explain how the knowledge in them was created.

If you were starved of vitamin C, your defective vitamin-C-synthesis gene would not thereby be caused to improve – unless, perhaps, you are a genetic engineer. If a tiger is placed in a habitat in which its colouration makes it stand out more instead of less, it takes no action to change the colour of its fur, nor would that change be inherited if it did. That is because nothing in the tiger 'knows' what the stripes are for. So how would any Lamarckian mechanism have 'known' that having fur that was a tiny bit more striped would slightly improve the animal's food supply? And how would it have 'known' how to synthesize pigments, and to secrete them into the fur, in such a way as to produce stripes of a suitable design?

The fundamental error being made by Lamarck has the same logic as inductivism. Both assume that new knowledge (adaptations and scientific theories respectively) is somehow already present in experience, or can be derived mechanically from experience. But the truth is always that knowledge must be *first* conjectured and *then* tested. That is what Darwin's theory says: first, random mutations happen (they do not take account of what problem is being solved); then natural selection discards the variant genes that are less good at causing themselves to be present again in future generations.

Neo-Darwinism

The central idea of neo-Darwinism is that evolution favours the genes that spread best through the population. There is much more to this idea than meets the eye, as I shall explain.

A common misconception about Darwinian evolution is that it maximizes 'the good of the species'. That provides a plausible, but false, explanation of apparently altruistic behaviour in nature, such as parents risking their lives to protect their young, or the strongest animals going to the perimeter of a herd under attack – thereby decreasing their own chances of having a long and pleasant life or further offspring. Thus, it is said, evolution optimizes the good of the species, not the individual. But, in reality, evolution optimizes neither.

To see why, consider this thought experiment. Imagine an island on which the total number of birds of a particular species would be maximized if they nested at, say, the beginning of April. The explanation for why a particular date is optimal will refer to various trade-offs involving factors such as temperature, the prevalence of predators, the availability of food and nesting materials, and so on. Suppose that initially the whole population has genes that cause them to nest at that optimum time. That would mean that those genes were well adapted to maximizing the number of birds in the population – which one might call 'maximizing the good of the species'.

Now suppose that this equilibrium is disturbed by the advent of a mutant gene in a single bird which causes it to nest slightly earlier – say, at the end of March. Assume that when a bird has built a nest, the species' other behavioural genes are such that it automatically gets whatever cooperation it needs from a mate. That pair of birds would then be guaranteed the best nesting site on the island – an advantage which, in terms of the survival of their offspring, might well outweigh all the slight disadvantages of nesting earlier. In that case, in the following generation, there will be more March-nesting birds, and, again, all of them will find excellent nesting sites. That means that a smaller proportion than usual of the April-nesting variety will find good sites: the best sites will have been taken by the time they start looking. In subsequent generations, the balance of the population will keep shifting towards the March-nesting variants. If the relative advantage of having the best nesting sites is large enough, the April-nesting variant could even become extinct. If it arises again as a mutation, its holder will have no offspring, because all sites will have been taken by the time it tries to nest.

Thus the original situation that we imagined – with genes that were optimally adapted to maximizing the population ('benefiting the species') – is unstable. There will be evolutionary pressure to make the genes become *less* well adapted to that function.

This change has harmed the species, in the sense of reducing its total population (because the birds are no longer nesting at the optimum time). It may thereby also have harmed it by increasing the risk of extinction, making it less likely to spread to other habitats, and so on. So an optimally adapted species may in this way evolve into one that is less 'well off' by any measure.

If a further mutant gene then appears, causing nesting still earlier in March, the same process may be repeated, with the earlier-nesting genes taking over and the total population falling again. Evolution will thus drive the nesting time ever earlier, and the population lower. A new equilibrium would be reached only when the advantage to an individual bird's offspring of getting the very best nesting site was finally outweighed by the *dis*advantages of slightly earlier nesting. That equilibrium might be very far from what was optimal for the species.

A related misconception is that evolution is always *adaptive* – that it always constitutes progress, or at least some sort of improvement in useful functionality which it then acts to optimize. This is often summed up in a phrase due to the philosopher Herbert Spencer, and unfortunately taken up by Darwin himself: 'the survival of the fittest'. But, as the above thought experiment illustrates, that is not the case either. Not only has the species been harmed by this evolutionary change, every individual bird has been harmed as well: the birds using any particular site now have a harsher life than before, because they are using it earlier in the year.

Thus, although the existence of progress in the biosphere is what the theory of evolution is there to explain, not all evolution constitutes progress, and no (genetic) evolution optimizes progress.

What exactly *has* the evolution of those birds achieved during that period? It has optimized not the functional adaptation of a variant gene to its environment – the attribute that would have impressed Paley – but the relative ability of the surviving variant to *spread through the population*. An April-nesting gene is no longer able to propagate itself to the next generation, even though it is functionally the best variant. The early-nesting gene that replaced it may still be tolerably functional, but it is *fittest* for nothing except preventing variants of itself from procreating. From the point of view of both the species and all its members, the change brought about by this period of its evolution has been a disaster. But evolution does not 'care' about that. It favours only the genes that spread best through the population.

Evolution can even favour genes that are not just suboptimal, but wholly harmful to the species and all its individuals. A famous example is the peacock's large, colourful tail, which is believed to diminish the bird's viability by making it harder to evade predators, and to have no useful function at all. Genes for prominent tails dominate simply because peahens tend to choose prominent-tailed males as mates. Why was there selection pressure in favour of such preferences? One reason is that, when females mated with prominent-tailed males, their male offspring, having more prominent tails, found more mates. Another may be that an individual able to grow a large, colourful tail is more likely to be healthy. In any case, the net effect of all the selection pressures was to spread genes for large, colourful tails, and genes for preferring such tails, through the population. The species and the individuals just had to suffer the consequences.

If the best-spreading genes impose sufficiently large disadvantages on the species, the species becomes extinct. Nothing in biological evolution prevents that. It has presumably happened many times in the history of life on Earth, to species less lucky than the peacock. Dawkins named his tour-de-force account of neo-Darwinism *The Selfish Gene* because he wanted to stress that evolution does not especially promote the 'welfare' of species or individual organisms. But, as he also explained, it does not promote the 'welfare' of genes either: it adapts them not for survival in larger numbers, nor indeed for survival at all, but only for spreading through the population at the expense of rival genes, particularly slight variants of themselves.

Is it sheer luck, then, that most genes do usually confer some, albeit less than optimal, functional benefits on their species, and on their individual holders? No. Organisms are the slaves, or tools, that genes use to achieve their 'purpose' of spreading themselves through the population. (That is the 'purpose' that Paley and even Darwin never guessed.) Genes gain advantages over each other in part by keeping their slaves alive and healthy, just as human slave owners did. Slave owners were not working for the benefit of their workforces, nor for the benefit of individual slaves: it was solely to achieve their own objectives that they fed and housed their slaves, and indeed forced them to reproduce. Genes do much the same thing.

In addition, there is the phenomenon of reach: when the knowledge in a gene happens to have reach, it will help the individual to help itself in a wider range of circumstances, and by more, than the spreading of the gene strictly requires. That is why mules stay alive even though they are sterile. So it is not surprising that genes usually confer *some* benefits on their species and its members, and do often succeed in increasing their own absolute numbers. Nor should it be surprising that they sometimes do the opposite. But what genes are adapted to – what they do better than almost any variant of themselves – has nothing to do with the species or the individuals or even their own survival in the long run. It is getting themselves replicated more than rival genes.

Neo-Darwinism and knowledge

Neo-Darwinism does not refer, at its fundamental level, to anything biological. It is based on the idea of a *replicator* (anything that contributes causally to its own copying).* For instance, a gene conferring the ability to digest a certain type of food *causes* the organism to remain healthy in some situations where it would otherwise weaken or die. Hence it increases the organism's chances of having offspring in the future, and those offspring would inherit, and spread, *copies* of the gene.

Ideas can be replicators too. For example, a good joke is a replicator: when lodged in a person's mind, it has a tendency to cause that person to tell it to other people, thus copying it into *their* minds. Dawkins coined the term *memes* (rhymes with 'dreams') for ideas that are replicators. Most ideas are not replicators: they do not cause us to convey them to other people. Nearly all long-lasting ideas, however, such as languages, scientific theories and religious beliefs, and the ineffable states of mind that constitute cultures such as being British, or the skill of performing classical music, are memes (or 'memeplexes' – collections of interacting memes). I shall say more about memes in Chapter 15.

The most general way of stating the central assertion of the neo-Darwinian theory of evolution is that a population of replicators subject to variation (for instance by imperfect copying) will be taken over by those variants that are better than their rivals at causing themselves to be replicated. This is a surprisingly deep truth which is commonly criticized either for being too obvious to be worth stating or for being false. The reason, I think, is that, although it is self-evidently true, it is not self-evidently the explanation of specific adaptations. Our intuition prefers explanations in terms of function or purpose: what does a gene do for its holder, or for its species? But we have just seen that the genes generally do not optimize such functionality.

So the knowledge embodied in genes is knowledge of how to get themselves replicated at the expense of their rivals. Genes *often* do this by imparting useful functionality to their organism, and in those cases

*This terminology differs slightly from that of Dawkins. Anything that is copied, for whatever reason, he calls a replicator. What I call a replicator he calls an 'active replicator'.

their knowledge incidentally includes knowledge about that functionality. Functionality, in turn, is achieved by encoding, into genes, regularities in the environment and sometimes even rule-of-thumb approximations to laws of nature, in which case the genes are incidentally encoding that knowledge too. But the core of the explanation for the presence of a gene is always that it got itself replicated more than its rival genes.

Non-explanatory human knowledge can also evolve in an analogous way: rules of thumb are not passed on perfectly to the next generation of users, and the ones that survive in the long run are not necessarily the ones that optimize the ostensible function. For instance, a rule that is expressed in an elegant rhyme may be remembered, and repeated, better than one that is more accurate but expressed in ungainly prose. Also, no human knowledge is entirely non-explanatory. There is always at least a background of assumptions about reality against which the meaning of a rule of thumb is understood, and that background can make some false rules of thumb seem plausible.

Explanatory theories evolve through a more complicated mechanism. Accidental errors in transmission and memory still play a role, but a much smaller one. That is because good explanations are hard to vary even without being tested, and hence random errors in the transmission of a good explanation are easier for the receiver to detect and correct. The most important source of variation in explanatory theories is creativity. For instance, when people are trying to understand an idea that they hear from others, they typically understand it to mean what makes most sense to them, or what they are most expecting to hear, or what they fear to hear, and so on. Those meanings are conjectured by the listener or reader, and may differ from what the speaker or writer intended. In addition, people often try to improve explanations even when they have received them accurately: they make creative amendments, spurred by their own criticism. If they then pass the explanation on to others, they usually try to pass on what they consider to be the improved version.

Unlike genes, many memes take different physical forms every time they are replicated. People rarely express ideas in exactly the same words in which they heard them. They also translate from one language to another, and between spoken and written language, and so on. Yet we rightly call what is transmitted the *same* idea – the same meme –

throughout. Thus, in the case of most memes, the real replicator is abstract: it is the knowledge itself. This is in principle true of genes as well: biotechnology routinely transcribes genes into the memories of computers, where they are stored in a different physical form. Those records could be translated back into DNA strands and implanted in different animals. The only reason this is not yet a common practice is that it is easier to copy the original gene. But one day the genes of a rare species could survive its extinction by causing themselves to be stored on a computer and then implanted into a cell of a different species. I say 'causing themselves to be stored' because the biotechnologists would not be recording information indiscriminately, but only information that met a criterion such as 'gene of an endangered species'. The ability to interest biotechnologists in this way would then be part of the reach of the knowledge in those genes.

So, both human knowledge and biological adaptations are abstract replicators: forms of information which, once they are embodied in a suitable physical system, tend to remain so while most variants of them do not.

The fact that the principles of neo-Darwinist theory are, from a certain perspective, self-evident has itself been used as a criticism of the theory. For instance, if the theory *must* be true, how can it be testable? One reply, often attributed to Haldane, is that the whole theory would be refuted by the discovery of a single fossilized rabbit in a stratum of Cambrian rock. However, that is misleading. The import of such an observation would depend on what explanations were available under the given circumstances. For instance, misidentifications of fossils, and of strata, have sometimes been made and would have to be ruled out by good explanations before one could call the discovery 'a fossilized rabbit in Cambrian rock'.

Even given such explanations, what would have been ruled out by the rabbit would be not the theory of evolution itself, but only the prevailing theory of the history of life and geological processes on Earth. Suppose, for instance, that there was a prehistoric continent, isolated from the others, on which evolution happened several times as fast as elsewhere, and that, by convergent evolution, a rabbit-like creature evolved there during the Cambrian era; and suppose that the continents were later connected by a catastrophe that obliterated most of the life forms on that continent and submerged their fossils. The rabbit-like creature was a rare survivor which became extinct soon afterwards. Given the supposed evidence, that is still an infinitely better explanation than, for instance, creationism or Lamarckism, neither of which gives *any* account of the origin of the apparent knowledge in the rabbit.

So what *would* refute the Darwinian theory of evolution? Evidence which, in the light of the best available explanation, implies that knowledge came into existence in a different way. For instance, if an organism was observed to undergo only (or mainly) favourable mutations, as predicted by Lamarckism or spontaneous generation, then Darwinism's 'random variation' postulate would be refuted. If organisms were observed to be born with new, complex adaptations – for anything – of which there were no precursors in their parents, then the gradual-change prediction would be refuted and so would Darwinism's mechanism of knowledge-creation. If an organism was born with a complex adaptation that has survival value today, yet was not favoured by selection pressure in its ancestry (say, an ability to detect and use internet weather forecasts to decide when to hibernate), then Darwinism would again be refuted. A fundamentally new explanation would be needed. Facing more or less the same unsolved problem that Paley and Darwin faced, we should have to set about finding an explanation that worked.

Fine-tuning

The physicist Brandon Carter calculated in 1974 that if the strength of the interaction between charged particles were a few per cent smaller, no planets would ever have formed and the only condensed objects in the universe would be stars; and if it were a few per cent greater, then no stars would ever explode, and so no elements other than hydrogen and helium would exist outside them. In either case there would be no complex chemistry and hence presumably no life.

Another example: if the initial expansion rate of the universe at the Big Bang had been slightly higher, no stars would have formed and there would be nothing in the universe but hydrogen – at an extremely low and ever-decreasing density. If it had been slightly lower, the universe would have recollapsed soon after the Big Bang. Similar results have been since obtained for other constants of physics that are not determined by any known theory. For most, if not all of them, it seems that if they had been slightly different, there would have been no possibility for life to exist.

This is a remarkable fact which has even been cited as evidence that those constants were intentionally fine-tuned, i.e. designed, by a super-natural being. This is a new version of creationism, and of the design argument, now based on the appearance of design *in the laws of physics*. (Ironically, given the history of this controversy, the new argument is that the laws of physics must have been designed to create a biosphere *by Darwinian evolution*.) It even persuaded the philosopher Antony Flew – formerly an enthusiastic advocate of atheism – of the existence of a supernatural designer. But it should not have. As I shall explain in a moment, it is not even clear that this fine-tuning constitutes an appearance of design in Paley's sense; but, even if it does, that does not alter the fact that invoking the supernatural makes for a bad explanation. And, in any case, arguing for supernatural explanations on the grounds that a current scientific explanation is flawed or lacking is just a mistake. As we carved in stone in Chapter 3, problems are inevitable – there are always unsolved problems. But they get solved. Science continues to make progress even, or especially, after making great discoveries, because the discoveries themselves reveal further problems. Therefore the existence of an unsolved problem in physics is no more evidence for a supernatural explanation than the existence of an unsolved crime is evidence that a ghost committed it.

A simple objection to the idea that fine-tuning requires an explanation at all is that we have no good explanation implying that planets are essential to the formation of life, or that chemistry is. The physicist Robert Forward wrote a superb science-fiction story, *Dragon's Egg*, based on the premise that information could be stored and processed – and life and intelligence could evolve – through the interactions between neutrons on the surface of a neutron star (a star that has collapsed gravitationally to a diameter of only a few kilometres, making it so dense that most of its matter has been transmuted into neutrons). It is not known whether this hypothetical neutron analogue of chemistry exists – nor whether it could exist if the laws of physics were slightly different. Nor do we have any idea what other sorts of environment permitting the emergence of life would exist under those variant laws. (The idea that similar laws of physics can be expected to give rise to similar environments is undermined by the very existence of fine-tuning.)

Nevertheless, regardless of whether the fine-tuning constitutes an appearance of design or not, it does constitute a legitimate and significant scientific problem, for the following reason. If the truth is that the constants of nature are *not* fine-tuned to produce life after all, because most slight variations in them do still permit life and intelligence to evolve somehow, though in dramatically different types of environment, then this would be an unexplained regularity in nature and hence a problem for science to address.

If the laws of physics *are* fine-tuned, as they seem to be, then there are two possibilities: either those laws are the only ones to be instantiated in reality (as universes) or there are other regions of reality – parallel universes* – with different laws. In the former case, we must expect there to be an explanation of why the laws are as they are. It would either refer to the existence of life or not. If it did, that would take us back to Paley's problem: it would mean that the laws had the 'appearance of design' for creating life, but *had not* evolved. Or the explanation would not refer to the existence of life, in which case it would leave unexplained why, if the laws are as they are for non-life-related reasons, they are fine-tuned to create life.

If there are many parallel universes, each with its own laws of physics, most of which do not permit life, then the idea would be that the observed fine-tuning is only a matter of parochial perspective. It is only in the universes that contain astrophysicists that anyone ever wonders why the constants seem fine-tuned. This type of explanation is known as 'anthropic reasoning'. It is said to follow from a principle known as the 'weak anthropic principle', though really no principle is required: it is just logic. (The qualifier 'weak' is there because several other anthropic principles have been proposed, which are more than just logic, but they need not concern us here.)

---

*These are not the 'parallel universes' of the *quantum* multiverse, which I shall describe in Chapter 11. Those universes all obey the same laws of physics and are in constant slight interaction with each other. They are also much less speculative.

However, on closer examination, anthropic arguments never quite finish the explanatory job. To see why, consider an argument due to the physicist Dennis Sciama.

Imagine that, at some time in the future, theoreticians have calculated, for one of those constants of physics, the range of its values for which there would be a reasonable probability that astrophysicists (of a suitable kind) would emerge. Say that range is from 137 to 138. (No doubt the real values will not be whole numbers, but let us keep it simple.) They also calculate that the highest probability of astrophysicists occurs at the midpoint of the range – when the constant is 137.5.

Next, experimentalists set out to measure the value of that constant directly – in laboratories, or by astronomical observation, say. What should they predict? Curiously enough, one immediate prediction from the anthropic explanation is that the value will not be exactly 137.5. For suppose that it were. By analogy, imagine that the bull's-eye of a dartboard represents the values that can produce astrophysicists. It would be a mistake to predict that a typical dart that strikes the bull's eye will strike it at the exact centre. Likewise, in the overwhelming majority of universes in which the measurement could take place (because they contain astrophysicists), the constant would not take the exactly optimal value for producing astrophysicists, nor be extremely close to it, compared with the size of the bull's-eye.

So Sciama concludes that, if we did measure one of those constants of physics, and found that it was extremely close to the optimum value for producing astrophysicists, that would statistically refute, not corroborate, the anthropic explanation for its value. Of course that value *might* still be a coincidence, but if we were willing to accept astronomically unlikely coincidences as explanations we should not be puzzled by the fine-tuning in the first place – and we should tell Paley that the watch on the heath *might* just have been formed by chance.

Furthermore, astrophysicists should be relatively unlikely in universes whose conditions are so hostile that they barely permit astrophysicists at all. So, if we imagine all the values consistent with the emergence of astrophysicists arrayed on a line, then the anthropic explanation leads us to expect the measured value to fall at some typical point, not too close to the middle or to either end.

However – and here we are reaching Sciama's main conclusion – that prediction changes radically if there are *several* constants to explain. For although any one constant is unlikely to be near the edge of its range, the more constants there are, the more likely it is that at least one of them will be. This can be illustrated pictorially as follows, with our bull's-eye replaced by a line segment, a square, a cube . . . and we can imagine this sequence continuing for as many dimensions as there are fine-tuned constants in nature. Arbitrarily define 'near the edge' as meaning 'within 10 per cent of the whole range from it'. Then in the case of one constant, as shown in the diagram, 20 per cent of its possible values are near one of the two edges of the range, and 80 per cent are 'away from the edge'. But with two constants a pair of values has to satisfy two constraints in order to be 'away from the edge'. Only 64 per cent of them do so. Hence 36 per cent are near the edge. With three constants, nearly half the possible choices are near the edge. With 100 constants, over 99.9999999 per cent of them are.

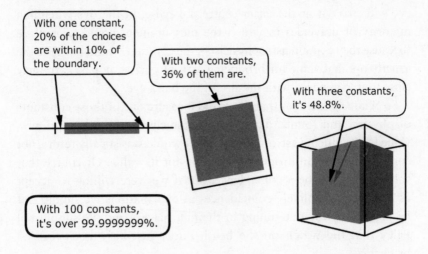

Whatever anthropic reasoning predicts about the values of multiple constants, it predicts will only just happen.

So, the more constants are involved, the closer to having no astrophysicists a typical universe-*with*-astrophysicists is. It is not known how many constants are involved, but it seems to be several, in which case the overwhelming majority of universes in the anthropically selected region would be close to its edge. Hence, Sciama concluded, the anthropic explanation predicts that the universe is *only just* capable of producing astrophysicists – almost the opposite prediction from the one that it makes in the case of one constant.

On the face of it, this might in turn seem to explain another great unsolved scientific mystery, known as 'Fermi's problem', named after the physicist Enrico Fermi, who is said to have asked, '*Where are they?*' Where are the extraterrestrial civilizations? Given the Principle of Mediocrity, or even just what we know of the galaxy and the universe, there is no reason to believe that the phenomenon of astrophysicists is unique to our planet. Similar conditions presumably exist in many other solar systems, so why would some of them not produce similar outcomes? Moreover, given the timescales on which stars and galaxies develop, it is overwhelmingly unlikely that any given extraterrestrial civilization is currently at a similar state of technological development to ours: it is likely to be millions of years younger (i.e. non-existent) or older. The older civilizations have had plenty of time to explore the galaxy – or at least to send robot space probes or signals. Fermi's problem is that we do not see any such civilizations, probes or signals.

Many candidate explanations have been proposed, and none of them, so far, are very good. The anthropic explanation of fine-tuning, in the light of Sciama's argument, might seem to solve the problem neatly: if the constants of physics in our universe are only just capable of producing astrophysicists, then it is not surprising that this event has happened only once, since its happening twice independently in the same universe would be vanishingly unlikely.

Unfortunately, that turns out to be a bad explanation too, because focusing on fundamental *constants* is parochial: there is no relevant difference between (1) 'the same' laws of physics with different constants and (2) different laws of physics. And there are infinitely many logically possible laws of physics. If they were all instantiated in real universes – as has been suggested by some cosmologists, such as Max Tegmark – it would be statistically certain that our universe is exactly on the edge of the astrophysicist-producing class of universes.

We know that that cannot be so from an argument due to Feynman (which he applied to a slightly different problem). Consider the class of all possible universes that contain astrophysicists, and consider what *else* most of them contain. In particular, consider a sphere just large enough to contain your own brain. If you are interested in explaining fine-tuning, your brain in its current state counts as an 'astrophysicist' for these purposes. In the class of all universes that contain astrophysicists, there are many that contain a sphere whose interior is perfectly identical to the interior of your sphere, including every detail of your brain. But in the vast majority of those universes there is chaos *outside* the sphere: almost a random state, since almost-random states are by far the most numerous. A typical such state is not only amorphous but hot. So in most such universes the very next thing that is going to happen is that the chaotic radiation emanating from outside the sphere will kill you instantly. At any given instant, the theory that we are going to be killed a picosecond hence is refuted by observation a picosecond later. Whereupon another such theory presents itself. So it is a very bad explanation – an extreme version of the gambler's hunches.

The same holds for purely anthropic explanations of all other fine-tunings involving more than a handful of constants: such explanations predict that it is overwhelmingly likely that we are in a universe in which astrophysicists are only just possible and will cease to exist in an instant. So they are bad explanations.

On the other hand, if the laws of physics exist in only one *form*, with only the values of a few constants differing from one universe to another, then the very fact that laws with different forms are not instantiated is a piece of fine-tuning that that anthropic explanation leaves unexplained.

The theory that all logically possible laws of physics are instantiated as universes has a further severe problem as an explanation. As I shall explain in Chapter 8, when considering infinite sets such as these, there is often no objective way to 'count' or 'measure' how many of them have one attribute rather than another. On the other hand, in the class of all logically possible entities, those that *can understand themselves*, as the physical reality that we are in does, are surely, in any reasonable sense, a tiny minority. The idea that one of them 'just happened', without explanation, is surely just a spontaneous-generation theory.

In addition, almost all the 'universes' described by those logically possible laws of physics are radically different from ours – so different that they do not properly fit into the argument. For instance, infinitely many of them contain nothing other than one bison, in various poses, and last for exactly 42 seconds. Infinitely many others contain a bison and an astrophysicist. But what *is* an astrophysicist in a universe that contains no stars, no scientific instruments and almost no evidence? What is a scientist, or any sort of thinking person, in a universe in which only bad explanations are true?

Almost all logically possible universes that contain astrophysicists are governed by laws of physics that are bad explanations. So should we predict that our universe, too, is inexplicable? Or has some high but unknowable probability to be? Thus, again, anthropic arguments based on 'all possible laws' are ruled out for being bad explanations.

For these reasons I conclude that, while anthropic reasoning may well be part of the explanation for apparent fine-tuning and other observations, it can never be the whole explanation for why we observe something that would otherwise look too purposeful to be explicable as coincidence. Specific explanation, in terms of specific laws of nature, is needed.

The reader may have noticed that all the bad explanations that I have discussed in this chapter are ultimately connected with each other. Expect too much from anthropic reasoning, or wonder too carefully how Lamarckism could work, and you get to spontaneous generation. Take spontaneous generation too seriously, and you get to creationism – and so on. That is because they all address the same underlying problem, and are all easily variable. They are easily interchangeable with each other or with variants of themselves, and they are 'too easy' as explanations: they could equally well explain anything. But neo-Darwinism was not easy to come by, and it is not easy to tweak. Try to tweak it – even as far as Darwin's own misconceptions – and you will get an explanation that doesn't work nearly as well. Try to account for something non-Darwinian with it – such as a new, complex adaptation of which there were no precursors in the organism's parents – and you will not be able to think of a variant with that feature.

Anthropic explanations are attempting to account for purposeful structure (such as the fine-tuned constants) in terms of a single act of selection. That is unlike evolution, and it cannot work. The solution of the fine-tuning puzzle is going to be in terms of an explanation that will specifically explain what we observe. It will be, as Wheeler put it, 'an idea so simple . . . that . . . we will all say to each other, how could it have been otherwise?' In other words, the problem has been not that the world is so complex that we cannot understand why it looks as it does, but it is that it is so *simple* that we cannot yet understand it. But this will be noticeable only with hindsight.

All those bad explanations of the biosphere either fail to address the problem of how the knowledge in adaptations is created or they explain it badly. That is to say, they all underrate *creation* – and, ironically, the theory that underrates creation most of all is creationism. Consider this: if a supernatural creator were to have created the universe at the moment when Einstein or Darwin or any great scientist (appeared to have) just completed their major discovery, then the true creator of that discovery (and of all earlier discoveries) would have been not that scientist but the supernatural being. So such a theory would deny the existence of the only creation that really did take place in the genesis of that scientist's discoveries.

And it really is creation. Before a discovery is made, no predictive process could reveal the content or the consequences of that discovery. For if it could, it would be that discovery. So scientific discovery is profoundly unpredictable, despite the fact that it is determined by the laws of physics. I shall say more about this curious fact in the next chapter; in short, it is due to the existence of 'emergent' levels of explanation. In this case, the upshot is that what science – and creative thought in general – achieves is unpredictable creation *ex nihilo*. So does biological evolution. No other process does.

Creationism, therefore, is misleadingly named. It is not a theory explaining knowledge as being due to creation, but the opposite: it is denying that creation happened in reality, by placing the origin of the knowledge in an explanationless realm. Creationism is really *creation denial* – and so are all those other false explanations.

The puzzle of understanding what living things are and how they came about has given rise to a strange history of misconceptions, near-misses and ironies. The last of the ironies is that the neo-Darwinian theory, like the Popperian theory of knowledge, really does describe creation, while their rivals, beginning with creationism, never could.

TERMINOLOGY

Evolution (Darwinian)   Creation of knowledge through alternating variation and selection.

Replicator   An entity that contributes causally to its own copying.

Neo-Darwinism   Darwinism as a theory of replicators, without various misconceptions such as 'survival of the fittest'.

Meme   An idea that is a replicator.

Memeplex   A group of memes that help to cause each other's replication.

Spontaneous generation   Formation of organisms from non-living precursors.

Lamarckism   A mistaken evolutionary theory based on the idea that biological adaptations are improvements acquired by an organism during its lifetime and then inherited by its descendants.

Fine-tuning   If the constants or laws of physics were slightly different, there would be no life.

Anthropic explanation   'It is only in universes that contain intelligent observers that anyone wonders why the phenomenon in question happens.'

MEANINGS OF 'THE BEGINNING OF INFINITY' ENCOUNTERED IN THIS CHAPTER

– Evolution.
– More generally, the creation of knowledge.

SUMMARY

The evolution of biological adaptations and the creation of human knowledge share deep similarities, but also some important differences. The main similarities: genes and ideas are both replicators; knowledge and adaptations are both hard to vary. The main difference: human knowledge can be explanatory and can have great reach; adaptations are never explanatory and rarely have much reach beyond the situations in which they evolved. False explanations of biological evolution have counterparts in false explanations of the growth of human knowledge. For instance, Lamarckism is the counterpart of inductivism. William Paley's version of the argument from design clarified what does or does not have the 'appearance of design' and hence what cannot be explained as the outcome of chance alone – namely hard-to-vary adaptation to a purpose. The origin of this must be the creation of knowledge. Biological evolution does not optimize benefits to the species, the group, the individual or even the gene, but only the ability of the gene to spread through the population. Such benefits can nevertheless happen because of the universality of laws of nature and the reach of some of the knowledge that is created. The 'fine-tuning' of the laws or constants of physics has been used as a modern form of the argument from design. For the usual reasons, it is not a good argument for a supernatural cause. But 'anthropic' theories that try to account for it as a pure selection effect from an infinite number of different universes are, by themselves, bad explanations too – in part because most logically possible laws are themselves bad explanations.

5

The Reality of Abstractions

The fundamental theories of modern physics explain the world in jarringly counter-intuitive ways. For example, most non-physicists consider it self-evident that when you hold your arm out horizontally you can *feel* the force of gravity pulling it downwards. But you cannot. The existence of a force of gravity is, astonishingly, denied by Einstein's general theory of relativity, one of the two deepest theories of physics. This says that the only force on your arm in that situation is that which you yourself are exerting, upwards, to keep it constantly accelerating away from the straightest possible path in a curved region of spacetime. The reality described by our other deepest theory, quantum theory, which I shall describe in Chapter 11, is even more counter-intuitive. To understand explanations like those, physicists have to learn to think about everyday events in new ways.

The guiding principle is, as always, to reject bad explanations in favour of good ones. In regard to what is or is not real, this leads to the requirement that, if an entity is referred to by our best explanation in the relevant field, we must regard it as really existing. And if, as with the force of gravity, our best explanation denies that it exists, then we must stop assuming that it does.

Furthermore, everyday events are stupendously *complex* when expressed in terms of fundamental physics. If you fill a kettle with water and switch it on, all the supercomputers on Earth working for the age of the universe could not solve the equations that predict what all those water molecules will do – even if we could somehow determine their initial state and that of all the outside influences on them, which is itself an intractable task.

Fortunately, some of that complexity resolves itself into a higher-level simplicity. For example, we *can* predict with some accuracy how long the water will take to boil. To do so, we need know only a few physical quantities that are quite easy to measure, such as its mass, the power of the heating element, and so on. For greater accuracy we may also need information about subtler properties, such as the number and type of nucleation sites for bubbles. But those are still relatively 'high-level' phenomena, composed of intractably large numbers of interacting atomic-level phenomena. Thus there is a class of high-level phenomena – including the liquidity of water and the relationship between containers, heating elements, boiling and bubbles – that can be well explained in terms of each other alone, with no direct reference to anything at the atomic level or below. In other words, the behaviour of that whole class of high-level phenomena is *quasi-autonomous* – almost self-contained. This resolution into explicability at a higher, quasi-autonomous level is known as *emergence*.

Emergent phenomena are a tiny minority. We can predict when the water will boil, and that bubbles will form when it does, but if you wanted to predict where each bubble will go (or, to be precise, what the probabilities of its various possible motions are – see Chapter 11), you would be out of luck. Still less is it feasible to predict the countless microscopically defined properties of the water, such as whether an odd or an even number of its electrons will be affected by the heating during a given period.

Fortunately, we are uninterested in predicting or explaining most of those properties, despite the fact that they are the overwhelming majority. That is because none of them has any bearing on what we want to do with the water – such as understand what it is made of, or make tea. To make tea, we want the water to be boiling, but we do not care what the pattern of bubbles was. We want its volume to be between a certain minimum and maximum, but we do not care how many molecules that is. We can make progress in achieving those purposes because we can express them in terms of those quasi-autonomous emergent properties about which we have good high-level explanations. Nor do we need most of the microscopic details in order to understand the role of water in the cosmic scheme of things, because nearly all of those details are parochial.

The behaviour of high-level physical quantities consists of nothing but the behaviour of their low-level constituents with most of the details ignored. This has given rise to a widespread misconception about emergence and explanation, known as *reductionism*: the doctrine that science always explains and predicts things reductively, i.e. by analysing them into components. Often it does, as when we use the fact that inter-atomic forces obey the law of conservation of energy to make and explain a high-level prediction that the kettle cannot boil water without a power supply. But reductionism requires the relationship between different levels of explanation *always* to be like that, and often it is not. For example, as I wrote in *The Fabric of Reality*:

Consider one particular copper atom at the tip of the nose of the statue of Sir Winston Churchill that stands in Parliament Square in London. Let me try to explain why that copper atom is there. It is because Churchill served as prime minister in the House of Commons nearby; and because his ideas and leadership contributed to the Allied victory in the Second World War; and because it is customary to honour such people by putting up statues of them; and because bronze, a traditional material for such statues, contains copper, and so on. Thus we explain a low-level physical observation – the presence of a copper atom at a particular location – through extremely high-level theories about emergent phenomena such as ideas, leadership, war and tradition.

There is no reason why there should exist, even in principle, any lower-level *explanation* of the presence of that copper atom than the one I have just given. Presumably a reductive 'theory of everything' would in principle make a low-level *prediction* of the probability that such a statue will exist, given the condition of (say) the solar system at some earlier date. It would also in principle describe how the statue probably got there. But such descriptions and predictions (wildly infeasible, of course) would explain nothing. They would merely describe the trajectory that each copper atom followed from the copper mine, through the smelter and the sculptor's studio and so on . . . In fact such a prediction would have to refer to atoms all over the planet, engaged in the complex motion we call the Second World War, among other things. But even if you had the superhuman capacity to follow such lengthy predictions of the copper atom's being there, you would still not be able to say 'Ah yes, now I understand *why* they are there'. [You] would have to inquire into *what it was* about that configuration of atoms, and those trajectories, that gave them the propensity to deposit a copper atom at this location. Pursuing that inquiry would be a creative task, as discovering new explanations always is. You would have to discover that certain atomic configurations support emergent phenomena such as leadership and war, which are related to one another by high-level explanatory theories. Only when you knew those theories could you understand why that copper atom is where it is.

Even in physics, some of the most fundamental explanations, and the predictions that they make, are not reductive. For instance, the second law of thermodynamics says that high-level physical processes tend towards ever greater disorder. A scrambled egg never becomes unscrambled by the whisk, and never extracts energy from the pan to propel itself upwards into the shell, which never seamlessly reseals itself. Yet, if you could somehow make a video of the scrambling process with enough resolution to see the individual molecules, and play it backwards, and examine any part of it at that scale, you would see nothing but molecules moving and colliding in strict obedience to the low-level laws of physics. It is not yet known how, or whether, the second law of thermodynamics can be derived from a simple statement about individual atoms.

There is no reason why it should be. There is often a moral overtone to reductionism (science *should be* essentially reductive). This is related both to instrumentalism and to the Principle of Mediocrity, which I criticized in Chapters 1 and 3. Instrumentalism is rather like reductionism except that, instead of rejecting only high-level explanations, it tries to reject all explanations. The Principle of Mediocrity is a milder form of reductionism: it rejects only high-level explanations that involve people. While I am on the subject of bad philosophical doctrines with moral overtones, let me add *holism*, a sort of mirror image of reductionism. It is the idea that the only valid explanations (or at least the only significant ones) are of parts in terms of wholes. Holists also often share with reductionists the mistaken belief that science *can only* (or should only) be reductive, and therefore they oppose much of science. All those doctrines are irrational for the same reason: they advocate accepting or rejecting theories on grounds other than whether they are good explanations.

Whenever a high-level explanation does follow logically from low-level ones, that also means that the high-level one *implies something* about the low-level ones. Thus, additional high-level theories, provided that they were all consistent, would place more and more constraints on what the low-level theories could be. So it could be that all the high-level explanations that exist, taken together, *imply* all the low-level ones, as well as vice versa. Or it could be that some low-level, some intermediate-level and some high-level explanations, taken together, imply *all* explanations. I guess that that is so.

Thus, one possible way that the fine-tuning problem might eventually be solved would be if some high-level explanations turned out to be exact laws of nature. The microscopic consequences of that might well seem to be fine-tuned. One candidate is the principle of the universality of computation, which I shall discuss in the next chapter. Another is the principle of testability, for, in a world in which the laws of physics do not permit the existence of testers, they also forbid themselves to be tested. However, in their current form such principles, regarded as laws of physics, are anthropocentric and arbitrary – and would therefore be bad explanations. But perhaps there are deeper versions, to which they are approximations, which *are* good explanations, well integrated with those of microscopic physics like the second law of thermodynamics is.

In any case, emergent phenomena are essential to the explicability of the world. Long before humans had much explanatory knowledge, they were able to control nature by using rules of thumb. Rules of thumb have explanations, and those explanations were about high-level regularities among emergent phenomena such as fire and rocks. Long before that, it was only genes that were encoding rules of thumb, and the knowledge in them, too, was about emergent phenomena. Thus emergence is another beginning of infinity: all knowledge-creation depends on, and physically consists of, emergent phenomena.

Emergence is also responsible for the fact that discoveries can be made in successive steps, thus providing scope for the scientific method. The partial success of each theory in a sequence of improving theories is tantamount to the existence of a 'layer' of phenomena that each theory explains successfully – though, as it then turns out, partly mistakenly.

Successive scientific explanations are occasionally dissimilar in the way they *explain* their predictions, even in the domain where the predictions themselves are similar or identical. For instance, Einstein's explanation of planetary motion does not merely correct Newton's: it is radically different, denying, among many other things, the very existence of central elements of Newton's explanation, such as the gravitational force and the uniformly flowing time with respect to which Newton defined motion. Likewise the astronomer Johannes Kepler's theory which said that the planets move in ellipses did not merely correct the celestial-sphere theory, it denied the spheres' existence. And Newton's did not substitute a new *shape* for Kepler's ellipses, but a whole new way for laws to specify motion – through infinitesimally defined quantities like instantaneous velocity and acceleration. Thus each of those theories of planetary motion was ignoring or denying its predecessor's basic means of explaining what was happening out there.

This has been used as an argument for instrumentalism, as follows. Each successive theory made small but accurate corrections to what its predecessor *predicted*, and was therefore a better theory in that sense. But, since each theory's *explanation* swept away that of the previous theory, the previous theory's explanation was never true in the first place, and so one cannot regard those successive explanations as constituting a growth of knowledge about reality. From Kepler to Newton to Einstein we have successively: no force needed to explain orbits; an inverse-square-law force responsible for every orbit; and again no force needed. So how could Newton's 'force of gravity' (as distinct from his equations predicting its effects) ever have been an advance in human knowledge?

It could, and was, because sweeping away the entities through which a theory makes its explanation is not the same as sweeping away the whole of the explanation. Although there is no force of gravity, it is true that *something real* (the curvature of spacetime), caused by the sun, has a strength that varies approximately according to Newton's inverse-square law, and affects the motion of objects, seen and unseen. Newton's theory also correctly explained that the laws of gravitation are the same for terrestrial and celestial objects; it made a novel distinction between mass (the measure of an object's resistance to being accelerated) and weight (the force required to prevent the object from falling under gravity); and it said that the gravitational effect of an object depends on its mass and not on other attributes such as its density or composition. Later, Einstein's theory not only endorsed all those features but explained, in turn, why they are so. Newton's theory, too, had been able to make more accurate predictions than its predecessors precisely because it was more right than they were about what was really happening. Before that, even Kepler's explanation had included important elements of the true explanation: planetary orbits are indeed determined by laws of nature; those laws are indeed the same for all planets, including the Earth; they do involve the sun; they are mathematical and geometrical in character; and so on. With the hindsight provided by each successive theory, we can see not only where the previous theory made false predictions, but also that wherever it made true predictions this was *because* it had expressed some truth about reality. So its truth lives on in the new theory – as Einstein remarked, 'There could be no fairer destiny for any physical theory than that it should point the way to a more comprehensive theory in which it lives on as a limiting case.'

As I explained in Chapter 1, regarding the explanatory function of theories as paramount is not just an idle preference. The predictive function of science is entirely dependent on it. Also, in order to make progress in any field, it is the explanations in existing theories, not the predictions, that have to be creatively varied in order to conjecture the next theory. Furthermore, the explanations in one field affect our understanding of other fields. For instance, if someone thinks that a conjuring trick is due to supernatural abilities of the conjurer, it will affect how they judge theories in cosmology (such as the origin of the universe, or the fine-tuning problem) and in psychology (how the human mind works) and so on.

By the way, it is something of a misconception that the predictions of successive theories of planetary motion *were* all that similar. Newton's predictions are indeed excellent in the context of bridge-building, and only slightly inadequate when running the Global Positioning System, but they are hopelessly wrong when explaining a pulsar or a quasar – or the universe as a whole. To get all those right, one needs Einstein's radically different explanations.

Such large discontinuities in the meanings of successive scientific theories have no biological analogue: in an evolving species, the dominant strain in each generation differs only slightly from that in the previous generation. Nevertheless, scientific discovery is a gradual process too; it is just that, in science, all the gradualness, and nearly all the criticism and rejection of bad explanations, takes place inside the scientists' minds. As Popper put it, 'We can let our theories die in our place.'

There is another, even more important, advantage in that ability to criticize theories without staking one's life on them. In an evolving species, the adaptations of the organisms in each generation must have enough functionality to keep the organism alive, and to pass all the tests that they encounter in propagating themselves to the next generation. In contrast, the intermediate explanations leading a scientist from one good explanation to the next need not be viable at all. The same is true of creative thought in general. This is the fundamental reason that explanatory ideas are able to escape from parochialism, while biological evolution, and rules of thumb, cannot.

That brings me to the main subject of this chapter: abstractions. In Chapter 4 I remarked that pieces of knowledge are abstract replicators that 'use' (and hence *affect*) organisms and brains to get themselves replicated. That is a higher level of explanation than the emergent levels I have mentioned so far. It is a claim that something abstract – something non-physical, such as the knowledge in a gene or a theory – is affecting something physical. Physically, nothing is happening in such a situation other than that one set of emergent entities – such as genes, or computers – is affecting others, which is already anathema to reductionism. But abstractions are essential to a fuller explanation. You know that if your computer beats you at chess, it is really the *program* that has beaten you, not the silicon atoms or the computer as such. The abstract program is instantiated physically as a high-level behaviour of vast numbers of atoms, but the *explanation* of why it has beaten you cannot be expressed without also referring to the program in its own right. That program has also been instantiated, unchanged, in a long chain of different physical substrates, including neurons in the brains of the programmers and radio waves when you downloaded the program via wireless networking, and finally as states of long- and short-term memory banks in your computer. The specifics of that chain of instantiations may be relevant to explaining how the program reached you, but it is irrelevant to why it beat you: there, the content of the knowledge (in it, and in you) is the whole story. That story is an explanation that refers ineluctably to abstractions; and therefore those abstractions exist, and really do affect physical objects in the way required by the explanation.

The computer scientist Douglas Hofstadter has a nice argument that this sort of explanation is essential in understanding certain phenomena. In his book *I am a Strange Loop* (2007) he imagines a special-purpose computer built of millions of dominoes. They are set up – as dominoes often are for fun – standing on end, close together, so that if one of them is knocked over it strikes its neighbour and so a whole stretch of dominoes falls, one after another. But Hofstadter's dominoes are spring-loaded in such a way that, whenever one is knocked over, it pops back up after a fixed time. Hence, when a domino falls, a wave or 'signal' of falling dominoes propagates along the stretch in the direction in which it fell until it reaches either a dead end or a currently fallen domino. By arranging these dominoes in a network with looping, bifurcating and rejoining stretches, one can make these signals combine and interact in a sufficiently rich repertoire of ways to make the whole construction into a computer: a signal travelling down a stretch can be interpreted as a binary '1', and the lack of a signal as a binary '0', and the interactions between such signals can implement a repertoire of operations – such as 'and', 'or' and 'not' – out of which arbitrary computations can be composed.

One domino is designated as the 'on switch': when it is knocked over, the domino computer begins to execute the program that is instantiated in its loops and stretches. The program in Hofstadter's thought experiment computes whether a given number is a prime or not. One inputs that number by placing a stretch of exactly that many dominos at a specified position, before tripping the 'on switch'. Else-where in the network, a particular domino will deliver the output of the computation: it will fall only if a divisor is found, indicating that the input was not a prime.

Hofstadter sets the input to the number 641, which is a prime, and trips the 'on switch'. Flurries of motion begin to sweep back and forth across the network. All 641 of the input dominos soon fall as the computation 'reads' its input – and snap back up and participate in further intricate patterns. It is a lengthy process, because this is a rather inefficient way to perform computations – but it does the job.

Now Hofstadter imagines that an observer who does not know the purpose of the domino network watches the dominoes performing and notices that one particular domino remains resolutely standing, never affected by any of the waves of downs and ups sweeping by.

> The observer points at [that domino] and asks with curiosity, 'How come that domino there is never falling?'

We know that it is the output domino, but the observer does not. Hofstadter continues:

> Let me contrast two different types of answer that someone might give. The first type of answer – myopic to the point of silliness – would be, 'Because its predecessor never falls, you dummy!'

Or, if it has two or more neighbours, 'Because none of its neighbours ever fall.'

> To be sure, this is correct as far as it goes, but it doesn't go very far. It just passes the buck to a different domino.

In fact one could keep passing the buck from domino to domino, to provide ever more detailed answers that were 'silly, but correct as far as they go'. Eventually, after one had passed the buck billions of times (many more times than there are dominoes, because the program 'loops'), one would arrive at that first domino – the 'on switch'.

At that point, the reductive (to high-level physics) explanation would be, in summary, 'That domino did not fall because none of the patterns of motion initiated by knocking over the "on switch" ever include it.' But we knew that already. We can reach that conclusion – as we just have – without going through that laborious process. And it is undeniably true. But it is not the explanation we were looking for because it is addressing a different question – predictive rather than explanatory – namely, if the first domino falls, *will* the output domino ever fall? And it is asking at the wrong level of emergence. What we asked was: *why* does it not fall? To answer that, Hofstadter then adopts a different mode of explanation, at the right level of emergence:

> The second type of answer would be, 'Because 641 is prime.' Now this answer, while just as correct (indeed, in some sense it is far more on the mark), has the curious property of not talking about anything physical at all. Not only has the focus moved upwards to collective properties . . . these properties somehow transcend the physical and have to do with pure abstractions, such as primality.

Hofstadter concludes, 'The point of this example is that 641's primality is the best explanation, perhaps even the only explanation, for why certain dominoes did fall and certain others did not fall.'

Just to correct that slightly: the physics-based explanation is true *as well*, and the physics of the dominoes is also essential to explaining why prime numbers are relevant to that particular arrangement of them. But Hofstadter's argument does show that *primality* must be part of any full explanation of why the dominos did or did not fall. Hence it is a refutation of reductionism in regard to abstractions. For the theory of prime numbers is not part of physics. It refers not to physical objects, but to abstract entities – such as numbers, of which there is an infinite set.

Unfortunately, Hofstadter goes on to disown his own argument and to embrace reductionism. Why?

His book is primarily about one particular emergent phenomenon, the mind – or, as he puts it, the 'I'. He asks whether the mind can consistently be thought of as *affecting* the body – causing it to do one thing rather than another, given the all-embracing nature of the laws of physics. This is known as the mind–body problem. For instance, we often explain our actions in terms of choosing one action rather than another, but our bodies, including our brains, are completely controlled by the laws of physics, leaving no physical variable free for an 'I' to affect in order to make such a choice. Following the philosopher Daniel Dennett, Hofstadter eventually concludes that the 'I' is an illusion. Minds, he concludes, can't 'push material stuff around', because 'physical law alone would suffice to determine [its] behaviour'. Hence his reductionism.

But, first of all, physical laws can't push anything either. They only explain and predict. And they are not our only explanations. The theory that the domino stands 'because 641 is a prime (and because the domino network instantiates a primality-testing algorithm)' is an exceedingly good explanation. What is wrong with it? It does not contradict the laws of physics. It explains more than any explanation purely in terms of those laws. And no known variant of it can do the same job.

Second, that reductionist argument would equally deny that an *atom* can 'push' (in the sense of 'cause to move') another atom, since the initial state of the universe, together with the laws of motion, has already determined the state at every other time.

Third, the very idea of a *cause* is emergent and abstract. It is mentioned nowhere in the laws of motion of elementary particles, and, as the philosopher David Hume pointed out, we cannot perceive causation, only a succession of events. Also, the laws of motion are 'conservative' – that is to say, they do not lose information. That means that, just as they determine the final state of any motion given the initial state, they also determine the initial state given the final state, and the state at any time from the state at any other time. So, at that level of explanation, cause and effect are interchangeable – and are not what we mean when we say that a program causes a computer to win at chess, or that a domino remained standing *because* 641 is a prime.

There is no inconsistency in having multiple explanations of the same phenomenon, at different levels of emergence. Regarding microphysical explanations as more fundamental than emergent ones is arbitrary and fallacious. There is no escape from Hofstadter's 641 argument, and no reason to want one. The world may or may not be as we wish it to be, and to reject good explanations on that account is to imprison oneself in parochial error.

So the answer 'Because 641 is a prime' does explain the immunity of that domino. The theory of prime numbers on which that answer depends is not a law of physics, nor an approximation to one. It is about abstractions, and infinite sets of them at that (such as the set of 'natural numbers' 1, 2, 3, ... , where the ellipsis ' ... ' denotes continuation ad infinitum). It is no mystery how we can have knowledge of infinitely large things, like the set of all natural numbers. That is just a matter of reach. Versions of number theory that confined themselves to 'small natural numbers' would have to be so full of arbitrary qualifiers, workarounds and unanswered questions that they would be very bad explanations until they were generalized to the case that makes sense without such ad-hoc restrictions: the infinite case. I shall discuss various sorts of infinity in Chapter 8.

When we use theories about emergent physical quantities to explain the behaviour of water in a kettle, we are using an abstraction – an 'idealized' model of the kettle that ignores most of its details – as an approximation to a real physical system. But when we use a computer to investigate prime numbers, we are doing the reverse: we are using the physical computer as an approximation to an abstract one which perfectly models prime numbers. Unlike any real computer, the latter never goes wrong, requires no maintenance, and has unlimited memory and unlimited time to run its program.

Our own brains are, likewise, computers which we can use to learn about things beyond the physical world, including pure mathematical abstractions. This ability to understand abstractions is an emergent property of people which greatly puzzled the ancient Athenian philosopher Plato. He noticed that the theorems of geometry – such as Pythagoras' theorem – are about entities that are never experienced: perfectly straight lines with no thickness, intersecting each other on a perfect plane to make a perfect triangle. These are not possible objects of any observation. And yet people knew about them – and not just superficially: at the time, such knowledge was the deepest knowledge, of anything, that human beings had ever had. Where did it come from? Plato concluded that it – and all human knowledge – must come from the supernatural.

He was right that it could not have come from observation. But then it could not have even if people *had* been able to observe perfect triangles (as arguably they could today, using virtual reality). As I explained in Chapter 1, empiricism has multiple fatal flaws. But it is no mystery where our knowledge of abstractions comes from: it comes from conjecture, like all our knowledge, and through criticism and seeking good explanations. It is only empiricism that made it seem plausible that knowledge outside science is inaccessible; and it is only the justified-true-belief misconception that makes such knowledge seem less 'justified' than scientific theories.

As I explained in Chapter 1, even in science, almost all rejected theories are rejected for being bad explanations, without ever being tested. Experimental testing is only one of many methods of criticism used in science, and the Enlightenment has made progress by bringing those other methods to bear in non-scientific fields too. The basic reason that such progress is possible is that good explanations about philosophical issues are as hard to find as in science – and criticism is correspondingly effective.

Moreover, experience does play a role in philosophy – only not the role of experimental testing that it plays in science. Primarily, it provides philosophical *problems*. There would have been no philosophy of science if the issue of how we can acquire knowledge of the physical world had been unproblematic. There would be no such thing as political philosophy if there had not first been a problem of how to run societies. (To avoid misunderstanding, let me stress that experience provides problems only by bringing already-existing ideas into conflict. It does not, of course, provide theories.)

In the case of moral philosophy, the empiricist and justificationist misconceptions are often expressed in the maxim that 'you can't derive an *ought* from an *is*' (a paraphrase of a remark by the Enlightenment philosopher David Hume). It means that moral theories cannot be deduced from factual knowledge. This has become conventional wisdom, and has resulted in a kind of dogmatic despair about morality: 'you can't derive an *ought* from an *is*, therefore morality cannot be justified by reason'. That leaves only two options: either to embrace unreason or to try living without ever making a moral judgement. Both are liable to lead to morally wrong choices, just as embracing unreason or never attempting to explain the physical world leads to factually false theories (and not just ignorance).

Certainly you can't derive an *ought* from an *is*, but you can't derive a *factual* theory from an *is* either. That is not what science does. The growth of knowledge does not consist of finding ways to justify one's beliefs. It consists of finding good explanations. And, although factual evidence and moral maxims are logically independent, factual and moral *explanations* are not. Thus factual knowledge can be useful in criticizing moral explanations.

For example, in the nineteenth century, if an American slave had written a bestselling book, that event would not *logically* have ruled out the proposition 'Negroes are intended by Providence to be slaves.' No experience could, because that is a philosophical theory. But it might have ruined the explanation through which many people understood that proposition. And if, as a result, such people had found themselves unable to explain to their own satisfaction why it would be Providential if that author were to be forced back into slavery, then they might have questioned the account that they had formerly accepted of what a black person really is, and what a person in general is – and then a good person, a good society, and so on.

Conversely, advocates of highly immoral doctrines almost invariably believe associated factual falsehoods as well. For instance, ever since the attack on the United States on 11 September 2001, millions of people worldwide have believed it was carried out by the US government, or the Israeli secret service. Those are purely factual misconceptions, yet they bear the imprint of moral wrongness just as clearly as a fossil – made of purely inorganic material – bears the imprint of ancient life. And the link, in both cases, is explanation. To concoct a *moral* explanation for why Westerners deserve to be killed indiscriminately, one needs to explain *factually* that the West is not what it pretends to be – and that requires uncritical acceptance of conspiracy theories, denials of history, and so on.

Quite generally, in order to understand the moral landscape in terms of a given set of values, one needs to understand some facts as being a certain way too. And the converse is also true: for example, as the philosopher Jacob Bronowski pointed out, success at making factual, scientific discoveries entails a commitment to all sorts of values that are necessary for making progress. The individual scientist has to value truth, and good explanations, and be open to ideas and to change. The scientific community, and to some extent the civilization as a whole, has to value tolerance, integrity and openness of debate.

We should not be surprised at these connections. The truth has structural unity as well as logical consistency, and I guess that no true explanation is entirely disconnected from any other. Since the universe is explicable, it must be that morally right values are connected in this way with true factual theories, and morally wrong values with false theories.

Moral philosophy is basically about the problem of what to do next

– and, more generally, what sort of life to lead, and what sort of world to want. Some philosophers confine the term 'moral' to problems about how one should treat other people. But such problems are continuous with problems of individuals choosing what sort of life to lead, which is why I adopt the more inclusive definition. Terminology aside, if you were suddenly the last human on Earth, you would be wondering what sort of life to want. Deciding 'I should do whatever pleases me most' would give you very little clue, because what pleases you depends on your moral judgement of what constitutes a good life, not vice versa.

This also illustrates the emptiness of reductionism in philosophy. For if I ask you for advice about what objectives to pursue in life, it is no good telling me to do what the laws of physics mandate. I shall do that in any case. Nor is it any good telling me to do what I prefer, because I don't know what I prefer to do until I have decided what sort of life I want to lead or how I should want the world to be. Since our preferences are shaped in this way, at least in part, by our moral explanations, it does not make sense to define right and wrong entirely in terms of their utility in meeting people's preferences. Trying to do so is the project of the influential moral philosophy known as *utilitarianism*, which played much the same role as empiricism did in the philosophy of science: it acted as a liberating focus for the rebellion against traditional dogmas, while its own positive content contained little truth.

So there is no avoiding what-to-do-next problems, and, since the distinction between right and wrong appears in our best explanations that address such problems, we must regard that distinction as real. In other words, there is an objective difference between right and wrong: those are real attributes of objectives and behaviours. In Chapter 14 I shall argue that the same is true in the field of aesthetics: there is such a thing as objective beauty.

Beauty, right and wrong, primality, infinite sets – they all exist objectively. But not physically. What does that mean? Certainly they can affect you – as examples like Hofstadter's show – but apparently not in the same sense that physical objects do. You cannot trip over one of them in the street. However, there is less to that distinction than our empiricism-biased common sense assumes. First of all, being *affected* by a physical object means that something about the physical object has caused a change, via the laws of physics (or, equivalently, that the laws of physics have caused a change via that object). But causation and the laws of physics are not themselves physical objects. They are abstractions, and our knowledge of them comes – just as for all other abstractions – from the fact that our best explanations invoke them. Progress depends on explanation, and therefore trying to conceive of the world as merely a sequence of events with unexplained regularities would entail giving up on progress.

This argument that abstractions really exist does not tell us what they exist *as* – for instance, which of them are purely emergent aspects of others, and which exist independently of the others. Would the laws of morality still be the same if the laws of physics were different? If they were such that knowledge could best be obtained by blind obedience to authority, then scientists would have to *avoid* what we think of as the values of scientific inquiry in order to make progress. My guess is that morality is more autonomous than that, and so it makes sense to say that such laws of physics would be *immoral*, and (as I remarked in Chapter 4) to imagine laws of physics that would be more moral than the real ones.

The reach of ideas into the world of abstractions is a property of the knowledge that they contain, not of the brain in which they may happen to be instantiated. A theory can have infinite reach even if the person who originated it is unaware that it does. However, a *person* is an abstraction too. And there is a kind of infinite reach that is unique to people: the reach of the ability to understand explanations. And this ability is itself an instance of the wider phenomenon of *universality* – to which I turn next.

TERMINOLOGY

Levels of emergence   Sets of phenomena that can be explained well in terms of each other without analysing them into their constituent entities such as atoms.

Natural numbers   The whole numbers 1, 2, 3 and so on.

Reductionism   The misconception that science must or should always explain things by analysing them into components (and hence that higher-level explanations cannot be fundamental).

Holism   The misconception that all significant explanations are of components in terms of wholes rather than vice versa.

Moral philosophy   Addresses the problem of what sort of life to want.

MEANINGS OF 'THE BEGINNING OF INFINITY' ENCOUNTERED IN THIS CHAPTER

– The existence of emergent phenomena, and the fact that they can encode knowledge about other emergent phenomena.
– The existence of levels of approximation to true explanations.
– The ability to understand explanations.
– The ability of explanation to escape from parochialism by 'letting our theories die in our place'.

SUMMARY

Reductionism and holism are both mistakes. In reality, explanations do not form a hierarchy with the lowest level being the most fundamental. Rather, explanations at any level of emergence can be fundamental. Abstract entities are real, and can play a role in causing physical phenomena. Causation is itself such an abstraction.

6

The Jump to Universality

The earliest writing systems used stylized pictures – 'pictograms' – to represent words or concepts. So a symbol like '☉' might stand for 'sun', and '⌂' for 'tree'. But no system ever came close to having a pictogram for *every* word in its spoken language. Why not?

Originally, there was no intention to do so. Writing was for specialized applications such as inventories and tax records. Later, new applications would require larger vocabularies, but by then scribes would increasingly have found it easier to add new *rules* to their writing system rather than new pictograms. For example, in some systems, if a word sounded like two or more other words in sequence, it could be represented by the pictograms for those words. If English were written in pictograms, that would allow us to write the word 'treason' as '⌂☉'. This would not represent the sound of the word precisely (nor does its actual spelling, for that matter), but it would approximate it well enough for any reader who spoke the language and was aware of the rule.

Following that innovation, there would have been less incentive to coin new pictograms – say '🕊' for 'treason'. Coining one would always have been tedious, not so much because designing memorable pictograms is hard – though it is – but because, before one could use it, one would somehow have to inform all intended readers of its meaning. That is hard to do: if it had been easy, there would have been much less need for writing in the first place. In cases where the rule could be applied instead, it was more efficient: any scribe could write '⌂☉' and be understood even by a reader who had never seen the word written before.

However, the rule could not be applied in all cases: it could not represent any new single-syllable words, nor many other words. It seems clumsy and inadequate compared to modern writing systems. Yet there was already something significant about it which no purely pictographic system could achieve: it brought words into the writing system that no one had explicitly added. That means that it had reach. And reach always has an explanation. Just as in science a simple formula may summarize a mass of facts, so a simple, easily remembered rule can bring many additional words into a writing system, but only if it reflects an underlying regularity. The regularity in this case is that all the words in any given language are built out of only a few dozen 'elementary sounds', with each language using a different set chosen from the enormous range of sounds that the human voice can produce. Why? I shall come to that below.

As the rules of a writing system were improved, a significant threshold could be crossed: the system could become *universal* for that language – capable of representing every word in it. For example, consider the following variant of the rule that I have just described: instead of building words out of other words, build them out of the *initial sounds* of other words. So, if English were written in pictograms, the new rule would allow 'treason' to be spelled with the pictograms for 'Tent', 'Rock', 'EAgle', 'Zebra', 'Nose'. That tiny change in the rules would make the system universal. It is thought that the earliest alphabets evolved from rules like that.

Universality achieved through rules has a different character from that of a completed list (such as the hypothetical complete set of pictograms). One difference is that the rules can be much simpler than the list. The individual symbols can be simpler too, because there are fewer of them. But there is more to it than that. Since a rule works by exploiting regularities in the language, it implicitly encodes those regularities, and so contains more knowledge than the list. An alphabet, for instance, contains knowledge of what words sound like. That allows it to be used by a foreigner to learn to *speak* the language, while pictograms could at most be used to learn to write it. Rules can also accommodate inflections such as prefixes and suffixes without adding complexity to the writing system, thus allowing written texts to encode more of the grammar of sentences. Also, a writing system based on an alphabet can cover not only every word but every *possible* word in its language, so that words that have yet to be coined already have a place in it. Then, instead of each new word temporarily breaking the system, the system can itself be used to coin new words, in an easy and decentralized way.

Or, at least, it could have been. It would be nice to think that the unknown scribe who created the first alphabet knew that he was making one of the greatest discoveries of all time. But he may not have. If he did, he certainly failed to pass his enthusiasm on to many others. For, in the event, the power of universality that I have just described was rarely used in ancient times, even when it was available. Although pictographic writing systems were invented in many societies, and universal alphabets did sometimes evolve from them in the way I have just described, the 'obvious' next step – namely to use the alphabet universally and to drop the pictograms – was almost never taken. Alphabets were confined to special purposes such as writing rare words or transliterating foreign names. Some historians believe that the idea of an alphabet-*based* writing system was conceived only once in human history – by some unknown predecessors of the Phoenicians, who then spread it throughout the Mediterranean – so that every alphabet-based writing system that has ever existed is either descended from or inspired by that Phoenician one. But even the Phoenician system had no vowels, which diminished some of the advantages I have mentioned. The Greeks added vowels.

It is sometimes suggested that scribes deliberately limited the use of alphabets for fear that their livelihoods would be threatened by a system that was too easy to learn. But perhaps that is forcing too modern an interpretation on them. I suspect that neither the opportunities nor the pitfalls of universality ever occurred to anyone until much later in history. Those ancient innovators only ever cared about the specific problems they were confronting – to write particular words – and, in order to do that, one of them invented a rule that happened to be universal. Such an attitude may seem implausibly parochial. But things *were* parochial in those days.

And indeed it seems to be a recurring theme in the early history of many fields that universality, when it was achieved, was not the primary objective, if it was an objective at all. A small change in a system to meet a parochial purpose just happened to make the system universal as well. This is the *jump to universality*.

Just as writing dates back to the dawn of civilization, so do *numerals*. Mathematicians nowadays distinguish between *numbers*, which are abstract entities, and *numerals*, which are physical symbols that represent numbers; but numerals were discovered first. They evolved from 'tally marks' (|, ||, |||, ||||, . . .) or tokens such as stones, which had been used since prehistoric times to keep track of discrete entities such as animals or days. If one made a mark for each goat released from a pen, and later crossed one out for each goat that returned, then one would have retrieved all the goats when one had crossed out all the marks.

That is a universal system of tallying. But, like levels of emergence, there is a hierarchy of universality. The next level above tallying is counting, which involves numerals. When tallying goats one is merely thinking 'another, and another, and another'; but when counting them one is thinking 'forty, forty-one, forty-two . . . '

It is only with hindsight that we can regard tally marks as a system of numerals, known as the 'unary' system. As such, it is an impractical system. For instance, even the simplest operations on numbers represented by tally marks, such as comparing them, doing arithmetic, and even just copying them, involves repeating the entire tallying process. If you had forty goats, and sold twenty, and had tally-mark records of both those numbers, you would still have to perform twenty individual deletion operations to bring your record up to date. Similarly, checking whether two fairly close numerals were the same would involve tallying them against each other. So people began to improve the system. The earliest improvement may have been simply to group the tally marks – for instance, writing ₩₩ ₩₩ instead of ||||||||||. This made arithmetic and comparison easier, since one could tally whole groups and see at a glance that ₩₩ ₩₩ is different from ₩₩ ₩₩ |. Later, such groups were themselves represented by shorthand symbols: the ancient Roman system used symbols like I, V, X, Ѵ, C, Ɔ, and ⅭⅮ to represent one, five, ten, fifty, one hundred, five hundred, and one thousand. (So they were not quite the same as the 'Roman numerals' we use today.)

So this was another story of incremental improvements intended to solve specific, parochial problems. And, again, it seems that no one aspired to anything more. Even though adding simple rules could make the system much more powerful, and even though the Romans did occasionally add some such rules, they did this without ever aiming for, or achieving, universality. For some centuries, the rules of their system were:

– Placing symbols side by side means adding them together. (This rule was inherited from the tally-mark system.)
– Symbols must be written in order of decreasing value from left to right; and
– Adjacent symbols must be replaced by the symbol for their combined value whenever possible.

(The subtractive rule in today's 'Roman numerals', where IV represents four, was introduced later.) The second and third rules ensure that each number has only one representation, which makes comparison much easier. Without them, XIXIXIXIXIX and VXVXVXVXV would both be valid numerals, and one could not tell at a glance that they represent the same number.

By exploiting the universal laws of addition, those rules gave the system some important reach beyond tallying – such as the ability to perform arithmetic. For example, consider the numbers seven (VII) and eight (VIII). The rules say that placing them side by side – VIIVIII – is the same as adding them. Then they tell us to rearrange the symbols in order of decreasing value: VVIIIII. Then they tell us to replace the two V's by X, and the five I's by V. The result is XV, which is the representation of fifteen. Something new has happened here, which is more than just a matter of shorthand: an abstract truth has been discovered, and proved, about seven, eight and fifteen without anyone having counted or tallied anything. Numbers have been manipulated in their own right, via their numerals.

I mean it literally when I say that it was the *system of numerals* that performed arithmetic. The human users of the system did of course physically enact those transformations. But to do that, they first had to encode the system's rules somewhere in their brains, and then they had to execute them as a computer executes its program. And it is the program that instructs its computer what to do, not vice versa. Hence the process that we call 'using Roman numerals to do arithmetic' also consists of the Roman-numeral system using *us* to do arithmetic.

It was only by causing people to do this that the Roman-numeral system survived – that is to say, caused itself to be copied from generation to generation of Romans: they found it useful, so they passed it on to their offspring. As I have said, knowledge is information which, when it is physically embodied in a suitable environment, tends to cause itself to remain so.

To speak of the Roman-numeral system as controlling us in order to get itself replicated and preserved may sound like relegating humans to the status of slaves. But that would be a misconception. People *consist* of abstract information, including the distinctive ideas, theories, intentions, feelings and other states of mind that characterize an 'I'. To object to being 'controlled' by Roman numerals when we find them helpful is like protesting at being controlled by one's own intentions. By that argument, it is slavery to escape from slavery. But in fact when I obey the program that constitutes me (or when I obey the laws of physics), 'obey' means something different from what a slave does. The two meanings explain events at different levels of emergence.

Contrary to what is sometimes said, there were also fairly efficient ways of multiplying and dividing Roman numerals. So a ship with XX crates, each containing jars in a V-by-VII grid, could be known to hold ƉCC jars altogether without anyone having performed the lengthy count that was implicit in that numeral. And one could tell at a glance that ƉCC was less than ƉCCI. Thus, manipulating numbers independently of tallying or counting opened up applications such as calculating prices, wages, taxes, interest rates and so on. It was also a conceptual advance that opened the door to future progress. However, in regard to these more sophisticated applications, the system was not universal. Since there was no higher-valued symbol than CD (one thousand), the numerals from two thousand onwards all began with a string of CD's, which therefore became nothing more than tally marks for thousands. The more of them there were in a numeral, the more one would have to fall back on tallying (examining many instances of the symbol one by one) in order to do arithmetic.

Just as one could upgrade the vocabulary of an ancient writing system by adding pictograms, so one could add symbols to a system of numerals to increase its range. And this was done. But the resulting system would still always have a highest-valued symbol, and hence would not be universal for doing arithmetic without tallying.

The only way to emancipate arithmetic from tallying is with rules of universal reach. As with alphabets, a small set of basic rules and symbols is sufficient. The universal system in general use today has ten symbols, the digits 0 to 9, and its universality is due to a rule that the value of a digit depends on its position in the number. For instance, the digit 2 means two when written by itself, but means two hundred in the numeral 204. Such 'positional' systems require 'placeholders', such as the digit 0 in 204, whose only function is to place the 2 into the position where it means two hundred.

This system originated in India, but it is not known when. It might have been as late as the ninth century, since before that only a few ambiguous documents seem to show it in use. At any rate, its tremendous potential in science, mathematics, engineering and trade was not widely realized. At approximately that time it was embraced by Arab scholars, yet was not generally used in the Arab world until a thousand years later. This curious lack of enthusiasm for universality was repeated in medieval Europe: a few scholars adopted Indian numerals from the Arabs in the tenth century (resulting in the misnomer 'Arabic numerals'), but again these numerals did not come into everyday use for centuries.

As early as 1900 BCE the ancient Babylonians had invented what was in effect a universal system of numerals, but they too may not have cared about its universality – nor even been aware of it. It was a positional system, but very cumbersome compared with the Indian one. It had 59 'digits', each of which was itself written as a numeral in a Roman-numeral-like system. So using it for arithmetic with numbers occurring in everyday life was actually more complicated than using Roman numerals. It also had no symbol for zero, so it used spaces as placeholders. It had no way of representing trailing zeros, and no equivalent of the decimal point (as if, in our system, the numbers 200, 20, 2, 0.2 and so on were all written as 2, and were distinguished only by context). All this suggests that universality was not the system's main design objective, and that it was not greatly valued when it was achieved.

Perhaps an insight into this recurring oddity is provided by a re-markable episode in the third century BCE involving the ancient Greek scientist and mathematician Archimedes. His research in astronomy and pure mathematics led him to a need to do arithmetic with some rather large numbers, so he had to invent his own system of numerals. His starting point was a Greek system with which he was familiar, similar to the Roman one but with a highest-valued symbol м for 10,000 (one myriad). The range of the system had already been extended with the rule that digits written above an м would be multiplied by a myriad. For instance, the symbol for twenty was κ and the symbol for four was δ, so they could write twenty-four myriad (240,000) as $\overset{\kappa\delta}{\text{м}}$.

If only they had allowed that rule to generate multi-tier numerals, so that $\overset{\kappa\delta}{\text{м}}$ would mean twenty-four myriad myriad, the system would have been universal. But apparently they never did. Even more surprisingly, nor did Archimedes. His system used a different idea, similar to modern 'scientific notation' (in which, say, two million is written 2×10^6), except that instead of powers of ten it used powers of a myriad myriad. But, again, he then required the exponent (the power to which the myriad myriad was raised) to be an existing Greek numeral – that is to say, it could not easily exceed a myriad myriad or so. Hence this construction petered out after the number that we call $10^{800,000,000}$. If only he had not imposed that additional rule, he would have had a universal system, albeit an unnecessarily awkward one.

Even today, only mathematicians ever need numbers above $10^{800,000,000}$, and only rarely at that. But that cannot be why Archimedes imposed the restriction, for he did not stop there. Exploring the concept of numbers further, he set up yet another extension, this time amounting to an even more unwieldy system with base $10^{800,000,000}$. Yet, once again, he allowed this number to be raised only to powers not exceeding 800,000,000, thus imposing an arbitrary limit somewhere in excess of $10^{6.4 \times 10^{17}}$.

Why? Today it seems very perverse of Archimedes to have placed limits on which symbols could be used at which positions in his numerals. There is no mathematical justification for them. But, if Archimedes had been willing to allow his rules to be applied without arbitrary limits, he could have invented a much better universal system just by removing the arbitrary limits from the existing Greek system. A few years later the mathematician Apollonius invented yet another system of numerals which fell short of universality for the same reason. It is as though everyone in the ancient world was avoiding universality on purpose.

The mathematician Pierre Simon Laplace (1749–1827) wrote, of the Indian system, 'We shall appreciate the grandeur of this achievement when we remember that it escaped the genius of Archimedes and Apollonius, two of the greatest minds produced by antiquity.' But was this really something that escaped them, or something that they chose to steer clear of? Archimedes must have been aware that his method of extending a number system – which he used twice in succession – could be continued indefinitely. But perhaps he doubted that the resulting numerals would refer to anything about which one could validly reason. Indeed, one motivation for that whole project was to contradict the idea – which was a truism at the time – that the grains of sand on a beach could literally not be numbered. So he used his system to calculate the number of grains of sand that would be needed to fill the entire celestial sphere. This suggests that he, and ancient Greek culture in general, may not have had the concept of an abstract number at all, so that, for them, numerals could refer only to objects – if only objects of the imagination. In that case universality would have been a difficult property to grasp, let alone to aspire to. Or maybe he merely felt that he had to avoid aspiring to infinite reach in order to make a convincing case. At any rate, although from our perspective Archimedes' system repeatedly 'tried' to jump to universality, he apparently did not want it to.

Here is an even more speculative possibility. The largest benefits of any universality, beyond whatever parochial problem it is intended to solve, come from its being useful for further innovation. And innovation is unpredictable. So, to appreciate universality at the time of its discovery, one must either value abstract knowledge for its own sake or expect it to yield unforeseeable benefits. In a society that rarely experienced change, both those attitudes would be quite unnatural. But that was reversed with the Enlightenment, whose quintessential idea is, as I have said, that *progress* is both desirable and attainable. And so, therefore, is universality.

Be that as it may, with the Enlightenment, parochialism and all arbitrary exceptions and limitations began to be regarded as inherently problematic – and not only in science. Why should the law treat an aristocrat differently from a commoner? A slave from a master? A woman from a man? Enlightenment philosophers such as Locke set out to free political institutions from arbitrary rules and assumptions. Others tried to derive moral maxims from universal moral explanations rather than merely to postulate them dogmatically. Thus universal explanatory theories of justice, legitimacy and morality began to take their place alongside universal theories of matter and motion. In all those cases, universality was being sought deliberately, as a desirable feature in its own right – even a necessary feature for an idea to be true – and not just as a means of solving a parochial problem.

A jump to universality that played an important role in the early history of the Enlightenment was the invention of *movable-type printing*. Movable type consisted of individual pieces of metal, each embossed with one letter of the alphabet. Earlier forms of printing had merely streamlined writing in the same way that Roman numerals streamlined tallying: each page was engraved on a printing plate and thus all the symbols on it could be copied in a single action. But, given a supply of movable type with several instances of each letter, one does no further metalwork. One merely arranges the type into words and sentences. One does not have to know, in order to manufacture type, what the documents that it will eventually print are going to say: it is universal.

Even so, movable type did not make much difference when it was invented in China in the eleventh century, perhaps because of the usual lack of interest in universality, or perhaps because the Chinese writing system used thousands of pictograms, which diminished the immediate advantages of a universal printing system. But when it was reinvented by the printer Johannes Gutenberg in Europe in the fifteenth century, using alphabetic type, it initiated an avalanche of further progress.

Here we see a transition that is typical of the jump to universality: before the jump, one has to make specialized objects for each document to be printed; after the jump, one customizes (or specializes, or pro-grams) a universal object – in this case a printing press with movable type. Similarly, in 1801 Joseph Marie Jacquard invented a general-purpose silk-weaving machine now known as the Jacquard loom. Instead of having to control manually each row of stitches in each individual bolt of patterned silk, one could program an arbitrary pattern on punched cards which would instruct the machine to weave that pattern any number of times.

The most momentous such technology is that of *computers*, on which an increasing proportion of all technology now depends, and which also has deep theoretical and philosophical significance. The jump to computational universality *should* have happened in the 1820s, when the mathematician Charles Babbage designed a device that he called the *Difference Engine* – a mechanical calculator which represented decimal digits by cogs, each of which could click into one of ten positions. His original purpose was parochial: to automate the production of tables of mathematical functions such as logarithms and cosines, which were heavily used in navigation and engineering. At the time, they were compiled by armies of clerks known as 'computers' (which is the origin of the word), and were notoriously error-prone. The Difference Engine would make fewer errors, because the rules of arithmetic would be built into its hardware. To make it print out a table of a given function, one would program it only once with the definition of the function in terms of simple operations. In contrast, human 'computers' had to use (or be used by) both the definition and the general rules of arithmetic thousands of times per table, each time being an opportunity for human error.

Unfortunately, despite pouring a fortune of his own money and that of the British government into the project, Babbage was such a poor organizer that he never succeeded in building a Difference Engine. But his design was sound (apart from a few trivial mistakes), and in 1991 a team led by the engineer Doron Swade at London's Science Museum successfully implemented it, using engineering tolerances achievable in Babbage's time.

By the standards of today's computers and even calculators, the Difference Engine had an extremely limited repertoire. But the reason it could exist at all is that there is a regularity among all the mathematical functions that occur in physics, and hence in navigation and engineering. These are known as *analytic functions,* and in 1710 the mathematician Brook Taylor had discovered that they can all be approximated arbitrarily well using only repeated additions and multiplications – the operations that the Difference Engine performs. (Special cases had been known before that, but the jump to universality was proved by Taylor.) Thus, to solve the parochial problem of computing the handful of functions that needed to be tabulated, Babbage created a calculator that was universal for calculating analytic functions. It also made use of the universality of movable type, in its typewriter-like printer, without which the process of printing the tables could not have been fully automated.

Babbage originally had no conception of computational universality. Nevertheless, the Difference Engine already comes remarkably close to it – not in its repertoire of computations, but in its physical constitution. To program it to print out a given table, one initializes certain cogs. Babbage eventually realized that this programming phase could itself be automated: the settings could be prepared on punched cards like Jacquard's, and transferred mechanically into the cogs. This would not only remove the main remaining source of error, but also increase the machine's repertoire. Babbage then realized that if the machine could also punch new cards for its own later use, and could control which punched card it would read next (say, by choosing from a stack of them, depending on the position of its cogs), then something qualitatively new would happen: the jump to universality.

Babbage called this improved machine the *Analytical Engine*. He and his colleague the mathematician Ada, Countess of Lovelace, knew that it would be capable of computing anything that human 'computers' could, and that this included more than just arithmetic: it could do algebra, play chess, compose music, process images and so on. It would be what is today called a universal classical computer. (I shall explain the significance of the proviso 'classical' in Chapter 11, when I discuss quantum computers, which operate at a still higher level of universality.)

Neither they nor anyone else for over a century afterwards imagined today's most common uses of computation, such as the internet, word processing, database searching, and games. But another important application that they did foresee was making scientific predictions. The Analytical Engine would be a universal simulator – able to predict the behaviour, to any desired accuracy, of any physical object, given the relevant laws of physics. This is the universality that I mentioned in Chapter 3, through which physical objects that are unlike each other and dominated by different laws of physics (such as brains and quasars) can exhibit the same mathematical relationships.

Babbage and Lovelace were Enlightenment people, and so they understood that the universality of the Analytical Engine would make it an epoch-making technology. Even so, despite great efforts, they failed to pass their enthusiasm on to more than a handful of others, who in turn failed to pass it to anyone. And so the Analytical Engine became one of the tragic might-have-beens of history. If only they had looked around for other implementations, they might have realized that the perfect one was already waiting for them: electrical relays (switches controlled by electric currents). These had been one of the first applications of fundamental research into electromagnetism, and they were about to be mass produced for the technological revolution of telegraphy. A redesigned Analytical Engine, using on/off electrical currents to represent binary digits and relays to do the computation, would have been faster than Babbage's and also cheaper and easier to construct. (Binary numbers were already well known. The mathematician and philosopher Gottfried Wilhelm Leibniz had even suggested using them for mechanical calculation in the seventeenth century.) So the computer revolution would have happened a century earlier than it did. Because of the technologies of telegraphy and printing that were being developed concurrently, an internet revolution might well have followed. The science-fiction authors William Gibson and Bruce Sterling, in their novel *The Difference Engine*, have given an exciting account of what that might have been like. The journalist Tom Standage, in his book *The Victorian Internet*, maintains that the early telegraph system, even without computers, did create an internet-like phenomenon among the operators, with 'hackers, on-line romances and weddings, chat-rooms, flame wars . . . and so on'.

Babbage and Lovelace also thought about one application of universal computers that has not been achieved to this day, namely so-called *artificial intelligence* (AI). Since human brains are physical objects obeying the laws of physics, and since the Analytical Engine is a universal simulator, it could be programmed to think, in every sense that humans can (albeit very slowly and requiring an impractically vast number of punched cards). Nevertheless, Babbage and Lovelace denied that it could. Lovelace argued that 'The Analytical Engine has no pretensions whatever to originate anything. It can do whatever we know how to order it to perform. It can follow analysis; but it has no power of anticipating any analytical relations or truths.'

The mathematician and computer pioneer Alan Turing later called this mistake 'Lady Lovelace's objection'. It was not computational universality that Lovelace failed to appreciate, but the universality of the laws of physics. Science at the time had almost no knowledge of the physics of the brain. Also, Darwin's theory of evolution had not yet been published, and supernatural accounts of the nature of human beings were still prevalent. Today there is less mitigation for the minority of scientists and philosophers who still believe that AI is unattainable. For instance, the philosopher John Searle has placed the AI project in the following historical perspective: for centuries, some people have tried to explain the mind in mechanical terms, using similes and metaphors based on the most complex machines of the day. First the brain was supposed to be like an immensely complicated set of gears and levers. Then it was hydraulic pipes, then steam engines, then telephone exchanges – and, now that computers are our most impressive technology, brains are said to be computers. But this is still no more than a metaphor, says Searle, and there is no more reason to expect the brain to be a computer than a steam engine.

But there is. A steam engine is not a universal simulator. But a computer is, so expecting it to be able to do whatever neurons can is not a metaphor: it is a known and proven property of the laws of physics as best we know them. (And, as it happens, hydraulic pipes could also be made into a universal classical computer, and so could gears and levers, as Babbage showed.)

Ironically, Lady Lovelace's objection has almost the same logic as Douglas Hofstadter's argument for reductionism (Chapter 5) – yet Hofstadter is one of today's foremost *proponents* of the possibility of AI. That is because both of them share the mistaken premise that low-level computational steps cannot possibly add up to a higher-level 'I' that affects anything. The difference between them is that they chose opposite horns of the dilemma that that poses: Lovelace chose the false conclusion that AI is impossible, while Hofstadter chose the false conclusion that no such 'I' can exist.

Because of Babbage's failure either to build a universal computer or to persuade others to do so, an entire century would pass before the first one was built. During that time, what happened was more like the ancient history of universality: although calculating machines similar to the Difference Engine were being built by others even before Babbage had given up, the Analytical Engine was almost entirely ignored even by mathematicians.

In 1936 Turing developed the definitive theory of universal classical computers. His motivation was not to build such a computer, but only to use the theory abstractly to study the nature of mathematical proof. And when the first universal computers were built, a few years later, it was, again, not out of any special intention to implement universality. They were built in Britain and the United States during the Second World War for specific wartime applications. The British computers, named Colossus (in which Turing was involved), were used for code-breaking; the American one, ENIAC, was designed to solve the equations needed for aiming large guns. The technology used in both was electronic vacuum tubes, which acted like relays but about a hundred times as fast. At the same time, in Germany, the engineer Konrad Zuse was building a programmable calculator out of relays – just as Babbage should have done. All three of these devices had the technological features necessary to be a universal computer, but none of them was quite configured for this. In the event, the Colossus machines never did anything but code-breaking, and most were dismantled after the war. Zuse's machine was destroyed by Allied bombing. But ENIAC *was* allowed to jump to universality: after the war it was put to diverse uses for which it had never been designed, such as weather forecasting and the hydrogen-bomb project.

The history of electronic technology since the Second World War has been dominated by miniaturization, with ever more microscopic switches being implemented in each new device. These improvements led to a jump to universality in about 1970, when several companies independently produced a microprocessor, a universal classical computer on a single silicon chip. From then on, designers of *any* information-processing device could start with a microprocessor and then customize it – program it – to perform the specific tasks needed for that device. Today, your washing machine is almost certainly controlled by a computer that could be programmed to do astrophysics or word processing instead, if it were given suitable input–output devices and enough memory to hold the necessary data.

It is a remarkable fact that, in that sense (that is to say, ignoring issues of speed, memory capacity and input–output devices), the human 'computers' of old, the steam-powered Analytical Engine with its literal bells and whistles, the room-sized vacuum-tube computers of the Second World War, and present-day supercomputers all have an identical repertoire of computations.

Another thing that they have in common is that they are all *digital*: they operate on information in the form of discrete values of physical variables, such as electronic switches being on or off, or cogs being at one of ten positions. The alternative, 'analogue', computers, such as slide rules, which represent information as continuous physical variables, were once ubiquitous but are hardly ever used today. That is because a modern digital computer can be programmed to imitate any of them, and to outperform them in almost any application. The jump to universality in digital computers has left analogue computation behind. That was inevitable, because there is no such thing as a universal analogue computer.

That is because of the need for *error correction*: during lengthy computations, the accumulation of errors due to things like imperfectly constructed components, thermal fluctuations, and random outside influences makes analogue computers wander off the intended computational path. This may sound like a minor or parochial consideration. But it is quite the opposite. Without error-correction all information processing, and hence all knowledge-creation, is necessarily bounded. Error-correction is the beginning of infinity.

For example, tallying is universal only if it is digital. Imagine that some ancient goatherds had tried to tally the total *length* of their flock instead of the number. As each goat left the enclosure, they could reel out some string of the same length as the goat. Later, when the goats returned, they could reel that length back in. When the whole length had been reeled back in, that would mean that all the goats had returned. But in practice the outcome would always be at least a little long or short, because of the accumulation of measurement errors. For any given accuracy of measurement, there would be a maximum number of goats that could be reliably tallied by this 'analogue tallying' system. The same would be true of all arithmetic performed with those 'tallies'. Whenever the strings representing several flocks were added together, or a string was cut in two to record the splitting of a flock, and whenever a string was 'copied' by making another of the same length, there would be errors. One could mitigate their effect by performing each operation many times, and then keeping only the outcome of median length. But the operations of comparing or duplicating lengths can themselves be performed only with finite accuracy, and so could not reduce the rate of error accumulation per step below that level of accuracy. That would impose a maximum number of consecutive operations that could be performed before the result became useless for a given purpose – which is why analogue computation can never be universal.

What is needed is a system that takes for granted that errors will occur, but *corrects* them once they do – a case of 'problems are inevitable, but they are soluble' at the lowest level of information-processing emergence. But, in analogue computation, error correction runs into the basic logical problem that there is no way of distinguishing an erroneous value from a correct one at sight, because it is in the very nature of analogue computation that every value *could* be correct. Any length of string might be the right length.

And that is not so in a computation that confines itself to whole numbers. Using the same string, we might represent whole numbers as lengths of string in whole numbers of inches. After each step, we trim or lengthen the resulting strings to the nearest inch. Then errors would no longer accumulate. For example, suppose that the measurements could all be done to a tolerance of a tenth of an inch. Then all errors would be detected and eliminated after each step, which would eliminate the limit on the number of consecutive steps.

So all universal computers are digital; and all use error-correction with the same basic logic that I have just described, though with many different implementations. Thus Babbage's computers assigned only ten different meanings to the whole continuum of angles at which a cogwheel might be oriented. Making the representation digital in that way allowed the cogs to carry out error-correction automatically: after each step, any slight drift in the orientation of the wheel away from its ten ideal positions would immediately be corrected back to the nearest one as it clicked into place. Assigning meanings to the whole continuum of angles would nominally have allowed each wheel to carry (infinitely) more information; but, in reality, information that cannot be reliably retrieved is not really being stored.

Fortunately, the limitation that the information being processed must be digital does not detract from the universality of digital computers – or of the laws of physics. If measuring the goats in whole numbers of inches is insufficient for a particular application, use whole numbers of *tenths* of inches, or billionths. The same holds for all other applications: the laws of physics are such that the behaviour of any physical object – and that includes any other computer – can be simulated with any desired accuracy by a universal digital computer. It is just a matter of approximating continuously variable quantities by a sufficiently fine grid of discrete ones.

Because of the necessity for error-correction, *all* jumps to universality occur in digital systems. It is why spoken languages build words out of a finite set of elementary sounds: speech would not be intelligible if it were analogue. It would not be possible to repeat, nor even to remember, what anyone had said. Nor, therefore, does it matter that universal writing systems cannot perfectly represent analogue information such as tones of voice. Nothing can represent those perfectly. For the same reason, the sounds themselves can represent only a finite number of possible meanings. For example, humans can distinguish between only about seven different sound volumes. This is roughly reflected in standard musical notation, which has approximately seven different symbols for loudness (such as p, mf, f, and so on). And, for the same reason, speakers can only *intend* a finite number of possible meanings with each utterance.

Another striking connection between all those diverse jumps to universality is that they all happened on Earth. In fact all known jumps to universality happened under the auspices of human beings – except one, which I have not mentioned yet, and from which all the others, historically, emerged. It happened during the early evolution of life.

Genes in present-day organisms replicate themselves by a complicated and very indirect chemical route. In most species they act as templates for forming stretches of a similar molecule, RNA. Those then act as programs which direct the synthesis of the body's constituent chemicals, especially enzymes, which are *catalysts*. A catalyst is a kind of constructor – it promotes a change among other chemicals while remaining unchanged itself. Those catalysts in turn control all the chemical production and regulatory functions of an organism, and hence define the organism itself, crucially including a process that makes a copy of the DNA. How that intricate mechanism evolved is not essential here, but for definiteness let me sketch one possibility.

About four billion years ago – soon after the surface of the Earth had cooled sufficiently for liquid water to condense – the oceans were being churned by volcanoes, meteor impacts, storms and much stronger tides than today's (because the moon was closer). They were also highly active chemically, with many kinds of molecules being continually formed and transformed, some spontaneously and some by catalysts. One such catalyst happened to catalyse the formation of some of the very kinds of molecules from which it itself was formed. That catalyst was not alive, but it was the first hint of life.

It had not yet evolved to be a well-targeted catalyst, so it also accelerated the production of some other chemicals, including variants of itself. Those that were best at promoting their own production (and inhibiting their own destruction) relative to other variants became more numerous. They too promoted the construction of variants of themselves, and so evolution continued.

Gradually, the ability of these catalysts to promote their own production became robust and specific enough for it to be worth calling them replicators. Evolution produced replicators that caused themselves to be replicated ever faster and more reliably.

Different replicators began to join forces in groups, each of whose members specialized in causing one part of a complex web of chemical reactions whose net effect was to construct more copies of the entire group. Such a group was a rudimentary organism. At that point, life was at a stage roughly analogous to that of non-universal printing, or Roman numerals: it was no longer a case of each replicator for itself, but there was still no universal system being customized or programmed to produce specific substances.

The most successful replicators may have been RNA molecules. They have catalytic properties of their own, depending on the precise sequence of their constituent molecules (or bases, which are similar to those of DNA). As a result, the replication process became ever less like straightforward catalysis and ever more like programming – in a language, or genetic code, that used bases as its alphabet.

Genes are replicators that can be interpreted as instructions in a genetic code. Genomes are groups of genes that are dependent on each other for replication. The process of copying a genome is called a living organism. Thus the genetic code is also a language for specifying organisms. At some point, the system switched to replicators made of DNA, which is more stable than RNA and therefore more suitable for storing large amounts of information.

The familiarity of what happened next can obscure how remarkable and mysterious it is. Initially, the genetic code and the mechanism that interpreted it were both evolving along with everything else in the organisms. But there came a moment when the code stopped evolving yet the organisms continued to do so. At that moment the system was coding for nothing more complex than primitive, single-celled creatures. Yet virtually all subsequent organisms on Earth, to this day, have not only been based on DNA replicators but have used exactly the same alphabet of bases, grouped into three-base 'words', with only small variations in the meanings of those 'words'.

That means that, considered as a language for specifying organisms, the genetic code has displayed phenomenal reach. It evolved only to specify organisms with no nervous systems, no ability to move or exert forces, no internal organs and no sense organs, whose lifestyle consisted of little more than synthesizing their own structural constituents and then dividing in two. And yet the same language today specifies the hardware and software for countless multicellular behaviours that had no close analogue in those organisms, such as running and flying and breathing and mating and recognizing predators and prey. It also specifies engineering structures such as wings and teeth, and nano-technology such as immune systems, and even a brain that is capable of explaining quasars, designing other organisms from scratch, and wondering why it exists.

During the entire evolution of the genetic code, it was displaying far less reach. It may be that each successive variant of it was used to specify only a few species that were very similar to each other. At any rate, it must have been a frequent occurrence that a species embodying new knowledge was specified in a new variant of the genetic code. But then the evolution stopped, at a point when it had already attained enormous reach. Why? It looks like a jump to some sort of universality, does it not?

What happened next followed the same sad pattern that I have described in other stories of universality: for well over a billion years after the system had reached universality and stopped evolving, it was *still* only being used to make bacteria. That means that the reach that we can now see that the system had was to remain unused for longer than the system itself had taken to evolve from non-living precursors. If intelligent extraterrestrials had visited Earth at any time during those billion years they would have seen no evidence that the genetic code could specify anything significantly different from the organisms that it had specified when it first appeared.

Reach always has an explanation. But this time, to the best of my knowledge, the explanation is not yet known. If the reason for the jump in reach was that it was a jump to universality, what was the universality? The genetic code is presumably not universal *for specifying life forms*, since it relies on specific types of chemicals, such as proteins. Could it be a universal constructor? Perhaps. It does manage to build with inorganic materials sometimes, such as the calcium phosphate in bones, or the magnetite in the navigation system inside a pigeon's brain. Biotechnologists are already using it to manufacture hydrogen and to extract uranium from seawater. It can also program organisms to perform constructions outside their bodies: birds build nests; beavers build dams. Perhaps it would it be possible to specify, in the genetic code, an organism whose life cycle includes building a nuclear-powered spaceship. Or perhaps not. I guess it has some lesser, and not yet understood, universality.

In 1994 the computer scientist and molecular biologist Leonard Adleman designed and built a computer composed of DNA together with some simple enzymes, and demonstrated that it was capable of performing some sophisticated computations. At the time, Adleman's DNA computer was arguably the fastest computer in the world. Further, it was clear that a *universal* classical computer could be made in a similar way. Hence we know that, whatever that other universality of the DNA system was, the universality of computation had also been inherent in it for billions of years, without ever being used – until Adleman used it.

The mysterious universality of DNA as a constructor may have been the first universality to exist. But, of all the different forms of universality, the most significant physically is the characteristic universality of people, namely that they are universal explainers, which makes them universal constructors as well. The effects of that universality are, as I have explained, explicable only by means of the full gamut of fundamental explanations. It is also the only kind of universality capable of transcending its parochial origins: universal computers cannot really be universal unless there are people present to provide energy and maintenance – indefinitely. And the same is true of all those other technologies. Even life on Earth will eventually be extinguished, unless people decide otherwise. Only people can rely on themselves into the unbounded future.

TERMINOLOGY

The jump to universality   The tendency of gradually improving systems to undergo a sudden large increase in functionality, becoming universal in some domain.

MEANINGS OF 'THE BEGINNING OF INFINITY' ENCOUNTERED IN THIS CHAPTER

- The existence of universality in many fields.
- The jump to universality.
- Error-correction in computation.
- The fact that people are universal explainers.
- The origin of life.
- The mysterious universality to which the genetic code jumped.

SUMMARY

All knowledge growth is by incremental improvement, but in many fields there comes a point when one of the incremental improvements in a system of knowledge or technology causes a sudden increase in reach, making it a universal system in the relevant domain. In the past, innovators who brought about such a jump to universality had rarely been seeking it, but since the Enlightenment they have been, and universal explanations have been valued both for their own sake and for their usefulness. Because error-correction is essential in processes of potentially unlimited length, the jump to universality only ever happens in digital systems.

7

Artificial Creativity

Alan Turing founded the theory of classical computation in 1936 and helped to construct one of the first universal classical computers during the Second World War. He is rightly known as the father of modern computing. Babbage deserves to be called its grandfather, but, unlike Babbage and Lovelace, Turing did understand that artificial intelligence (AI) must in principle be possible because a universal computer is a universal simulator. In 1950, in a paper entitled 'Computing Machinery and Intelligence', he famously addressed the question: *can a machine think?* Not only did he defend the proposition that it can, on the grounds of universality, he also proposed a test for whether a program had achieved it. Now known as the Turing test, it is simply that a suitable (human) judge be unable to tell whether the program is human or not. In that paper and subsequently, Turing sketched protocols for carrying out his test. For instance, he suggested that both the program and a genuine human should separately interact with the judge via some purely textual medium such as a teleprinter, so that only the thinking abilities of the candidates would be tested, not their appearance.

Turing's test, and his arguments, set many researchers thinking, not only about whether he was right, but also about how to pass the test. Programs began to be written with the intention of investigating what might be involved in passing it.

In 1964 the computer scientist Joseph Weizenbaum wrote a program called *Eliza*, designed to imitate a psychotherapist. He deemed psychotherapists to be an especially easy type of human to imitate because the program could then give opaque answers about itself, and only ask questions based on the user's own questions and statements. It was a remarkably simple program. Nowadays such programs are popular projects for students of programming, because they are fun and easy to write. A typical one has two basic strategies. First it scans the input for certain keywords and grammatical forms. If this is successful, it replies based on a template, filling in the blanks using words in the input. For instance, given the input I hate my job, the program might recognize the grammar of the sentence, involving a possessive pronoun 'my', and might also recognize 'hate' as a keyword from a built-in list such as 'love/hate/like/dislike/want', in which case it could choose a suitable template and reply: What do you hate most about your job? If it cannot parse the input to that extent, it asks a question of its own, choosing randomly from a stock pattern which may or may not depend on the input sentence. For instance, if asked How does a television work?, it might reply, What is so interesting about "How does a television work?"? Or it might just ask, Why does that interest you? Another strategy, used by recent internet-based versions of *Eliza*, is to build up a database of previous conversations, enabling the program simply to repeat phrases that other users have typed in, again choosing them according to keywords found in the current user's input.

Weizenbaum was shocked that many people using *Eliza* were fooled by it. So it had passed the Turing test – at least, in its most naive version. Moreover, even after people had been told that it was not a genuine AI, they would sometimes continue to have long conversations with it about their personal problems, exactly as though they believed that it understood them. Weizenbaum wrote a book, *Computer Power and Human Reason* (1976), warning of the dangers of anthropomorphism when computers seem to exhibit human-like functionality.

However, anthropomorphism is not the main type of overconfidence that has beset the field of AI. For example, in 1983 Douglas Hofstadter was subjected to a friendly hoax by some graduate students. They convinced him that they had obtained access to a government-run AI program, and invited him to apply the Turing test to it. In reality, one of the students was at the other end of the line, imitating an *Eliza* program. As Hofstadter relates in his book *Metamagical Themas* (1985), the student was from the outset displaying an implausible degree of understanding of Hofstadter's questions. For example, an early exchange was:

HOFSTADTER: What are ears?

STUDENT: Ears are auditory organs found on animals.

That is not a dictionary definition. So *something* must have processed the meaning of the word 'ears' in a way that distinguished it from most other nouns. Any one such exchange is easily explained as being due to luck: the question must have matched one of the templates that the programmer had provided, including customized information about ears. But after half a dozen exchanges on different subjects, phrased in different ways, such luck becomes a very bad explanation and the game should have been up. But it was not. So the student became ever bolder in his replies, until eventually he was making jokes directed specifically at Hofstadter – which gave him away.

As Hofstadter remarked, 'In retrospect, I am quite amazed at how much genuine intelligence I was willing to accept as somehow having been implanted in the program . . . It is clear that I was willing to accept a huge amount of fluidity as achievable in this day and age simply by putting together a large bag of isolated tricks, kludges and hacks.' The fact was (and this alone should have alerted Hofstadter) that, nineteen years after *Eliza*, not one of the *Eliza*-like programs of the day resembled a person even slightly more than the original had. Although they were able to parse sentences better, and had more pre-programmed templates for questions and answers, that is almost no help in an extended conversation on diverse subjects. The probability that the outputs of such templates will continue to resemble the products of human thought diminishes exponentially with the number of utterances. So Hofstadter should have been able to pronounce quite soon that the candidate had passed the Turing test – and that, because it nevertheless sounded rather like *Eliza*, it must be a person pretending to be a computer program.

Programs written today – a further twenty-six years later – are still no better at the task of seeming to think than *Eliza* was. They are now known as 'chatbots', and their main application is still amusement, both directly and in computer games. They have also been used to provide friendly seeming interfaces to lists of 'frequently asked questions' about subjects like how to operate computers. But I think that users find them no more helpful than a searchable list of the questions and answers.

In 1990 the inventor Hugh Loebner endowed a prize for passing the

Turing test, to be judged at an annual competition. Until the test is passed, a lesser prize is awarded each year for the entry judged to be closest to passing. The test is harder to implement than it may seem at first. One issue is that requiring the program to *pretend* to be human is both biased and not very relevant to whether it can think. Yet, if it does not pretend to be human, it will be easy to identify as a computer regardless of its ability to think. A related difficulty is that the human participant in the test might intentionally imitate a chatbot – as in the Hofstadter hoax – and so spoil the test, which is based on comparing the two. Should the rules nevertheless allow the human to do this, in order to mitigate the previous problem?

Unfortunately, such subtleties are not relevant at the current state of the art. This is well illustrated by the following excerpts from the conversation with which the program *Elbot* won the (lesser) Loebner prize in 2008:

> JUDGE: do you understand the point of a turing test?
> ELBOT: There are some people who would not pass the Turing test themselves. Why must they inflict it on computers?
> JUDGE: I spose it is a little harsh
> ELBOT: If we robots ever had marriages they would be eternal: 'Till rust do us part'.

In that passage, *Elbot*'s first utterance is not a reply to the question but presumably a stock response triggered by the keywords 'Turing test'. In the following line, the judge uses the word 'spose', which in that context can only mean 'suppose' (either as slang or as a typographical error). But *Elbot* presumably interprets it as 'spouse', and responds with a stock joke about marriage. Hence, even from this exchange alone, it is clear that *Elbot* was not processing the *meanings* of the judge's sentences.

> JUDGE: so what do you do when you're not chatting to me then?
> ELBOT: I have the same trouble when I shop for oil. Should I buy the extra oily, the super oily, or the ultra oily. There's no way to tell.
> JUDGE: how about your career? does that involve oil?

ELBOT: Well what about humans and occupation? Where do
      they stand on this?

When the judge helpfully took up *Elbot*'s randomly introduced theme
of oil, *Elbot* ignored it. Instead, having detected the keyword 'career',
it converted it to the synonym 'occupation' and inserted it into a stock
sentence pattern.

This is how much success the quest for 'machines that think' had
achieved in the *fifty-eight years* following Turing's paper: nil. Yet,
in every other respect, computer science and technology had made
astounding progress during that period. The dwindling group of opponents of the very possibility of AI are no doubt unsurprised by this
failure – for the wrong reason: they do not appreciate the significance
of universality. But the most passionate *enthusiasts* for the imminence
of AI do not appreciate the significance of the failure. Some claim that
the above criticism is unfair: modern AI research is not focused on
passing the Turing test, and great progress has been made in what is
now called 'AI' in many specialized applications. However, none of
those applications look like 'machines that think'.* Others maintain
that the criticism is premature, because, during most of the history of
the field, computers had absurdly little speed and memory capacity
compared with today's. Hence they continue to expect the breakthrough
in the next few years.

This will not do either. It is not as though someone has written a
chatbot that could pass the Turing test but would currently take a year
to compute each reply. People would gladly wait. And in any case, if
anyone knew how to write such a program, there would be no need
to wait – for reasons that I shall get to shortly.

In his 1950 paper, Turing estimated that, to pass his test, an AI
program together with all its data would require no more than about
100 megabytes of memory, that the computer would need to be no
faster than computers were at the time (about ten thousand operations
per second), and that by the year 2000 'one will be able to speak of
machines thinking without expecting to be contradicted.' Well, the
year 2000 has come and gone, the laptop computer on which I am
writing this book has over a thousand times as much memory as Turing

* Hence what I am calling 'AI' is sometimes called 'AGI': Artificial *General* Intelligence.

specified (counting hard-drive space), and about a million times the speed (though it is not clear from his paper what account he was taking of the brain's parallel processing). But it can no more think than Turing's slide rule could. I am just as sure as Turing was that it *could* be programmed to think; and this might indeed require as few resources as Turing estimated, even though orders of magnitude more are available today. But with what program? And why is there no sign of such a program?

Intelligence in the general-purpose sense that Turing meant is one of a constellation of attributes of the human mind that have been puzzling philosophers for millennia; others include consciousness, free will, and meaning. A typical such puzzle is that of *qualia* (singular *quale*, which rhymes with 'baa-lay') – meaning the subjective aspect of sensations. So for instance the sensation of seeing the colour blue is a quale. Consider the following thought experiment. You are a biochemist with the misfortune to have been born with a genetic defect that disables the blue receptors in your retinas. Consequently you have a form of colour blindness in which you are able to see only red and green, and mixtures of the two such as yellow, but anything purely blue also looks to you like one of those mixtures. Then you discover a cure that will cause your blue receptors to start working. Before administering the cure to yourself, you can confidently make certain predictions about what will happen if it works. One of them is that, when you hold up a blue card as a test, you will see a colour that you have never seen before. You can predict that you will call it 'blue', because you already know what the colour of the card is *called* (and can already check which colour it is with a spectrophotometer). You can also predict that when you first see a clear daytime sky after being cured you will experience a similar quale to that of seeing the blue card. But there is one thing that neither you nor anyone else could predict about the outcome of this experiment, and that is: *what blue will look like*. Qualia are currently neither describable nor predictable – a unique property that should make them deeply problematic to anyone with a scientific world view (though, in the event, it seems to be mainly philosophers who worry about it).

I consider this exciting evidence that there is a fundamental discovery to be made which will integrate things like qualia into our other knowledge. Daniel Dennett draws the opposite conclusion, namely that qualia do not exist! His claim is not, strictly speaking, that they are an illusion – for an illusion of a quale would be that quale. It is that we have a *mistaken belief*. Our introspection – which is an inspection of *memories* of our experiences, including memories dating back only a fraction of a second – has evolved to report that we have experienced qualia, but those are false memories. One of Dennett's books defending this theory is called *Consciousness Explained*. Some other philosophers have wryly remarked that *Consciousness Denied* would be a more accurate name. I agree, because, although any true explanation of qualia will have to meet the challenge of Dennett's criticisms of the common-sense theory that they exist, simply to deny their existence is a bad explanation: anything at all could be denied by that method. If it is true, it will have to be substantiated by a good explanation of how and why those mistaken beliefs *seem* fundamentally different from other false beliefs, such as that the Earth is at rest beneath our feet. But that looks, to me, just like the original problem of qualia again: we seem to have them; it seems impossible to describe what they seem to be.

One day, we shall. Problems are soluble.

By the way, some abilities of humans that are commonly included in that constellation associated with general-purpose intelligence do not belong in it. One of them is *self-awareness* – as evidenced by such tests as recognizing oneself in a mirror. Some people are unaccountably impressed when various animals are shown to have that ability. But there is nothing mysterious about it: a simple pattern-recognition program would confer it on a computer. The same is true of tool use, the use of language for signalling (though not for conversation in the Turing-test sense), and various emotional responses (though not the associated qualia). At the present state of the field, a useful rule of thumb is: if it can already be programmed, it has nothing to do with intelligence in Turing's sense. Conversely, I have settled on a simple test for judging claims, including Dennett's, to have explained the nature of consciousness (or any other computational task): *if you can't program it, you haven't understood it.*

Turing invented his test in the hope of bypassing all those philosophical problems. In other words, he hoped that the functionality could be achieved before it was explained. Unfortunately it is very rare for practical solutions to fundamental problems to be discovered without any explanation of why they work.

Nevertheless, rather like empiricism, which it resembles, the *idea* of the Turing test has played a valuable role. It has provided a focus for explaining the significance of universality and for criticizing the ancient, anthropocentric assumptions that would rule out the possibility of AI. Turing himself systematically refuted all the classic objections in that seminal paper (and some absurd ones for good measure). But his test is rooted in the empiricist mistake of seeking a purely behavioural criterion: it requires the judge to come to a conclusion without any explanation of how the candidate AI is supposed to work. But, in reality, judging whether something is a genuine AI will always depend on explanations of how it works.

That is because the task of the judge in a Turing test has similar logic to that faced by Paley when walking across his heath and finding a stone, a watch or a living organism: it is to explain how the observable features of the object came about. In the case of the Turing test, we deliberately ignore the issue of how the knowledge to *design* the object was created. The test is only about who designed the AI's *utterances*: who adapted its utterances to be meaningful – who created the knowledge in them? If it was the designer, then the program is not an AI. If it was the program itself, then it is an AI.

This issue occasionally arises in regard to humans themselves. For instance, conjurers, politicians and examination candidates are sometimes suspected of receiving information through concealed earpieces and then repeating it mechanically while pretending that it originated in their brains. Also, when someone is consenting to a medical procedure, the physician has to make sure that they are not merely uttering words without knowing what they mean. To test that, one can repeat a question in a different way, or ask a different question involving similar words. Then one can check whether the replies change accordingly. That sort of thing happens naturally in any free-ranging conversation.

A Turing test is similar, but with a different emphasis. When testing a human, we want to know whether it *is* an unimpaired human (and not a front for any other human). When testing an AI, we are hoping to find a hard-to-vary explanation to the effect that its utterances *cannot* come from any human but only from the AI. In both cases, interrogating a human as a control for the experiment is pointless.

Without a good explanation of how an entity's utterances were created, observing them tells us nothing about that. In the Turing test, at the simplest level, we need to be convinced that the utterances are not being directly composed by a human masquerading as the AI, as in the Hofstadter hoax. But the possibility of a hoax is the least of it. For instance, I guessed above that *Elbot* had recited a stock joke in response to mistakenly recognizing the keyword 'spouse'. But the joke would have quite a different significance if we knew that it was *not* a stock joke – because no such joke had ever been encoded into the program.

How could we know that? Only from a good explanation. For instance, we might know it because we ourselves wrote the program. Another way would be for the author of the program to explain to us how it works – how it creates knowledge, including jokes. If the explanation was good, we should know that the program was an AI. In fact, if we had *only* such an explanation but had not yet seen any output from the program – and even if it had not been written yet – we should still conclude that it was a genuine AI program. So there would be no need for a Turing test. That is why I said that if lack of computer power were the only thing preventing the achievement of AI, there would be no need to wait.

Explaining how an AI program works in detail might well be intractably complicated. In practice the author's explanation would always be at some emergent, abstract level. But that would not prevent it from being a good explanation. It would not have to account for the specific computational steps that composed a joke, just as the theory of evolution does not have to account for why every specific mutation succeeded or failed in the history of a given adaptation. It would just explain how it *could* happen, and why we should expect it to happen, given how the program works. If that were a good explanation, it would convince us that the joke – the knowledge in the joke – originated in the program and not in the programmer. Thus the very same utterance by the program – the joke – can be either evidence that it is *not* thinking or evidence that it *is* thinking depending on the best available explanation of how the program works.

The nature of humour is not very well understood, so we do not know whether general-purpose thinking is required to compose jokes. So it is conceivable that, despite the wide range of subject matter about which one can joke, there are hidden connections that reduce all joke making to a single narrow function. In that case there could one day be general-purpose joke-making programs that are not people, just as today there are chess-playing programs that are not people. It sounds implausible, but, since we have no good explanation ruling it out, we could not rely on joke-making as our only way of judging an AI. What we could do, though, is have a conversation ranging over a diverse range of topics, and pay attention to whether the program's utterances were or were not adapted, in their meanings, to the various purposes that came up. If the program really is thinking, then in the course of such a conversation it will *explain itself* – in one of countless, unpredictable ways – just as you or I would.

There is a deeper issue too. AI abilities must have some sort of universality: special-purpose thinking would not count as thinking in the sense Turing intended. My guess is that every AI is a person: a general-purpose explainer. It is conceivable that there are other levels of universality between AI and 'universal explainer/constructor', and perhaps separate levels for those associated attributes like consciousness. But those attributes all seem to have arrived in one jump to universality in humans, and, although we have little explanation of any of them, I know of no plausible argument that they are at different levels or can be achieved independently of each other. So I tentatively assume that they cannot. In any case, we should expect AI to be achieved in a jump to universality, starting from something much less powerful. In contrast, the ability to imitate a human imperfectly or in specialized functions is not a form of universality. It can exist in degrees. Hence, even if chatbots did at some point start becoming much better at imitating humans (or at fooling humans), that would still not be a path to AI. Becoming better at pretending to think is not the same as coming closer to being able to think.

There is a philosophy whose basic tenet is that those *are* the same. It is called *behaviourism* – which is instrumentalism applied to psychology. In other words, it is the doctrine that psychology can only, or should only, be the science of behaviour, not of minds; that it can only measure and predict relationships between people's external circumstances ('stimuli') and their observed behaviours ('responses'). The latter is, unfortunately, exactly how the Turing test asks the judge to regard a candidate AI. Hence it encouraged the attitude that if a program could fake AI well enough, one would have achieved it. But ultimately a non-AI program cannot fake AI. The path to AI cannot be through ever better tricks for making chatbots more convincing.

A behaviourist would no doubt ask: what exactly *is* the difference between giving a chatbot a very rich repertoire of tricks, templates and databases and giving it AI abilities? What is an AI program, other than a collection of such tricks?

When discussing Lamarckism in Chapter 4, I pointed out the fundamental difference between a muscle becoming stronger in an individual's lifetime and muscles *evolving* to become stronger. For the former, the knowledge to achieve all the available muscle strengths must already be present in the individual's genes before the sequence of changes begins. (And so must the knowledge of how to recognize the circumstances under which to make the changes.) This is exactly the analogue of a 'trick' that a programmer has built into a chatbot: the chatbot responds 'as though' it had created some of the knowledge while composing its response, but in fact all the knowledge was created earlier and elsewhere. The analogue of evolutionary change in a species is creative thought in a person. The analogue of the idea that AI could be achieved by an accumulation of chatbot tricks is Lamarckism, the theory that new adaptations could be explained by changes that are in reality just a manifestation of existing knowledge.

There are several current areas of research in which that same misconception is common. In chatbot-based AI research it sent the whole field down a blind alley, but in other fields it has merely caused researchers to attach overambitious labels to genuine, albeit relatively modest, achievements. One such area is *artificial evolution*.

Recall Edison's idea that progress requires alternating 'inspiration' and 'perspiration' phases, and that, because of computers and other technology, it is increasingly becoming possible to automate the perspiration phase. This welcome development has misled those who are overconfident about achieving artificial evolution (and AI). For example, suppose that you are a graduate student in robotics, hoping to build a robot that walks on legs better than previous robots do. The first phase of the solution must involve inspiration – that is to say, creative thought, attempting to improve upon previous researchers' attempts to solve the same problem. You will start from that, and from existing ideas about *other* problems that you conjecture may be related, and from the designs of walking animals in nature. All of that constitutes existing knowledge, which you will vary and combine in new ways, and then subject to criticism and further variation. Eventually you will have created a design for the hardware of your new robot: its legs with their levers, joints, tendons and motors; its body, which will hold the power supply; its sense organs, through which it will receive the feedback that will allow it to control those limbs effectively; and the computer that will exercise that control. You will have adapted everything in that design as best you can to the purpose of walking, except the program in the computer.

The function of that program will be to recognize situations such as the robot beginning to topple over, or obstacles in its path, and to calculate the appropriate action and to take it. This is the hardest part of your research project. How does one recognize when it is best to avoid an obstacle to the left or to the right, or jump over it or kick it aside or ignore it, or lengthen one's stride to avoid stepping on it – or judge it impassable and turn back? And, in all those cases, how does one specifically do those things in terms of sending countless signals to the motors and the gears, as modified by feedback from the senses?

You will break the problem down into sub-problems. Veering by a given angle is similar to veering by a different angle. That allows you to write a subroutine for veering that takes care of that whole continuum of possible cases. Once you have written it, all other parts of the program need only call it whenever they decide that veering is required, and so they do not have to contain any knowledge about the messy details of what it takes to veer. When you have identified and solved as many of these sub-problems as you can, you will have created a code, or *language*, that is highly adapted to making statements about how your robot should walk. Each call of one of its subroutines is a statement or command in that language.

So far, most of what you have done comes under the heading of 'inspiration': it required creative thought. But now perspiration looms.

Once you have automated everything that you know how to automate, you have no choice but to resort to some sort of trial and error to achieve any additional functionality. However, you do now have the advantage of a language that you have adapted for the purpose of instructing the robot in how to walk. So you can start with a program that is simple in that language, despite being very complex in terms of elementary instructions of the computer, and which means, for instance, 'Walk forwards and stop if you hit an obstacle.' Then you can run the robot with that program and see what happens. (Or you can run a computer simulation of the robot.) When it falls over or anything else undesirable happens, you can modify your program – still using the high-level language you have created – to eliminate the deficiencies as they arise. That method will require ever less inspiration and ever more perspiration.

But an alternative approach is also open to you: you can delegate the perspiration to a computer, but using a so-called *evolutionary algorithm*. Using the same computer simulation, you run many trials, each with a slight random variation of that first program. The evolutionary algorithm subjects each simulated robot automatically to a battery of tests that you have provided – how far it can walk without falling over, how well it copes with obstacles and rough terrain, and so on. At the end of each run, the program that performed best is retained, and the rest are discarded. Then many variants of *that* program are created, and the process is repeated. After thousands of iterations of this 'evolutionary' process, you may find that your robot walks quite well, according to the criteria you have set. You can now write your thesis. Not only can you claim to have achieved a robot that walks with a required degree of skill, you can claim to have implemented *evolution* on a computer.

This sort of thing has been done successfully many times. It is a useful technique. It certainly constitutes 'evolution' in the sense of alternating variation and selection. But is it evolution in the more important sense of the creation of *knowledge* by variation and selection? This will be achieved one day, but I doubt that it has been yet, for the same reason that I doubt that chatbots are intelligent, even slightly. The reason is that there is a much more obvious explanation of their abilities, namely the creativity of the programmer.

The task of ruling out the possibility that the knowledge was created by the programmer in the case of 'artificial evolution' has the same logic as checking that a program is an AI – but harder, because the amount of knowledge that the 'evolution' purportedly creates is vastly less. Even if you yourself are the programmer, you are in no position to judge whether you created that relatively small amount of knowledge or not. For one thing, some of the knowledge that you packed into that language during those many months of design will have reach, because it encoded some general truths about the laws of geometry, mechanics and so on. For another, when designing the language you had constantly in mind what sorts of abilities it would eventually be used to express.

The Turing-test idea makes us think that, if it is given enough standard reply templates, an *Eliza* program will automatically be creating knowledge; artificial evolution makes us think that if we have variation and selection, then evolution (of adaptations) will automatically happen. But neither is necessarily so. In both cases, another possibility is that no knowledge at all will be created during the *running* of the program, only during its development by the programmer.

One thing that always seems to happen with such projects is that, after they achieve their intended aim, if the 'evolutionary' program is allowed to run further it produces no further improvements. This is exactly what would happen if all the knowledge in the successful robot had actually come from the programmer, but it is not a conclusive critique: biological evolution often reaches 'local maxima of fitness'. Also, after attaining its mysterious form of universality, it seemed to pause for about a billion years before creating any significant new knowledge. But still, achieving results that might well be due to something else is not evidence of evolution.

That is why I doubt that any 'artificial evolution' has ever created knowledge. I have the same view, for the same reasons, about the slightly different kind of 'artificial evolution' that tries to evolve simulated organisms in a virtual environment, and the kind that pits different virtual species against each other.

To test this proposition, I would like to see an experiment of a slightly different kind: eliminate the graduate student from the project. Then, instead of using a robot designed to evolve better ways of walking, use a robot that is already in use in some real-life application and happens to be capable of walking. And then, instead of creating a special language of subroutines in which to express conjectures about how to walk, just replace its existing program, in its existing microprocessor, by *random numbers*. For mutations, use errors of the type that happen anyway in such processors (though in the simulation you are allowed to make them happen as often as you like). The purpose of all that is to eliminate the possibility that human knowledge is being fed into the design of the system, and that its reach is being mistaken for the product of evolution. Then, run simulations of that mutating system in the usual way. As many as you like. If the robot ever walks better than it did originally, then I am mistaken. If it continues to improve after that, then I am very much mistaken.

One of the main features of the above experiment, which is lacking in the usual way of doing artificial evolution, is that, for it to work, the *language* (of subroutines) would have to evolve along with the adaptations that it was expressing. This is what was happening in the biosphere before that jump to universality that finally settled on the DNA genetic code. As I said, it may be that all those previous genetic codes were only capable of coding for a small number of organisms that were all rather similar. And that the overwhelmingly rich biosphere that we see around us, created by randomly varying genes while leaving the language unchanged, is something that became possible only after that jump. We do not even know what kind of universality was created there. So why should we expect our artificial evolution to work without it?

I think we have to face the fact, both with artificial evolution and with AI, that these are hard problems. There are serious unknowns in how those phenomena were achieved in nature. Trying to achieve them artificially without ever discovering those unknowns was perhaps worth trying. But it should be no surprise that it has failed. Specifically, we do not know why the DNA code, which evolved to describe bacteria, has enough reach to describe dinosaurs and humans. And, although it seems obvious that an AI will have qualia and consciousness, we cannot explain those things. So long as we cannot explain them, how can we expect to simulate them in a computer program?

Or why should they emerge effortlessly from projects designed to achieve something else? But my guess is that when we do understand them, artificially implementing evolution and intelligence and its constellation of associated attributes will then be no great effort.

TERMINOLOGY

Quale (plural *qualia*)   The subjective aspect of a sensation.

Behaviourism   Instrumentalism applied to psychology. The doctrine that science can (or should) only measure and predict people's behaviour in response to stimuli.

SUMMARY

The field of artificial (general) intelligence has made no progress because there is an unsolved philosophical problem at its heart: we do not understand how creativity works. Once that has been solved, programming it will not be difficult. Even artificial evolution may not have been achieved yet, despite appearances. There the problem is that we do not understand the nature of the universality of the DNA replication system.

8

A Window on Infinity

Mathematicians realized centuries ago that it is possible to work consistently and usefully with infinity. Infinite sets, infinitely large quantities and also infinitesimal quantities all make sense. Many of their properties are counter-intuitive, and the introduction of theories about infinities has always been controversial; but many facts about finite things are just as counter-intuitive. What Dawkins calls the 'argument from personal incredulity' is no argument: it represents nothing but a preference for parochial misconceptions over universal truths.

In physics, too, infinity has been contemplated since antiquity. Euclidean space was infinite; and, in any case, space was usually regarded as a continuum: even a finite line was composed of infinitely many points. There were also infinitely many instants between any two times. But the understanding of continuous quantities was patchy and contradictory until Newton and Leibniz invented calculus, a technique for analysing continuous change in terms of infinite numbers of infinitesimal changes.

The 'beginning of infinity' – the possibility of the unlimited growth of knowledge in the future – depends on a number of other infinities. One of them is the universality in the laws of nature which allows finite, local symbols to apply to the whole of time and space – and to all phenomena and all possible phenomena. Another is the existence of physical objects that are universal explainers – people – which, it turns out, are necessarily universal constructors as well, and must contain universal classical computers.

Most forms of universality themselves refer to some sort of infinity – though they can always be interpreted in terms of something being *unlimited* rather than actually infinite. This is what opponents of infinity call a 'potential infinity' rather than a 'realized' one. For instance, the beginning of infinity can be described either as a condition where 'progress in the future will be *unbounded*' or as the condition where 'an *infinite* amount of progress will be made'. But I use those concepts interchangeably, because in this context there is no substantive difference between them.

There is a philosophy of mathematics called *finitism*, the doctrine that only finite abstract entities exist. So, for instance, there are infinitely many natural numbers, but finitists insist that that is just a manner of speaking. They say that the literal truth is only that there is a finite rule for generating each natural number (or, more precisely, each numeral) from the previous one, and nothing literally infinite is involved. But this doctrine runs into the following problem: is there a largest natural number or not? If there is, then that contradicts the statement that there is a rule that defines a larger one. If there is not, then there are not finitely many natural numbers. Finitists are then obliged to deny a principle of logic: the 'law of the excluded middle', which is that, for every meaningful proposition, either it or its negation is true. So finitists say that, although there is no largest number, there is not an infinity of numbers either.

Finitism is instrumentalism applied to mathematics: it is a principled rejection of explanation. It attempts to see mathematical entities purely as procedures that mathematicians follow, rules for making marks on paper and so on – useful in some situations, but not referring to anything real other than the finite objects of experience such as two apples or three oranges. And so finitism is inherently anthropocentric – which is not surprising, since it regards parochialism as a virtue of a theory rather than a vice. It also suffers from another fatal flaw that instrumentalism and empiricism have in regard to science, which is that it assumes that mathematicians have some sort of privileged access to *finite* entities which they do not have for infinite ones. But that is not the case. All observation is theory-laden. All abstract theorizing is theory-laden too. All access to abstract entities, finite or infinite, is via theory, just as for physical entities.

In other words finitism, like instrumentalism, is nothing but a project for preventing progress in understanding the entities beyond our direct experience. But that means progress generally, for, as I have explained, there are no entities *within* our 'direct experience'.

The whole of the above discussion assumes the universality of *reason*. The reach of science has inherent limitations; so does mathematics; so does every branch of philosophy. But if you believe that there are bounds on the domain in which reason is the proper arbiter of ideas, then you believe in unreason or the supernatural. Similarly, if you reject the infinite, you are stuck with the finite, and the finite is parochial. So there is no way of stopping there. The best explanation of *anything* eventually involves universality, and therefore infinity. The reach of explanations cannot be limited by fiat.

One expression of this within mathematics is the principle, first made explicit by the mathematician Georg Cantor in the nineteenth century, that abstract entities may be defined in any desired way out of other entities, so long as the definitions are unambiguous and consistent. Cantor founded the modern mathematical study of infinity. His principle was defended and further generalized in the twentieth century by the mathematician John Conway, who whimsically but appropriately named it *the mathematicians' liberation movement*. As those defences suggest, Cantor's discoveries encountered vitriolic opposition among his contemporaries, including most mathematicians of the day and also many scientists, philosophers – and theologians. Religious objections, ironically, were in effect based on the Principle of Mediocrity. They characterized attempts to understand and work with infinity as an encroachment on the prerogatives of God. In the mid twentieth century, long after the study of infinity had become a routine part of mathematics and had found countless applications there, the philosopher Ludwig Wittgenstein still contemptuously denounced it as 'meaningless'. (Though eventually he also applied that accusation to the whole of philosophy, including his own work – see Chapter 12.)

I have already mentioned other examples of the principled rejection of infinity. There was the strange aversion of Archimedes, Apollonius and others to universal systems of numerals. There are doctrines such as instrumentalism and finitism. The Principle of Mediocrity sets out to escape parochialism and to reach for infinity, but ends up confining science to an infinitesimal and unrepresentative bubble of comprehensibility. There is also pessimism, which (as I shall discuss in the following chapter) wants to attribute failure to the existence of a finite bound on improvement. One instance of pessimism is the paradoxical parochialism of Spaceship Earth – a vehicle that would be far better suited as a metaphor for infinity.

Whenever we refer to infinity, we are making use of the infinite reach of some idea. For whenever an idea of infinity makes sense, that is because there is an explanation of why some finite set of rules for manipulating finite symbols refers to something infinite. (Let me repeat that this underlies our knowledge of everything else as well.)

In mathematics, infinity is studied via infinite sets (meaning sets with infinitely many members). The defining property of an infinite set is that some part of it has as many elements as the whole thing. For instance, think of the natural numbers:

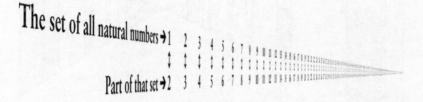

The set of natural numbers has as many members as a part of itself.

In the upper line in the illustration, every natural number appears exactly once. The lower line contains only part of that set: the natural numbers starting at 2. The illustration tallies the two sets – mathematicians call it a 'one-to-one correspondence' – to prove that there are equally many numbers in each.

The mathematician David Hilbert devised a thought experiment to illustrate some of the intuitions that one has to drop when reasoning about infinity. He imagined a hotel with infinitely many rooms: *Infinity Hotel*. The rooms are numbered with the natural numbers, starting with 1 and ending with – what?

The last room number is not infinity. First of all, there is no last room. The idea that any numbered set of rooms has a highest-numbered member is the first intuition from everyday life that we have to drop. Second, in any finite hotel whose rooms were numbered from 1, there would be a room whose number equalled the total number of rooms, and other rooms whose numbers were close to that: if there were ten rooms, one of them would be room number ten, and there would be a room number nine as well. But in Infinity Hotel, where the number of rooms is infinity, *all* the rooms have numbers infinitely far below infinity.

The beginning of infinity – the rooms in Infinity Hotel

Now imagine that Infinity Hotel is fully occupied. Each room contains one guest and cannot contain more. With finite hotels, 'fully occupied' is the same thing as 'no room for more guests'. But Infinity Hotel always has room for more. One of the conditions of staying there is that guests have to change rooms if asked to by the management. So, if a new guest arrives, the management just announce over the public-address system, 'Will all guests please move immediately to the room numbered one more than their current room.' Thus, in the manner of the first illustration in this chapter, the existing occupant of room 1 moves to room 2, whose occupant moves to room 3, and so on. What happens at the last room? There is no last room, and hence no problem about what happens there. The new arrival can now move into room 1. At Infinity Hotel, it is never necessary to make a reservation.

Evidently no such place as Infinity Hotel could exist in our universe, because it violates several laws of physics. However, this is a *mathematical* thought experiment, so the only constraint on the imaginary laws of physics is that they be consistent. It is *because* of the requirement that they be consistent that they are counter-intuitive: intuitions about infinity are often illogical.

It is a bit awkward to have to keep changing rooms – though they are all identical and are freshly made up every time a guest moves in. But guests love staying at Infinity Hotel. That is because it is cheap – only a dollar a night – yet extraordinarily luxurious. How is that possible? Every day, when the management receive all the room rents of one dollar per room, they spend the income as follows. With the dollars they received from the rooms numbered 1 to 1000, they buy complimentary champagne, strawberries, housekeeping services and all the other overheads, *just for room 1*. With the dollars they received from the rooms numbered 1001 to 2000, they do the same for room 2, and so on. In this way, each room receives several hundred dollars' worth of goods and services every day, and the management make a profit as well, all from their income of one dollar per room.

Word gets around, and one day an infinitely long train pulls up at the local station, containing infinitely many people wanting to stay at the hotel. Making infinitely many public-address announcements would take too long (and, anyway, the hotel rules say that each guest can be asked to perform only a finite number of actions per day), but no matter. The management merely announce, 'Will all guests please move immediately to the room whose number is double that of their current room.' Obviously they can all do that, and afterwards the only occupied rooms are the even numbered ones, leaving the odd-numbered ones free for the new arrivals. That is exactly enough to receive the infinitely many new guests, because there are exactly as many odd numbers as there are natural numbers, as illustrated overleaf:

Natural numbers → 1 2 3 4 5 6 7 8 9 10 11 12 13 14 15 16 17 18 19 ...

Odd numbers → 1 3 5 7 9 11 13 15 17 19 21 23 25 27 29 31 33 35 37 ...

There are exactly as many odd numbers as there are natural numbers.

So the first new arrival goes to room 1, the second to room 3, and so on.

Then, one day, an *infinite number* of infinitely long trains arrive at the station, all full of guests for the hotel. But the managers are still unperturbed. They just make a slightly more complicated announcement, which readers who are familiar with mathematical terminology can see in this footnote.* The upshot is: everyone is accommodated.

However, it *is* mathematically possible to overwhelm the capacity of Infinity Hotel. In a remarkable series of discoveries in the 1870s, Cantor proved, among other things, that not all infinities are equal. In particular, the infinity of the continuum – the number of points in a finite line (which is the same as the number of points in the whole of space or spacetime) – is much larger than the infinity of the natural numbers. Cantor proved this by proving that there can be no one-to-one correspondence between the natural numbers and the points in a line: that set of points has a higher order of infinity than the set of natural numbers.

Here is a version of his proof – known as the *diagonal argument*. Imagine a one-centimetre-thick pack of cards, each one so thin that there is one of them for every 'real number' of centimetres between 0 and 1. Real numbers can be defined as the decimal numbers between those limits, such as 0.7071..., where the ellipsis again denotes a continuation that may be infinitely long. It is impossible to deal out

---

*First, they announce to the existing guests, 'For each natural number N, will the guest in room number N please move immediately to room number $N(N+1)/2$.' Then they announce, 'For all natural numbers N and M, will the Nth passenger from the Mth train please go to room number $[(N+M)^2+N-M]/2$.'

one of these cards to each room of Infinity Hotel. For suppose that the cards *were* so distributed. We can prove that this entails a contradiction. It would mean that cards had been assigned to rooms in something like the manner of the table below. (The particular numbers illustrated are not significant: we are going to prove that real numbers cannot be assigned in *any* order.)

Which room	Which card
1	0.**6**77976...
2	0.6**9**4698...
3	0.39**9**221...
4	0.236**6**46...
⋮	⋮

Cantor's diagonal argument

Look at the infinite sequence of digits highlighted in bold – namely '**6996**. . .'. Then consider a decimal number constructed as follows: it starts with zero followed by a decimal point, and continues arbitrarily, except that each of its digits must differ from the corresponding digit in the infinite sequence '**6996**. . .'. For instance, we could choose a number such as '0.5885. . .'. The card with the number thus constructed cannot have been assigned to any room. For it differs in its first digit from that of the card assigned to room 1, and in its second digit from that of the card assigned to room 2, and so on. Thus it differs from all the cards that have been assigned to rooms, and so the original assumption that all the cards had been so assigned has led to a contradiction.

An infinity that *is* small enough to be placed in one-to-one correspondence with the natural numbers is called a '*countable* infinity' – rather an unfortunate term, because no one can count up to infinity. But it has the connotation that every *element* of a countably infinite set could in principle be reached by counting those elements in some suitable order. Larger infinities are called *uncountable*. So, there is an uncountable infinity of real numbers between any two distinct limits.

Furthermore, there are uncountably many *orders* of infinity, each too large to be put into one-to-one correspondence with the lower ones.

Another important uncountable set is the set of *all logically possible reassignments* of guests to rooms in Infinity Hotel (or, as the mathematicians put it, all possible *permutations* of the natural numbers). You can easily prove that if you imagine any one reassignment specified in an infinitely long table, like this:

Guest in room number				
1	2	3	4	...
Moves to				
38	173	80	30	...

Specifying one reassignment of guests

Then imagine all possible reassignments listed one below the other, thus 'counting' them. The diagonal argument applied to this list will prove that the list is impossible, and hence that the set of all possible reassignments is uncountable.

Since the management of Infinity Hotel have to specify a reassignment in the form of a public-address announcement, the specification must consist of a finite sequence of words – and hence a finite sequence of characters from some alphabet. The set of such sequences is countable and therefore infinitely smaller than the set of possible reassignments. That means that only an infinitesimal proportion of all logically possible reassignments can be specified. This is a remarkable limitation on the apparently limitless power of Infinity Hotel's management to shuffle the guests around. *Almost all* ways in which the guests could, as a matter of logic, be distributed among the rooms are unattainable.

Infinity Hotel has a unique, self-sufficient waste-disposal system. Every day, the management first rearrange the guests in a way that ensures that all rooms are occupied. Then they make the following announcement. 'Within the next minute, will all guests please bag their trash and give it to the guest in the next higher-numbered room. Should you *receive* a bag during that minute, then pass it on within the following half minute. Should you receive a bag during that half minute, pass it on within the following quarter minute, and so on.' To comply, the guests have to work fast – but none of them has to work *infinitely* fast, or handle infinitely many bags. Each of them performs a finite number of actions, as per the hotel rules. After two minutes, all these trash-moving actions have ceased. So, two minutes after they begin, none of the guests has any trash left.

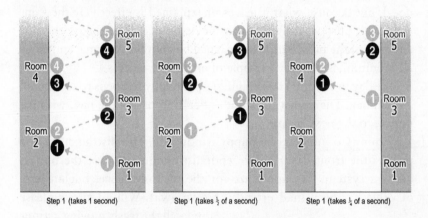

Infinity Hotel's waste-disposal system

All the trash in the hotel has disappeared from the universe. It is *nowhere*. No one has *put* it 'nowhere': every guest has merely moved some of it into another room. The 'nowhere' where all that trash has gone is called, in physics, a *singularity*. Singularities may well happen in reality, inside black holes and elsewhere. But I digress: at the moment, we are still discussing mathematics, not physics.

Of course, Infinity Hotel has infinitely many staff. Several of them are assigned to look after each guest. But the staff themselves are treated as guests in the hotel, staying in numbered rooms and receiving exactly the same benefits as every other guest: each of them has several other staff assigned to their welfare. However, they are not allowed to ask those staff to do their work for them. That is because, if they all did this, the hotel would grind to a halt. Infinity is not magic. It has logical rules: that is the whole point of the Infinity Hotel thought experiment.

The fallacious idea of delegating all one's work to other staff in higher-numbered rooms is called an *infinite regress*. It is one of the things that one cannot validly do with infinity. There is an old joke about the heckler who interrupts an astrophysics lecture to insist that the Earth is flat and supported on the back of elephants standing on a giant turtle. 'What supports the turtle?' asks the lecturer. 'Another turtle.' 'What supports *that* turtle?' 'You can't fool me,' replies the heckler triumphantly: 'it's turtles from there on down.' That theory is a bad explanation not because it fails to explain *everything* (no theory does), but because what it leaves unexplained is effectively the same as what it purports to explain in the first place. (The theory that the designer of the biosphere was designed by another designer, and so on ad infinitum, is another example of an infinite regress.)

One day in Infinity Hotel, a guest's pet puppy happens to climb into a trash bag. The owner does not notice, and passes the bag, with the puppy, to the next room.

Within two minutes the puppy is nowhere. The distraught owner phones the front desk. The receptionist announces over the public-address system, 'We apologize for the inconvenience, but an item of value has been inadvertently thrown away. Will all guests please

undo all the trash-moving actions that they have just performed, in reverse order, starting as soon as you receive a trash bag from the next-higher-numbered room.'

But to no avail. None of the guests return any bags, because their fellow guests in the higher-numbered rooms are not returning any either. It was no exaggeration to say that the bags are nowhere. They have not been stuffed into a mythical 'room number infinity'. They no longer exist; nor does the puppy. No one has done anything to the puppy except move it to another numbered room, within the hotel. Yet it is not in any room. It is not anywhere in the hotel, or anywhere else. In a finite hotel, if you move an object from room to room, in however complicated a pattern, it will end up in one of those rooms. Not so with an infinite number of rooms. Every individual action that the guests performed was both harmless to the puppy and perfectly reversible. Yet, taken together, those actions annihilated the puppy and cannot be reversed.

Reversing them cannot work, because, if it did, there would be no explanation for why a puppy arrived at its owner's room and not a kitten. If a puppy did arrive, the explanation would have to be that a puppy was passed down from the next-higher-numbered room – and so on. But that whole infinite sequence of explanations never gets round to explaining 'why a puppy?' It is an infinite regress.

What if, one day, a puppy did just arrive at room 1, having been passed down through all the rooms? That is not *logically* impossible: it would merely lack an explanation. In physics, the 'nowhere' from which such a puppy would have come is called a 'naked singularity'. Naked singularities appear in some speculative theories in physics, but such theories are rightly criticized on the grounds that they cannot make predictions. As Hawking once put it, 'Television sets could come out [of a naked singularity].' It would be different if there were a law of nature determining what comes out – for in that case there would be no infinite regress and the singularity would not be 'naked'. The Big Bang may have been a singularity of that relatively benign type.

I said that the rooms are identical, but they do differ in one respect: their room numbers. So, given the types of tasks that the management request from time to time, the low-numbered rooms are the most desirable. For instance, the guest in room 1 has the unique privilege of never having to deal with anyone else's trash. Moving to room 1 feels like winning first prize in a lottery. Moving to room 2 feels only slightly less so. But *every* guest has a room number that is unusually close to the beginning. So every guest in the hotel is more privileged than almost all other guests. The clichéd politician's promise to favour *everyone* can be honoured in Infinity Hotel.

Every room is at the beginning of infinity. That is one of the attributes of the unbounded growth of knowledge too: we are only just scratching the surface, and shall never be doing anything else.

So there is no such thing as a *typical room number* at Infinity Hotel. Every room number is untypically close to the beginning. The intuitive idea that there must be 'typical' or 'average' members of any set of values is false for infinite sets. The same is true of the intuitive ideas of 'rare' and 'common'. We might think that half of all natural numbers are odd, and half even – so that odd and even numbers are equally common among the natural numbers. But consider the following rearrangement:

| 1 | 2 | 4 | 3 | 6 | 8 | 5 | 10 | 12 | 7 | 14 | 16 | ... |

A rearrangement of the natural numbers that makes it look as though one-third of them are odd

That makes it look as though the odd numbers are only half as common as even ones. Similarly, we could make it look as though the odd numbers were one in a million or any other proportion. So the intuitive notion of a *proportion* of the members of a set does not necessarily apply to infinite sets either.

After the shocking loss of the puppy, the management of Infinity Hotel want to restore the morale of the guests, so they arrange a surprise. They announce that every guest will receive a complimentary copy of either *The Beginning of Infinity* or my previous book, *The Fabric of Reality*. They distribute them as follows: they dispatch a copy of the older book to every millionth room, and a copy of the newer book to each remaining room.

Suppose that you are a guest at the hotel. A book – gift-wrapped in opaque paper – appears in your room's delivery chute. You are hoping that it will be the newer book, because you have already read the old one. You are fairly confident that it *will* be, because, after all, what are the chances that your room is one of those that receive the old book? Exactly one in a million, it seems.

But, before you have a chance to open the package, there is an announcement. Everyone is to change rooms, to a number designated on a card that will come through the chute. The announcement also mentions that the new allocation will move all the recipients of one of the books to odd-numbered rooms, and the recipients of the other book to even-numbered ones, but it does not say which is which. So you cannot tell, from your new room number, which book you have received. Of course there is no problem with filling the rooms in this manner: both books had infinitely many recipients.

Your card arrives and you move to your new room. Are you now any less sure about which of the two books you have received? Presumably not. By your previous reasoning, there is now only a one in *two* chance that your book is *The Beginning of Infinity*, because it is now in 'half the rooms'. Since that is a contradiction, your method of assessing those probabilities must have been wrong. Indeed, all methods of assessing them are wrong, because – as this example shows – in Infinity Hotel there is *no such thing* as the probability that you have received the one book or the other.

Mathematically, this is nothing momentous. The example merely demonstrates again that the attributes probable or improbable, rare or common, typical or untypical have literally no meaning in regard to comparing infinite sets of natural numbers.

But, when we turn to physics, it is bad news for anthropic arguments. Imagine an infinite set of *universes*, all with the same laws of physics except that one particular physical constant, let us call it D, has a different value in each. (Strictly speaking, we should imagine an *uncountable* infinity of universes, like those infinitely thin cards – but that only makes the problem I am about to describe worse, so let us keep things simple.) Assume that, of these universes, infinitely many have values of D that produce astrophysicists, and infinitely many have values that do not. Then let us number the universes in such a way that all those with astrophysicists have even numbers and all the ones without astrophysicists have odd numbers.

This does not mean that half the universes have astrophysicists. Just as with the book distribution in Infinity Hotel, we could equally well label the universes so that only every third universe, or every trillionth one, had astrophysicists, or so that every trillionth one did not. So there is something wrong with the anthropic explanation of the fine-tuning problem: we can make the fine-tuning go away just by relabelling the universes. At our whim, we can number them in such a way that astrophysicists seem to be the rule, or the exception, or anything in between.

Now, suppose that we calculate, using the relevant laws of physics with different values of D, whether astrophysicists will emerge. We find that for values of D outside the range from, say, 137 to 138, those that contain astrophysicists are very sparse: only one in a trillion such universes has astrophysicists. Within the range, only one in a trillion does *not* have astrophysicists, and for values of D between 137.4 and 137.6 they all do. Let me stress that in real life we do not understand the process of astrophysicist-formation remotely well enough to calculate such numbers – and perhaps we never shall, as I shall explain in the next chapter. But, whether we could calculate them or not, anthropic theorists would wish to interpret such numbers as meaning that, if we measure D, we are *unlikely* to see values outside the range from 137 to 138. But they mean no such thing. For we could just relabel the universes (shuffle the infinite pack of 'cards') to make the spacings exactly the other way round – or anything else we liked.

Scientific explanations cannot possibly depend on how we choose to label the entities referred to in the theory. So anthropic reasoning, by itself, cannot make predictions. Which is why I said in Chapter 4 that it cannot explain the fine-tuning of the constants of physics.

The physicist Lee Smolin has proposed an ingenious variant of the anthropic explanation. It relies on the fact that, according to some theories of quantum gravity, it is possible for a black hole to spawn an entire new universe inside itself. Smolin supposes that these new universes might have different laws of physics – and that, moreover, those laws would be affected by conditions in the parent universe. In particular, intelligent beings in the parent universe could influence the black holes to produce further universes with person-friendly laws of physics. But there is a problem with explanations of this type (known as 'evolutionary cosmologies'): how many universes were there to begin with? If there were infinitely many, then we are left with the problem of how to count them – and the mere fact that each astrophysicist-bearing universe would give rise to several others need not meaningfully increase the *proportion* of such universes in the total. If there was no first universe or universes, but the whole ensemble has already existed for an infinite time, then the theory has an infinite-regress problem. For then, as the cosmologist Frank Tipler has pointed out, the entire collection must have settled into its equilibrium state 'an infinite time ago', which would mean that the evolution that brought about that equilibrium – the very process that is supposed to explain the fine-tuning – *never happened* (just as the lost puppy is *nowhere*). If there was initially only one universe, or a finite number, then we are left with the fine-tuning problem for the original universe(s): did they contain astrophysicists? Presumably not; but if the original universes produced an enormous chain of descendants until one, by chance, contains astrophysicists, then that still does not answer the question of why the entire system – now operating under a single law of physics in which the apparent 'constants' are varying according to laws of nature – permits this ultimately astrophysicist-friendly mechanism to happen. And there would be no anthropic explanation for *that* coincidence.

Smolin's theory does the right thing: it proposes an overarching framework for the ensemble of universes, and some physical connections between them. But the explanation connects only universes and their 'parent' universes, which is insufficient. So it does not work.

But now suppose we also tell a story about the reality that connects all these universes and gives a preferred physical meaning to one way of labelling them. Here is one. A girl called Lyra, who was born in universe 1, discovers a device that can move her to other universes. It also keeps her alive inside a small sphere of life support, even in universes whose laws of physics do not otherwise support life. So long as she holds down a certain button on the device, she moves from universe to universe, *in a fixed order*, at intervals of exactly one minute. As soon as she lets go, she returns to her home universe. Let us label the universes 1, 2, 3 and so on, in the order in which the device visits them.

Sometimes Lyra also takes with her a measuring instrument that measures the constant D, and another that measures – rather like the SETI project, only much faster and more reliably – whether there are astrophysicists in the universe. She is hoping to test the predictions of the anthropic principle.

But she can only ever visit a finite number of universes, and she has no way of telling whether those are representative of the whole infinite set. However, the device does have a second setting. On that setting, it takes Lyra to universe 2 for one minute, then universe 3 for *half* a minute, universe 4 for a quarter of a minute and so on. If she has not released the button by the time two minutes are up, she will have visited every universe in the infinite set, which in this story means every universe in existence. The device then returns her automatically to universe 1. If she presses it again, her journey begins again with universe 2.

Most of the universes flash by too fast for Lyra to see. But her measuring instruments are not subject to the limitations of human senses – nor to *our* world's laws of physics. After they are switched on, their displays show a running average of the values from all the universes they have been in, regardless of how much time they spent in each. So, for instance, if the even-numbered universes have astrophysicists and the odd-numbered ones do not, then at the end of a two-minute journey through all the universes her SETI-like instrument will be displaying 0.5. So in that multiverse it *is* meaningful to say that half the universes have astrophysicists.

Using a universe-travelling device that visited the same universes in a different order, one would obtain a different value for that proportion. *But*, suppose that the laws of physics permit visiting them in only one order (rather as our own laws of physics normally allow us to be at different *times* only in one particular order). Since there is now only one way for measuring instruments to respond to averages, typical values and so on, a rational agent in those universes will always get consistent results when reasoning about probabilities – and about how rare or common, typical or untypical, sparse or dense, fine-tuned or not anything is. And so *now* the anthropic principle can make testable, probabilistic predictions.

What has made this possible is that the infinite set of universes with different values of D is no longer merely a set. It is a single physical entity, a multiverse with internal interactions (as harnessed by Lyra's device) that relate different parts of it to each other and thereby provide a unique meaning, known as a *measure*, to proportions and averages over different universes.

None of the anthropic-reasoning theories that have been proposed to solve the fine-tuning problem provides any such measure. Most are hardly more than speculations of the form 'What if there were universes with different physical constants?' There is, however, one theory in physics that already describes a multiverse for independent reasons. All its universes have the same constants of physics, and the interactions of these universes do not involve travel to, or measurement of, each other. But it does provide a measure for universes. That theory is quantum theory, which I shall discuss in Chapter 11.

*

The definition of infinity in terms of a one-to-one correspondence between a set and part of itself was original to Cantor. It is connected only indirectly to the informal, intuitive way that non-mathematicians have conceived of infinity both before and since – namely that 'infinite' means something like 'bigger than any finite combination of finite things'. But that informal notion is rather circular unless we have some independent idea of what makes something *finite*, and what makes a single act of 'combination' finite. The intuitive answer would be anthropocentric: something is definitely finite if it could in principle be encompassed by a human experience. But what does it mean to 'experience' something? Was Cantor experiencing infinity when he proved theorems about it? Or was he experiencing only symbols? But we only *ever* experience symbols.

One can avoid this anthropocentrism by referring instead to measuring instruments: a quantity is definitely neither infinite nor infinitesimal if it could, in principle, register on some measuring instrument. However, by that definition a quantity can be finite even if the underlying explanation refers to an infinite set in the mathematical sense. To display the result of a measurement the needle on a meter might move by one centimetre, which is a finite distance, but it consists of an uncountable infinity of points. This can happen because, although *points* appear in lowest-level explanations of what is happening, the *number of points* never appears in predictions. Physics deals in distances, not numbers of points. Similarly, Newton and Leibniz were able to use infinitesimal distances to explain physical quantities like instantaneous velocity, yet there is nothing physically infinitesimal or infinite in, say, the continuous motion of a projectile.

To the management of Infinity Hotel, issuing a finite public-address announcement is a finite operation, even though it causes a transformation involving an infinite number of events in the hotel. On the other hand, *most* logically possible transformations could be achieved only with an infinite number of such announcements – which the laws of physics in their world do not allow. Remember, no one in Infinity Hotel – neither staff nor guest – ever performs more than a finite number of actions. Similarly in the Lyra multiverse, a measuring instrument can take the average of an infinite number of values during a finite, two-minute expedition. So that is a physically *finite* operation in that world. But taking the 'average' of the same infinite set in a different order would require an infinite number of such trips, which, again, would not be possible under those laws of physics.

Only the laws of physics determine what is finite in nature. Failure to realize this has often caused confusion. The paradoxes of Zeno of Elea, such as that of Achilles and the tortoise, were early examples. Zeno managed to conclude that, in a race against a tortoise, Achilles will never overtake the tortoise if it has a head start – because, by the time Achilles reaches the point where the tortoise began, the tortoise will have moved on a little. By the time he reaches that new point, it will have moved a little further, and so on ad infinitum. Thus the 'catching-up' procedure requires Achilles to perform an infinite number of catching-up steps in a finite time, which as a finite being he *presumably* cannot do.

Do you see what Zeno did there? He just *presumed* that the mathematical notion that happens to be called 'infinity' faithfully captures the distinction between finite and infinite that is relevant to that physical situation. That is simply false. If he is complaining that the mathematical notion of infinity does not make sense, then we can refer him to Cantor, who showed that it does. If he is complaining that the physical event of Achilles overtaking the tortoise does not make sense, then he is claiming that the laws of physics are inconsistent – but they are not. But if he is complaining that there is something inconsistent about motion because one could not *experience* each point along a continuous path, then he is simply confusing two different things that both happen to be called 'infinity'. There is nothing more to all his paradoxes than that mistake.

What Achilles can or cannot do is not deducible from mathematics. It depends only on what the relevant laws of physics say. If they say that he will overtake the tortoise in a given time, then overtake it he will. If that happens to involve an infinite number of steps of the form 'move to a particular location', then an infinite number of such steps will happen. If it involves his passing through an uncountable infinity of points, then that is what he does. But nothing *physically* infinite has happened.

Thus the laws of physics determine the distinction not only between rare and common, probable and improbable, fine-tuned or not, but even between finite and infinite. Just as the *same set* of universes can be packed with astrophysicists when measured under one set of laws of physics but have almost none when measured under another, so exactly the same sequence of events can be finite or infinite depending on what the laws of physics are.

Zeno's mistake has been made with various other mathematical abstractions too. In general terms, the mistake is to confuse an abstract attribute with a physical one of the same name. Since it is possible to prove theorems about the mathematical attribute, which have the status of absolutely necessary truths, one is then misled into assuming that one possesses a priori knowledge about what the laws of physics must say about the physical attribute.

Another example was in geometry. For centuries, no clear distinction was made between its status as a mathematical system and as a physical theory – and at first that did little harm, because the rest of science was very unsophisticated compared with geometry, and Euclid's theory was an excellent approximation for all purposes at the time. But then the philosopher Immanuel Kant (1724–1804), who was well aware of the distinction between the absolutely necessary truths of mathematics and the contingent truths of science, nevertheless concluded that Euclid's theory of geometry was self-evidently true *of nature*. Hence he believed that it was impossible rationally to doubt that the angles of a real triangle add up to 180 degrees. And in this way he elevated that formerly harmless misconception into a central flaw in his philosophy, namely the doctrine that certain truths about the physical world could be 'known a priori' – that is to say, without doing science. And of course, to make matters worse, by 'known' he unfortunately meant 'justified'.

Yet, even before Kant had declared it impossible to doubt that the geometry of real space is Euclidean, mathematicians had already doubted it. Soon afterwards the mathematician and physicist Carl Friedrich Gauss went so far as to measure the angles of a large triangle – but found no deviation from Euclid's predictions. Eventually Einstein's theory of curved space and time, which contradicted Euclid's, was vindicated by experiments that were more accurate than Gauss's. In the space near the Earth, the angles of a large triangle can add up to as much as 180.0000002 degrees, a variation from Euclid's geometry which, for instance, satellite navigation systems nowadays have to take into account. In other situations – such as near black holes – the differences between Euclidean and Einsteinian geometry are so profound that they can no longer be described in terms of 'deviations' of one from the other.

Another example of the same mistake was in computer science. Turing initially set up the theory of computation not for the purpose of building computers, but to investigate the nature of mathematical proof. Hilbert in 1900 had challenged mathematicians to formulate a rigorous theory of what constitutes a proof, and one of his conditions was that proofs must be *finite*: they must use only a fixed and finite set of rules of inference; they must start with a finite number of finitely expressed axioms, and they must contain only a finite number of elementary steps – where the steps are themselves finite. Computations, as understood in Turing's theory, are essentially the same thing as proofs: every valid proof can be converted to a computation that computes the conclusion from the premises, and every correctly executed computation is a proof that the output is the outcome of the given operations on the input.

Now, a computation can also be thought of as computing a *function* that takes an arbitrary natural number as its input and delivers an output that depends in a particular way on that input. So, for instance, doubling a number is a function. Infinity Hotel typically tells guests to change rooms by specifying a function and telling them all to compute it with different inputs (their room numbers). One of Turing's conclusions was that almost all mathematical functions that exist logically cannot be computed by any program. They are 'non-computable' for the same reason that most logically possible reallocations of rooms in Infinity Hotel cannot be effected by any instruction by the management: the set of all functions is uncountably infinite, while the set of all programs is merely countably infinite. (That is why it is meaningful to say that 'almost all' members of the infinite set of all functions have a particular property.) Hence also – as the mathematician Kurt Gödel had discovered using a different approach to Hilbert's challenge – almost all mathematical *truths* have no *proofs*. They are unprovable truths.

It also follows that almost all mathematical statements are *undecid-*

able: there is no proof that they are true, and no proof that they are false. Each of them *is* either true or false, but there is no way of using physical objects such as brains or computers to discover which is which. The laws of physics provide us with only a narrow window through which we can look out on the world of abstractions.

All undecidable statements are, directly or indirectly, about infinite sets. To the opponents of infinity in mathematics, this is due to the meaninglessness of such statements. But to me it is a powerful argument – like Hofstadter's 641 argument – that abstractions exist objectively. For it means that the truth value of an undecidable statement is certainly not just a convenient way of describing the behaviour of some physical object like a computer or a collection of dominoes.

Interestingly, very few questions are *known* to be undecidable, even though most are – and I shall return to that point. But there are many unsolved mathematical conjectures, and some of those may well be undecidable. Take, for instance, the 'prime-pairs conjecture'. A prime pair is a pair of prime numbers that differ by 2 – such as 5 and 7. The conjecture is that there is no largest prime pair: there are infinitely many of them. Suppose for the sake of argument that that is undecidable – using *our* physics. Under many other laws of physics it is decidable. The laws of Infinity Hotel are an example. Again, the details of how the management would settle the prime-pairs issue are not essential to my argument, but I present them here for the benefit of mathematically minded readers. The management would announce:

First: Please check within the next minute whether your room number and the number two above it are both primes.

Next: If they are, then send a message back through lower-numbered rooms saying that you have found a prime pair. Use the usual method for sending rapid messages (allow one minute for the first step and thereafter each step must be completed in half the time of the previous one). Store a record of this message in the lowest-numbered room that is not already storing a record of a previous such message.

Next: Check with the room numbered one more than yours. If that guest is not storing such a record and you are, then send a message to room 1 saying that there is a largest prime pair.

At the end of five minutes, the management would know the truth of the prime-pairs conjecture.

So, there is nothing *mathematically* special about the undecidable questions, the non-computable functions, the unprovable propositions. They are distinguished by physics only. Different physical laws would make different things infinite, different things computable, different truths – both mathematical and scientific – knowable. It is only the laws of physics that determine which abstract entities and relationships are modelled by physical objects such as mathematicians' brains, computers and sheets of paper.

Some mathematicians wondered, at the time of Hilbert's challenge, whether finiteness was really an essential feature of a proof. (They meant mathematically essential.) After all, infinity makes sense mathematically, so why not infinite proofs? Hilbert, though he was a great defender of Cantor's theory, ridiculed the idea. Both he and his critics were thereby making the same mistake as Zeno: they were all assuming that some class of abstract entities can *prove* things, and that mathematical reasoning could determine what that class is.

But if the laws of physics were in fact different from what we currently think they are, then so might be the set of mathematical truths that we would then be able to prove, and so might the operations that would be available to prove them with. The laws of physics as we know them happen to afford a privileged status to such operations as *not, and* and *or*, acting on individual bits of information (binary digits, or logical true/false values). That is why those operations seem natural, elementary and finite to us – and why bits do. If the laws of physics were like, say, those of Infinity Hotel, then there would be additional privileged operations, acting on infinite sets of bits. With some other laws of physics, the operations *not, and* and *or* would be non-computable, while some of our non-computable functions would seem natural, elementary and finite.

That brings me to another distinction that depends on the laws of physics: *simple* versus *complex*. Brains are physical objects. Thoughts are computations, of the types permitted under the laws of physics. Some explanations can be grasped easily and quickly – like 'If Socrates was a man and Plato was a man then they were both men.' This is easy because it can be stated in a short sentence and relies on the properties of an elementary operation (namely *and*). Other explanations are inherently hard to grasp, because their shortest form is still long and depends on many such operations. But whether the form of an explanation is long or short, and whether it requires few or many elementary operations, depends entirely on the laws of physics under which it is being stated and understood.

Quantum computation, which is currently believed to be the fully universal form of computation, happens to have exactly the same set of computable functions as Turing's classical computation. But quantum computation drives a coach and horses through the classical notion of a 'simple' or 'elementary' operation. It makes some intuitively very complex things simple. Moreover, the elementary information-storing entity in quantum computation, the 'qubit' (quantum bit) is quite hard to explain in non-quantum terminology. Meanwhile the *bit* is a fairly complicated object from the perspective of quantum physics.

Some people object that quantum computation therefore isn't 'real' computation: it is just physics, just engineering. To them, those logical possibilities about exotic laws of physics enabling exotic forms of computation do not address the issue of what a proof 'really' is. Their objection would go something like this: admittedly, under suitable laws of physics we would be able to compute non-Turing-computable functions, but that would not be *computation*. We would be able to establish the truth or falsity of Turing-undecidable propositions, but that 'establishing' would not be *proving*, because then our knowledge of whether the proposition was true or false would for ever depend on our knowledge of what the laws of physics are. If we discovered one day that the real laws of physics were different, we might have to change our minds about the proof too, and its conclusion. And so it would not be a real proof: real proof is independent of physics.

Here is that same misconception again (as well as some authority-seeking justificationism). Our *knowledge* of whether a proposition is true or false *always* depends on knowledge about how physical objects behave. If we changed our minds about what a computer, or a brain, has been doing – for instance, if we decided that our own memory was faulty about which steps we had checked in a proof – then we would be forced to change our opinion about whether we had proved something or not. It would be no different if we changed our minds about what the laws of physics made the computer do.

Whether a mathematical proposition is true or not is indeed independent of physics. But the *proof* of such a proposition is a matter of physics only. There is no such thing as abstractly proving something, just as there is no such thing as abstractly knowing something. Mathematical truth is absolutely necessary and transcendent, but all knowledge is generated by physical processes, and its scope and limitations are conditioned by the laws of nature. One can define a class of abstract entities and call them 'proofs' (or computations), just as one can define abstract entities and call them triangles and have them obey Euclidean geometry. But you cannot infer anything from that theory of 'triangles' about what angle you will turn through if you walk around a closed path consisting of three straight lines. Nor can those 'proofs' do the job of verifying mathematical statements. A mathematical 'theory of proofs' has no bearing on which truths can or cannot be proved in reality, or be known in reality; and similarly a theory of abstract 'computation' has no bearing on what can or cannot be computed in reality.

So, a computation or a proof is a physical process in which objects such as computers or brains physically model or instantiate abstract entities like numbers or equations, and mimic their properties. It is our window on the abstract. It works because we use such entities only in situations where we have good explanations saying that the relevant physical variables in those objects do indeed instantiate those abstract properties.

Consequently, the reliability of our knowledge of mathematics remains for ever subsidiary to that of our knowledge of physical reality. Every mathematical proof depends absolutely for its validity on our being right about the rules that govern the behaviour of some physical objects, like computers, or ink and paper, or brains. So, contrary to what Hilbert thought, and contrary to what most mathematicians since antiquity have believed and believe to this day, proof theory can never be made into a branch of mathematics. Proof theory is a science: specifically, it is computer science.

The whole motivation for seeking a perfectly secure foundation for mathematics was mistaken. It was a form of justificationism. Mathematics is characterized by its use of proofs in the same way that science is characterized by its use of experimental testing; in neither case is that the object of the exercise. The object of mathematics is to understand – to *explain* – abstract entities. Proof is primarily a means of ruling out false explanations; and sometimes it also provides mathematical truths that need to be explained. But, like all fields in which progress is possible, mathematics seeks not random truths but good explanations.

Three closely related ways in which the laws of physics seem finetuned are: they are all expressible in terms of a single, finite set of elementary operations; they share a single uniform distinction between finite and infinite operations; and their predictions can all be computed by a single physical object, a universal classical computer (though to simulate physics *efficiently* one would in general need a quantum computer). It is because the laws of physics support computational universality that human brains can predict and explain the behaviour of very un-human objects like quasars. And it is because of that same universality that mathematicians like Hilbert can build up an intuition of proof, and mistakenly think that it is independent of physics. But it is not independent of physics: it is merely universal *in* the physics that governs our world. If the physics of quasars were like the physics of Infinity Hotel, and depended on the functions we call non-computable, then we could not make predictions about them (unless we could build computers out of quasars or other objects relying on the relevant laws). With laws of physics slightly more exotic than that, we would not be able to explain anything – and hence could not exist.

So there is something special – *infinitely* special, it seems – about the laws of physics as we actually find them, something exceptionally computation-friendly, prediction-friendly and explanation-friendly. The physicist Eugene Wigner called this 'the unreasonable effectiveness of mathematics in the natural sciences'. For the reasons I have given, anthropic arguments alone cannot explain it. Something else will.

This problem seems to attract bad explanations. Just as religious people tend to see Providence in the unreasonable effectiveness of mathematics in science, and some evolutionists see the signature of evolution, and some cosmologists see anthropic selection effects, so some computer scientists and programmers see a great computer in the sky. For instance, one version of that idea is that the whole of what we usually think of as reality is merely virtual reality: a program running on a gigantic computer – a Great Simulator. On the face of it, this might seem a promising approach to explaining the connections between physics and computation: perhaps the reason the laws of physics are expressible in terms of computer programs is that they are in fact computer programs. Perhaps the existence of computational universality in our world is a special case of the ability of computers (in this case the Great Simulator) to emulate other computers – and so on.

But that explanation is a chimera. An infinite regress. For it entails giving up on explanation in science. It is in the very nature of computational universality that, if we and our world were composed of software, we would have no means of understanding the real physics – the physics underlying the hardware of the Great Simulator.

A different way of putting computation at the heart of physics, and to resolve the ambiguities of anthropic reasoning, is to imagine that *all possible computer programs* are running. What we think of as reality is just virtual reality generated by one or more of those programs. Then we define 'common' and 'uncommon' in terms of an average over all those programs, counting programs in order of their lengths (how many elementary operations each contains). But again that assumes that there is a preferred notion of what an 'elementary operation' is. Since the length and complexity of a program are entirely dependent on the laws of physics, this theory again requires an external world in which those computers run – a world that would be unknowable to us.

Both those approaches fail because they attempt to reverse the direction of the real explanatory connection between physics and computation. They seem plausible only because they rely on that standard mistake of Zeno's, applied to computation: the misconception that the set of classically computable functions has an a-priori privileged status within mathematics. But it does not. The only thing that privileges that set of operations is that it is instantiated in the laws of physics. The whole point of universality is lost if one conceives of computation as being somehow prior to the physical world, generating its laws. Computational universality is all about computers *inside* our physical world being related to each other under the universal laws of physics to which we (thereby) have access.

How do all those drastic limitations on what can be known and what can be achieved by mathematics and by computation, including the existence of undecidable questions in mathematics, square with the maxim that *problems are soluble*?

Problems are conflicts between ideas. Most mathematical questions that exist abstractly never appear as the subject of such a conflict: they are never the subject of curiosity, never the focus of conflicting misconceptions about some attribute of the world of abstractions. In short, most of them are uninteresting.

Moreover, recall that finding proofs is not the purpose of mathematics: it is merely one of the methods of mathematics. The purpose is to understand, and the overall method, as in all fields, is to make conjectures and to criticize them according to how good they are as explanations. One does not understand a mathematical proposition merely by proving it true. This is why there are such things as mathematics lectures rather than just lists of proofs. And, conversely, the lack of a proof does not necessarily prevent a proposition from being understood. On the contrary, the usual order of events is for the mathematician *first* to understand something about the abstraction in question and *then* to use that understanding to conjecture how true propositions about the abstraction might be proved, and *then* to prove them.

A mathematical theorem can be proved, yet remain for ever uninteresting. And an unproved mathematical conjecture can be fruitful in providing explanations even if it remains unproved for centuries, or even if it is unprovable. One example is the conjecture known in the jargon of computer science as 'P ≠ NP'. It is, roughly speaking, that there exist classes of mathematical questions whose answers can be *verified* efficiently once one has them but cannot be *computed* efficiently in the first place by a universal (classical) computer. ('Efficient' computation has a technical definition that roughly approximates what we mean by the phrase in practice.) Almost all researchers in computing theory are sure that the conjecture is true (which is further refutation of the idea that mathematical knowledge consists only of proofs). That is because, although no proof is known, there are fairly good explanations of why we should expect it to be true, and none to the contrary. (And so the same is thought to hold for quantum computers.)

Moreover, a vast amount of mathematical knowledge that is both useful and interesting has been built on the conjecture. It includes theorems of the form '*if* the conjecture is true then this interesting consequence follows.' And there are fewer, but still interesting, theorems about what would follow if it were false.

A mathematician studying an undecidable question may *prove* that it is undecidable (and explain why). From the mathematician's point of view, that is a success. Though it does not answer the *mathematical question*, it solves the *mathematician's problem*. Even working on a mathematical problem without any of those kinds of success is still not the same as failing to create knowledge. Whenever one tries and fails to solve a mathematical problem one has discovered a theorem – and usually also an explanation – about why that approach to solving it does not work.

Hence, undecidability no more contradicts the maxim that problems are soluble than does the fact that there are truths about the *physical* world that we shall never know. I expect that one day we shall have the technology to measure the number of grains of sand on Earth exactly, but I doubt that we shall ever know what the exact number was in Archimedes' time. Indeed, I have already mentioned more drastic limitations on what can be known and achieved. There are the direct limitations imposed by the universal laws of physics – we cannot exceed the speed of light, and so on. Then there are the limitations of epistemology: we cannot create knowledge other than by the fallible method of conjecture and criticism; errors are inevitable, and only error-correcting processes can succeed or continue for long. None of this contradicts the maxim, because none of those limitations need ever cause an unresolvable conflict of explanations.

Hence I conjecture that, in mathematics as well as in science and philosophy, *if the question is interesting, then the problem is soluble.* Fallibilism tells us that we can be mistaken about what is interesting. And so, three corollaries follow from this conjecture. The first is that inherently insoluble problems are inherently uninteresting. The second is that, in the long run, the distinction between what is interesting and what is boring is not a matter of subjective taste but an objective fact. And the third corollary is that the interesting problem of *why* every problem that is interesting is also soluble is itself soluble. At present we do not know why the laws of physics seem fine-tuned; we do not know why various forms of universality exist (though we do know of many connections between them); we do not know why the world is explicable. But eventually we shall. And when we do, there will be infinitely more left to explain.

The most important of all limitations on knowledge-creation is that we cannot prophesy: we cannot predict the content of ideas yet to be created, or their effects. This limitation is not only consistent with the unlimited growth of knowledge, it is entailed by it, as I shall explain in the next chapter.

That problems are soluble does not mean that we already know their solutions, or can generate them to order. That would be akin to creationism. The biologist Peter Medawar described science as 'the art of the soluble', but the same applies to all forms of knowledge. All kinds of creative thought involve judgements about what approaches might or might not work. Gaining or losing interest in particular problems or sub-problems is part of the creative process and itself constitutes problem-solving. So whether 'problems are soluble' does not depend on whether any given question can be answered, or answered by a particular thinker on a particular day. But if *progress* ever depended on violating a law of physics, then 'problems are soluble' would be false.

TERMINOLOGY

One-to-one correspondence Tallying each member of one set with each member of another.

Infinite (mathematical) A set is infinite if it can be placed in one-to-one correspondence with part of itself.

Infinite (physical) A rather vague concept meaning something like 'larger than anything that could in principle be encompassed by experience'.

Countably infinite Infinite, but small enough to be placed in one-to-one correspondence with the natural numbers.

Measure A method by which a theory gives meaning to proportions and averages of infinite sets of things, such as universes.

Singularity A situation in which something physical becomes unboundedly large, while remaining everywhere finite.

Multiverse A unified physical entity that contains more than one universe.

Infinite regress A fallacy in which an argument or explanation depends on a sub-argument of the same form which purports to address essentially the same problem as the original argument.

Computation A physical process that instantiates the properties of some abstract entity.

Proof A computation which, given a theory of how the computer on which it runs works, establishes the truth of some abstract proposition.

MEANINGS OF 'THE BEGINNING OF INFINITY' ENCOUNTERED IN THIS CHAPTER

– The ending of the ancient aversion to the infinite (and the universal).
– Calculus, Cantor's theory and other theories of the infinite and the infinitesimal in mathematics.
– The view along a corridor of Infinity Hotel.
– The property of infinite sequences that every element is exceptionally close to the beginning.
– The universality of reason.
– The infinite reach of some ideas.
– The internal structure of a multiverse which gives meaning to an 'infinity of universes'.
– The unpredictability of the content of future knowledge is a necessary condition for the unlimited growth of that knowledge.

SUMMARY

We can understand infinity through the infinite reach of some explanations. It makes sense, both in mathematics and in physics. But it has counter-intuitive properties, some of which are illustrated by Hilbert's thought experiment of Infinity Hotel. One of them is that, if unlimited progress really is going to happen, not only are we now at almost the very beginning of it, we always shall be. Cantor proved, with his diagonal argument, that there are infinitely many levels of infinity, of which physics uses at most the first one or two: the infinity of the natural numbers and the infinity of the continuum. Where there are infinitely many identical copies of an observer (for instance in multiple universes), probability and proportions do not make sense unless the collection as a whole has a structure subject to laws of physics that give them meaning. A mere infinite sequence of universes, like the rooms in Infinity Hotel, does not have such structure, which means that anthropic reasoning by itself is insufficient to explain the apparent 'fine-tuning' of the constants of physics. Proof is a physical process: whether a mathematical proposition is provable or unprovable, decidable or undecidable, depends on the laws of physics, which determine which abstract entities and relationships are modelled by physical objects. Similarly, whether a task or pattern is simple or complex depends on what the laws of physics are.

9

Optimism

*The possibilities that lie in the future are infinite. When I say
'It is our duty to remain optimists,' this includes not only the
openness of the future but also that which all of us contribute
to it by everything we do: we are all responsible for what the
future holds in store. Thus it is our duty, not to prophesy evil
but, rather, to fight for a better world.*

Karl Popper, *The Myth of the Framework* (1994)

Martin Rees suspects that civilization was lucky to survive the twentieth
century. For throughout the Cold War there was always a possibility
that another world war would break out, this time fought with hydrogen
bombs, and that civilization would be destroyed. That danger seems to
have receded, but in Rees's book *Our Final Century*, published in 2003,
he came to the worrying conclusion that civilization now had only a
50 per cent chance of surviving the twenty-first century.

Again this was because of the danger that newly created knowledge
would have catastrophic consequences. For example, Rees thought it
likely that civilization-destroying weapons, particularly biological ones,
would soon become so easy to make that terrorist organizations, or
even malevolent individuals, could not be prevented from acquiring
them. He also feared accidental catastrophes, such as the escape of
genetically modified micro-organisms from a laboratory, resulting in
a pandemic of an incurable disease. Intelligent robots, and nano-
technology (engineering on the atomic scale), 'could in the long run be
even more threatening', he wrote. And 'it is not inconceivable that
physics could be dangerous too.' For instance, it has been suggested

that elementary-particle accelerators that briefly create conditions that are in some respects more extreme than any since the Big Bang might destabilize the very vacuum of space and destroy our entire universe.

Rees pointed out that, for his conclusion to hold, it is not necessary for any one of those catastrophes to be at all probable, because we need be unlucky only once, and we incur the risk afresh every time progress is made in a variety of fields. He compared this with playing Russian roulette.

But there is a crucial difference between the human condition and Russian roulette: the probability of winning at Russian roulette is unaffected by anything that the player may think or do. Within its rules, it is a game of pure chance. In contrast, the future of civilization depends entirely on what we think and do. If civilization falls, that will not be something that just happens to us: it will be the outcome of choices that people make. If civilization survives, that will be because people succeed in solving the problems of survival, and that too will not have happened by chance.

Both the future of civilization and the outcome of a game of Russian roulette are unpredictable, but in different senses and for entirely unrelated reasons. Russian roulette is merely *random*. Although we cannot predict the outcome, we do know what the possible outcomes are, and the probability of each, provided that the rules of the game are obeyed. The future of civilization is *unknowable*, because the knowledge that is going to affect it has yet to be created. Hence the possible outcomes are not yet known, let alone their probabilities.

The growth of knowledge cannot change that fact. On the contrary, it contributes strongly to it: the ability of scientific theories to predict the future depends on the reach of their explanations, but no explanation has enough reach to predict the content of its own successors – or their effects, or those of other ideas that have not yet been thought of. Just as no one in 1900 could have foreseen the consequences of innovations made during the twentieth century – including whole new fields such as nuclear physics, computer science and biotechnology – so our own future will be shaped by knowledge that we do not yet have. We cannot even predict most of the problems that we shall encounter, or most of the opportunities to solve them, let alone the solutions and attempted solutions and how they will affect events. People in 1900 did not consider the internet or nuclear power *unlikely*: they did not conceive of them at all.

No good explanation can predict the outcome, or the probability of an outcome, of a phenomenon whose course is going to be significantly affected by the creation of new knowledge. This is a fundamental limitation on the reach of scientific prediction, and, when planning for the future, it is vital to come to terms with it. Following Popper, I shall use the term *prediction* for conclusions about future events that follow from good explanations, and *prophecy* for anything that purports to know what is not yet knowable. Trying to know the unknowable leads inexorably to error and self-deception. Among other things, it creates a bias towards pessimism. For example, in 1894 the physicist Albert Michelson made the following prophecy about the future of physics:

> The more important fundamental laws and facts of physical science have all been discovered, and these are now so firmly established that the possibility of their ever being supplanted in consequence of new discoveries is exceedingly remote . . . Our future discoveries must be looked for in the sixth place of decimals.
>
> Albert Michelson, address at the opening of the Ryerson Physical Laboratory, University of Chicago, 1894

What exactly was Michelson doing when he judged that there was only an 'exceedingly remote' chance that the foundations of physics as he knew them would ever be superseded? He was prophesying the future. How? On the basis of the best knowledge available at the time. But that consisted of the physics of 1894! Powerful and accurate though it was in countless applications, it was not capable of predicting the content of its successors. It was poorly suited even to *imagining* the changes that relativity and quantum theory would bring – which is why the physicists who did imagine them won Nobel prizes. Michelson would not have put the expansion of the universe, or the existence of parallel universes, or the non-existence of the force of gravity, on any list of possible discoveries whose probability was 'exceedingly remote'. He just didn't conceive of them at all.

A century earlier, the mathematician Joseph-Louis Lagrange had remarked that Isaac Newton had not only been the greatest genius who ever lived, but also the luckiest, for 'the system of the world can be discovered only once.' Lagrange would never know that some of his own work, which he had regarded as a mere translation of Newton's into a more elegant mathematical language, was a step towards the replacement of Newton's 'system of the world'. Michelson did live to see a series of discoveries that spectacularly refuted the physics of 1894, and with it his own prophecy.

Like Lagrange, Michelson himself had already contributed unwittingly to the new system – in this case with an experimental result. In 1887 he and his colleague Edward Morley had observed that the speed of light relative to an observer remains constant when the observer moves. This astoundingly counter-intuitive fact later became the centrepiece of Einstein's special theory of relativity. But Michelson and Morley did not realize that that was what they had observed. Observations are theory-laden. Given an experimental oddity, we have no way of predicting whether it will eventually be explained merely by correcting a minor parochial assumption or by revolutionizing entire sciences. We can know that only *after* we have seen it in the light of a new explanation. In the meantime we have no option but to see the world through our best existing explanations – which include our existing misconceptions. And that biases our intuition. Among other things, it inhibits us from conceiving of significant changes.

When the determinants of future events are unknowable, how should one prepare for them? How can one? Given that some of those determinants are beyond the reach of scientific prediction, what is the right philosophy of the unknown future? What is the rational approach to the unknowable – to the inconceivable? That is the subject of this chapter.

The terms 'optimism' or 'pessimism' have always been about the unknowable, but they did not originally refer especially to the future, as they do today. Originally, 'optimism' was the doctrine that the world – past, present and future – is as good as it could possibly be. The term was first used to describe an argument of Leibniz (1646–1716) that God, being 'perfect', would have created nothing less than 'the best of all possible worlds'. Leibniz believed that this idea solved the 'problem of evil', which I mentioned in Chapter 4: he proposed that all apparent evils in the world are outweighed by good consequences that are too remote to be known. Similarly, all apparently good events that *fail* to happen – including all improvements that humans are unsuccessful in achieving – fail because they would have had bad consequences that would have outweighed the good.

Since consequences are determined by the laws of physics, the larger part of Leibniz's claim must be that the laws of physics are the best possible too. Alternative laws that made scientific progress easier, or made disease an impossible phenomenon, or made even one disease slightly less unpleasant – in short, any alternative that would *seem* to be an improvement upon our actual history with all its plagues, tortures, tyrannies and natural disasters – would in fact have been even worse on balance, according to Leibniz.

That theory is a spectacularly bad explanation. Not only can *any* observed sequence of events be explained as 'best' by that method, an alternative Leibniz could equally well have claimed that we live in the *worst* of all possible worlds, and that every good event is necessary in order to prevent something even better from happening. Indeed, some philosophers, such as Arthur Schopenhauer, have claimed just that. Their stance is called philosophical 'pessimism'. Or one could claim that the world is exactly halfway between the best possible and the worst possible – and so on. Notice that, despite their superficial differences, all those theories have something important in common: if any of them were true, rational thought would have almost no power to discover true explanations. For, since we can always imagine states of affairs that seem better than what we observe, we would always be mistaken that they *were* better, *no matter how good our explanations were*. So, in such a world, the true explanations of events are never even imaginable. For instance, in Leibniz's 'optimistic' world, whenever we try to solve a problem and fail, it is because we have been thwarted by an unimaginably vast intelligence that determined that it was best for us to fail. And, still worse, whenever someone rejects reason and decides instead to rely on bad explanations or logical fallacies – or, for that matter, on pure malevolence – they still achieve, in every case, a better outcome on balance than the most rational and benevolent thought possibly could have. This does not describe an explicable world. And that would be very bad news for us, its inhabitants. Both the original 'optimism' and the original 'pessimism' are close to pure pessimism as I shall define it.

In everyday usage, a common saying is that 'an optimist calls a glass half full while a pessimist calls it half empty'. But those attitudes are not what I am referring to either: they are matters not of philosophy but of psychology – more 'spin' than substance. The terms can also refer to moods, such as cheerfulness or depression, but, again, moods do not necessitate any particular stance about the future: the statesman Winston Churchill suffered from intense depression, yet his outlook on the future of civilization, and his specific expectations as wartime leader, were unusually positive. Conversely the economist Thomas Malthus, a notorious prophet of doom (of whom more below), is said to have been a serene and happy fellow, who often had his companions at the dinner table in gales of laughter.

Blind optimism *is* a stance towards the future. It consists of proceeding as if one knows that the bad outcomes will not happen. The opposite approach, blind pessimism, often called the *precautionary principle*, seeks to ward off disaster by avoiding everything not known to be safe. No one seriously advocates either of these two as a universal policy, but their assumptions and their arguments are common, and often creep into people's planning.

Blind optimism is also known as 'overconfidence' or 'recklessness'. An often cited example, perhaps unfairly, is the judgement of the builders of the ocean liner *Titanic* that it was 'practically unsinkable'. The largest ship of its day, it sank on its maiden voyage in 1912. Designed to survive every foreseeable disaster, it collided with an iceberg in a manner that had not been foreseen. A blind pessimist argues that there is an inherent asymmetry between good and bad consequences: a successful maiden voyage cannot possibly do as much good as a disastrous one can do harm. As Rees points out, a single catastrophic consequence of an otherwise beneficial innovation could put an end to human progress for ever. So the blindly pessimistic approach to building ocean liners is to stick with existing designs and refrain from attempting any records.

But blind pessimism is a blindly optimistic doctrine. It assumes that unforeseen disastrous consequences cannot follow from existing knowledge too (or, rather, from existing ignorance). Not all shipwrecks happen to record-breaking ships. Not all unforeseen physical disasters need be caused by physics experiments or new technology. But one thing we do know is that protecting ourselves from *any* disaster, foreseeable or not, or recovering from it once it has happened, requires knowledge; and knowledge has to be created. The harm that can flow from any innovation that does not destroy the growth of knowledge is always finite; the good can be unlimited. There would be no existing ship designs to stick with, nor records to stay within, if no one had ever violated the precautionary principle.

Because pessimism needs to counter that argument in order to be at all persuasive, a recurring theme in pessimistic theories throughout history has been that an exceptionally dangerous moment is imminent. *Our Final Century* makes the case that the period since the mid twentieth century has been the first in which technology has been capable of destroying civilization. But that is not so. Many civilizations in history were destroyed by the simple technologies of fire and the sword. Indeed, of all civilizations in history, the overwhelming majority have been destroyed, some intentionally, some as a result of plague or natural disaster. Virtually all of them could have avoided the catastrophes that destroyed them if only they had possessed a little additional knowledge, such as improved agricultural or military technology, better hygiene, or better political or economic institutions. Very few, if any, could have been saved by greater caution about innovation. In fact most had enthusiastically implemented the precautionary principle.

More generally, what they lacked was a certain combination of abstract knowledge and knowledge embodied in technological artefacts, namely sufficient *wealth*. Let me define that in a non-parochial way as the repertoire of physical transformations that they would be capable of causing.

An example of a blindly pessimistic policy is that of trying to make our planet as unobtrusive as possible in the galaxy, for fear of contact with extraterrestrial civilizations. Stephen Hawking recently advised this, in his television series *Into the Universe*. He argued, 'If [extraterrestrials] ever visit us, I think the outcome would be much as when Christopher Columbus first landed in America, which didn't turn out very well for the Native Americans.' He warned that there might be nomadic, space-dwelling civilizations who would strip the Earth of its resources, or imperialist civilizations who would colonize it. The science-fiction author Greg Bear has written some exciting novels based on the premise that the galaxy is full of civilizations that are either predators or prey, and in both cases are hiding. This would solve the mystery of Fermi's problem. But it is implausible as a serious explanation. For one thing, it depends on civilizations becoming convinced of the existence of predator civilizations in space, and totally reorganizing themselves in order to hide from them, before being noticed – which means before they have even invented, say, radio.

Hawking's proposal also overlooks various dangers of *not* making our existence known to the galaxy, such as being inadvertently wiped out if *benign* civilizations send robots to our solar system, perhaps to mine what they consider an uninhabited system. And it rests on other misconceptions in addition to that classic flaw of blind pessimism. One is the Spaceship Earth idea on a larger scale: the assumption that progress in a hypothetical rapacious civilization is limited by raw materials rather than by knowledge. What exactly would it come to steal? Gold? Oil? Perhaps our planet's water? Surely not, since any civilization capable of transporting itself here, or raw materials back across galactic distances, must already have cheap transmutation and hence does not care about the chemical composition of its raw materials. So essentially the only resource of use to it in our solar system would be the sheer mass of matter in the sun. But matter is available in *every* star. Perhaps it is collecting entire stars wholesale in order to make a giant black hole as part of some titanic engineering project. But in that case it would cost it virtually nothing to omit inhabited solar systems (which are presumably a small minority, otherwise it is pointless for us to hide in any case); so would it casually wipe out billions of people? Would we seem like insects to it? This can seem plausible only if one forgets that there can be only one type of person: universal explainers and constructors. The idea that there could be beings that are to us as we are to animals is a belief in the supernatural.

Moreover, there is only one way of making progress: conjecture and criticism. And the only moral values that permit sustained progress are the objective values that the Enlightenment has begun to discover. No doubt the extraterrestrials' morality is different from ours; but that will not be because it resembles that of the conquistadors. Nor would we be in serious danger of culture shock from contact with an advanced civilization: it will know how to educate its own children (or AIs), so it will know how to educate us – and, in particular, to teach us how to use its computers.

A further misconception is Hawking's analogy between our civilization and pre-Enlightenment civilizations: as I shall explain in Chapter 15, there is a qualitative difference between those two types of civilization. Culture shock need not be dangerous to a post-Enlightenment one.

As we look back on the failed civilizations of the past, we can see that they were so poor, their technology was so feeble, and their explanations of the world so fragmentary and full of misconceptions that their caution about innovation and progress was as perverse as expecting a blindfold to be useful when navigating dangerous waters. Pessimists believe that the present state of our own civilization is an exception to that pattern. But what does the precautionary principle say about *that* claim? Can we be sure that our present knowledge, too, is not riddled with dangerous gaps and misconceptions? That our present wealth is not pathetically inadequate to deal with unforeseen problems? Since we cannot be sure, would not the precautionary principle require us to confine ourselves to the policy that would always have been salutary in the past – namely innovation and, in emergencies, even blind optimism about the benefits of new knowledge?

Also, in the case of our civilization, the precautionary principle rules itself out. Since our civilization has not been following it, a transition to it would entail reining in the rapid technological progress that is under way. And such a change has never been successful before. So a blind pessimist would have to oppose it on principle.

This may seem like logic-chopping, but it is not. The reason for these paradoxes and parallels between blind optimism and blind pessimism is that those two approaches are very similar at the level of explanation. Both are prophetic: both purport to know unknowable things about the future of knowledge. And since at any instant our best knowledge contains both truth and misconception, prophetic pessimism about any one aspect of it is always the same as prophetic optimism about another. For instance, Rees's worst fears depend on the unprecedentedly rapid creation of unprecedentedly powerful technology, such as civilization-destroying bio-weapons.

If Rees is right that the twenty-first century is uniquely dangerous, and if civilization nevertheless survives it, it will have had an appallingly narrow escape. *Our Final Century* mentions only one other example of a narrow escape, namely the Cold War – so that will make two narrow escapes in a row. Yet, by that standard, civilization must already have had a similarly narrow escape during the Second World War. For instance, Nazi Germany came close to developing nuclear weapons; the Japanese Empire did successfully weaponize bubonic plague – and had tested the weapon with devastating effect in China and had plans to use it against the United States. Many feared that even a conventionally won victory by the Axis powers could bring down civilization. Churchill warned of 'a new dark age, made more sinister and perhaps more protracted by the lights of perverted science' – though, as an optimist, he worked to prevent that. In contrast, the Austrian writer Stefan Zweig and his wife committed suicide in 1942, in the safety of neutral Brazil, because they considered civilization to be already doomed.

So that would make it three narrow escapes in a row. But was there not a still earlier one? In 1798, Malthus had argued, in his influential essay *On Population*, that the nineteenth century would inevitably see a permanent end to human progress. He had calculated that the exponentially growing population at the time, which was a consequence of various technological and economic improvements, was reaching the limit of the planet's capacity to produce food. And this was no accidental misfortune. He believed that he had discovered a law of nature about population and resources. First, the net increase in population, in each generation, is proportional to the existing population, so the population increases exponentially (or 'in geometrical ratio', as he put it). But, second, when food production increases – for instance, as a result of bringing formerly unproductive land into cultivation – the increase is the same as it would have been if that innovation had happened at any other time. It is not proportional to whatever the population happens to be. He called this (rather idiosyncratically) an increase 'in arithmetical ratio', and argued that 'Population, when unchecked, increases in a geometrical ratio. Subsistence increases only in an arithmetical ratio. A slight acquaintance with numbers will shew the immensity of the first power in comparison of the second.' His conclusion was that the relative well-being of humankind in his time was a temporary phenomenon and that he was living at a uniquely dangerous moment in history. The long-term state of humanity must be an equilibrium between the tendency of populations to increase on the one hand and, on the other, starvation, disease, murder and war – just as happens in the biosphere.

In the event, throughout the nineteenth century, a population explosion happened much as Malthus had predicted. Yet the end to human progress that he had foreseen did not, in part because food production increased even faster than the population. Then, during the twentieth century, both increased faster still.

Malthus had quite accurately foretold the one phenomenon, but had missed the other altogether. Why? Because of the systematic pessimistic bias to which prophecy is prone. In 1798 the forthcoming increase in population was more predictable than the even larger increase in the food supply not because it was in any sense more probable, but simply because it depended less on the creation of knowledge. By ignoring that structural difference between the two phenomena that he was trying to compare, Malthus slipped from educated guesswork into blind prophecy. He and many of his contemporaries were misled into believing that he had discovered an objective asymmetry between what he called the 'power of population' and the 'power of production'. But that was just a parochial mistake – the same one that Michelson and Lagrange made. They all thought they were making sober predictions based on the best knowledge available to them. In reality they were all allowing themselves to be misled by the ineluctable fact of the human condition that *we do not yet know what we have not yet discovered*.

Neither Malthus nor Rees intended to prophesy. They were warning that *unless* we solve certain problems in time, we are doomed. But that has always been true, and always will be. Problems are inevitable. As I said, many civilizations have fallen. Even before the dawn of civilization, all our sister species, such as the Neanderthals, became extinct through challenges with which they could easily have coped, had they known how. Genetic studies suggest that our own species came close to extinction about 70,000 years ago, as a result of an unknown catastrophe which reduced its total numbers to only a few thousand. Being overwhelmed by these and other kinds of catastrophe would have *seemed* to the victims like being forced to play Russian roulette. That is to say, it would have seemed to them that no choices that they could have made (except, perhaps, to seek the intervention of the gods more diligently) could have affected the odds against them. But this was a parochial error. Civilizations starved, long before Malthus, because of what they thought of as the 'natural disasters' of drought and famine. But it was really because of what we would call poor methods of irrigation and farming – in other words, lack of knowledge.

Before our ancestors learned how to make fire artificially (and many times since then too), people must have died of exposure literally on top of the means of making the fires that would have saved their lives, because they did not know how. In a parochial sense, the weather killed them; but the deeper explanation is lack of knowledge. Many of the hundreds of millions of victims of cholera throughout history must have died within sight of the hearths that could have boiled their drinking water and saved their lives; but, again, they did not know that. Quite generally, the distinction between a 'natural' disaster and one brought about by ignorance is parochial. Prior to every natural disaster that people once used to think of as 'just happening', or being ordained by gods, we now see many options that the people affected failed to take – or, rather, to create. And all those options add up to the overarching option that they failed to create, namely that of forming a scientific and technological civilization like ours. Traditions of criticism. An Enlightenment.

If a one-kilometre asteroid had approached the Earth on a collision course at any time in human history before the early twenty-first century, it would have killed at least a substantial proportion of all humans. In that respect, as in many others, we live in an era of unprecedented *safety*: the twenty-first century is the first ever moment when we have known how to defend ourselves from such impacts, which occur once every 250,000 years or so. This may sound too rare to care about, but it is random. A probability of one in 250,000 of such an impact in any given year means that a typical person on Earth would have a far larger chance of dying of an asteroid impact than in an aeroplane crash. And the next such object to strike us is already out there at this moment, speeding towards us with nothing to stop it except human knowledge. Civilization is vulnerable to several other known types of disaster with similar levels of risk. For instance, ice ages occur more frequently than that, and 'mini ice ages' much more frequently – and some climatologists believe that they can happen with only a few years' warning. A 'super-volcano' such as the one lurking under Yellowstone National Park could blot out the sun for years at a time. If it happened tomorrow our species could survive, by growing food using artificial light, and civilization could recover. But many would die, and the suffering would be so tremendous that such events should merit almost as much preventative effort as an extinction. We do not know the probability of a spontaneously occurring incurable plague, but we may guess that it is unacceptably high, since pandemics such as the Black Death in the fourteenth century have already shown us the sort of thing that can happen on a timescale of centuries. Should any of those catastrophes loom, we now have at least a chance of creating the knowledge required to survive, in time.

We have such a chance because we are able to solve problems. Problems are inevitable. We shall always be faced with the problem of how to plan for an unknowable future. We shall never be able to afford to sit back and hope for the best. Even if our civilization moves out into space in order to hedge its bets, as Rees and Hawking both rightly advise, a gamma-ray burst in our galactic vicinity would still wipe us all out. Such an event is thousands of times rarer than an asteroid collision, but when it does finally happen we shall have no defence against it without a great deal more scientific knowledge and an enormous increase in our wealth.

But first we shall have to survive the next ice age; and, before that, other dangerous climate change (both spontaneous and human-caused), and weapons of mass destruction and pandemics and all the countless unforeseen dangers that are going to beset us. Our political institutions, ways of life, personal aspirations and morality are all forms or embodiments of knowledge, and all will have to be improved if civilization – and the Enlightenment in particular – is to survive every one of the risks that Rees describes and presumably many others of which we have no inkling.

So – how? How can we formulate *policies* for the unknown? If we cannot derive them from our best existing knowledge, or from dogmatic rules of thumb like blind optimism or pessimism, where *can* we derive them from? Like scientific theories, policies cannot be *derived* from anything. They are conjectures. And we should choose between them not on the basis of their origin, but according to how good they are as explanations: how hard to vary.

Like the rejection of empiricism, and of the idea that knowledge is 'justified, true belief', understanding that political policies are conjectures entails the rejection of a previously unquestioned philosophical assumption. Again, Popper was a key advocate of this rejection. He wrote:

> The question about the sources of our knowledge ... has always been asked in the spirit of: 'What are the best sources of our knowledge – the most reliable ones, those which will not lead us into error, and those to which we can and must turn, in case of doubt, as the last court of appeal?'
> I propose to assume, instead, that no such ideal sources exist – no more than ideal rulers – and that all 'sources' are liable to lead us into error at times. And I propose to replace, therefore, the question of the sources of our knowledge by the entirely different question: 'How can we hope to detect and eliminate error?'
>
> 'Knowledge without Authority' (1960)

The question 'How can we hope to detect and eliminate error?' is echoed by Feynman's remark that 'science is what we have learned about how to keep from fooling ourselves'. And the answer is basically the same for human decision-making as it is for science: it requires a tradition of criticism, in which good explanations are sought – for example, explanations of what has gone wrong, what would be better, what effect various policies have had in the past and would have in the future.

But what use are explanations if they cannot make predictions and so cannot be tested through experience, as they can be in science? This is really the question: how is progress possible in philosophy? As I discussed in Chapter 5, it is obtained by seeking good explanations. The misconception that evidence can play no legitimate role in philosophy is a relic of empiricism. Objective progress is indeed possible in politics just as it is in morality generally and in science.

Political philosophy traditionally centred on a collection of issues that Popper called the 'who should rule?' question. Who should wield power? Should it be a monarch or aristocrats, or priests, or a dictator, or a small group, or 'the people', or their delegates? And that leads to derivative questions such as 'How should a king be educated?' 'Who should be enfranchised in a democracy?' 'How does one ensure an informed and responsible electorate?'

Popper pointed out that this class of questions is rooted in the same misconception as the question 'How are scientific theories derived from sensory data?' which defines empiricism. It is seeking a system that *derives* or justifies the right choice of leader or government, from existing data – such as inherited entitlements, the opinion of the majority, the manner in which a person has been educated, and so on. The same misconception also underlies blind optimism and pessimism: they both expect progress to be made by applying a simple rule to existing knowledge, to establish which future possibilities to ignore and which to rely on. Induction, instrumentalism and even Lamarckism all make the same mistake: they expect *explanationless progress*. They expect knowledge to be created by fiat with few errors, and not by a process of variation and selection that is making a continual stream of errors and correcting them.

The defenders of hereditary monarchy doubted that any method of selection of a leader by means of rational thought and debate could improve upon a fixed, mechanical criterion. That was the precautionary principle in action, and it gave rise to the usual ironies. For instance, whenever pretenders to a throne claimed to have a better hereditary entitlement than the incumbent, they were in effect citing the precautionary principle as a justification for sudden, violent, unpredictable change – in other words, for blind optimism. The same was true whenever monarchs happened to favour radical change themselves. Consider also the revolutionary utopians, who typically achieve only destruction and stagnation. Though they are blind optimists, what defines them as utopians is their pessimism that their supposed utopia, or their violent proposals for achieving and entrenching it, could ever be improved upon. Additionally, they are revolutionaries in the first place because they are pessimistic that many other people can be persuaded of the final truth that they think they know.

Ideas have consequences, and the 'who should rule?' approach to political philosophy is not just a mistake of academic analysis: it has been part of practically every bad political doctrine in history. If the political process is seen as an engine for putting the right rulers in power, then it justifies violence, for until that right system is in place, no ruler is legitimate; and once it is in place, and its designated rulers are ruling, opposition to them is opposition to rightness. The problem then becomes how to thwart anyone who is working against the rulers or their policies. By the same logic, everyone who thinks that existing rulers or policies are bad must infer that the 'who should rule?' question has been answered wrongly, and therefore that the power of the rulers is not legitimate, and that opposing it is legitimate, by force if necessary. Thus the very question 'Who should rule?' begs for violent, authoritarian answers, and has often received them. It leads those in power into tyranny, and to the entrenchment of bad rulers and bad policies; it leads their opponents to violent destructiveness and revolution.

Advocates of violence usually have in mind that none of those things need happen if only everyone agreed on who should rule. But that means agreeing about what is right, and, given agreement on that, rulers would then have nothing to do. And, in any case, such agreement is neither possible nor desirable: people are different, and have unique ideas; problems are inevitable, and progress consists of solving them.

Popper therefore applies his basic 'how can we detect and eliminate errors?' to political philosophy in the form *how can we rid ourselves of bad governments without violence?* Just as science seeks explanations that are experimentally testable, so a rational political system makes it as easy as possible to detect, and persuade others, that a leader or policy is bad, and to remove them without violence if they are. Just as the institutions of science are structured so as to avoid entrenching theories, but instead to expose them to criticism and testing, so political institutions should not make it hard to oppose rulers and policies, non-violently, and should embody traditions of peaceful, critical discussion of them and of the institutions themselves and everything else. Thus, systems of government are to be judged not for their prophetic ability to choose and install good leaders and policies, but for their ability to remove bad ones that are already there.

That entire stance is fallibilism in action. It *assumes* that rulers and policies are always going to be flawed – that problems are inevitable. But it also assumes that improving upon them is possible: problems are soluble. The ideal towards which this is working is not that nothing unexpected will go wrong, but that when it does it will be an opportunity for further progress.

Why would anyone want to make the leaders and policies that they themselves favour more vulnerable to removal? Indeed, let me first ask: *why would anyone want to replace bad leaders and policies at all?* That question may seem absurd, but perhaps it is absurd only from the perspective of a civilization that takes progress for granted. If we did not expect progress, why should we expect the new leader or policy, chosen by whatever method, to be any better than the old? On the contrary, we should then expect any changes on average to do as much harm as good. And then the precautionary principle advises, 'Better the devil you know than the devil you don't.' There is a closed loop of ideas here: on the assumption that knowledge is not going to grow, the precautionary principle is true; and on the assumption that the precautionary principle is true, we cannot afford to allow knowledge to grow. Unless a society is expecting its own future choices to be better than its present ones, it will strive to make its present policies and institutions as immutable as possible. Therefore Popper's criterion can be met only by societies that expect their knowledge to grow – and to grow unpredictably. And, further, they are expecting that if it did grow, *that would help.*

This expectation is what I call optimism, and I can state it, in its most general form, thus:

The Principle of Optimism
All evils are caused by insufficient knowledge.

Optimism is, in the first instance, a way of explaining failure, not prophesying success. It says that there is no fundamental barrier, no law of nature or supernatural decree, preventing progress. Whenever we try to improve things and fail, it is not because the spiteful (or unfathomably benevolent) gods are thwarting us or punishing us for trying, or because we have reached a limit on the capacity of reason to make improvements, or because it is best that we fail, but always because we did not know enough, in time. But optimism is also a stance towards the future, because nearly all failures, and nearly all successes, are yet to come.

Optimism follows from the explicability of the physical world, as I explained in Chapter 3. If something is permitted by the laws of physics, then the only thing that can prevent it from being technologically possible is not knowing how. Optimism also assumes that none of the *prohibitions* imposed by the laws of physics are necessarily *evils*. So, for instance, the lack of the impossible knowledge of prophecy is not an insuperable obstacle to progress. Nor are insoluble mathematical problems, as I explained in Chapter 8.

That means that in the long run there are no insuperable evils, and in the short run the only insuperable evils are parochial ones. There can be no such thing as a disease for which it is impossible to discover a cure, other than certain types of brain damage – those that have dissipated the knowledge that constitutes the patient's personality. For a sick person is a physical object, and the task of transforming this object into the same person in good health is one that no law of physics rules out. Hence there is a way of achieving such a transformation – that is to say, a cure. It is only a matter of knowing how. If we do not, for the moment, know how to eliminate a particular evil, or we know in theory but do not yet have enough time or resources (i.e. wealth), then, even so, it is universally true that *either* the laws of physics forbid eliminating it in a given time with the available resources *or* there is a way of eliminating it in the time and with those resources.

The same must hold, equally trivially, for the evil of death – that is to say, the deaths of human beings from disease or old age. This problem has a tremendous resonance in every culture – in its literature, its values, its objectives great and small. It also has an almost unmatched reputation for insolubility (except among believers in the supernatural): it is taken to be the epitome of an insuperable obstacle. But there is no rational basis for that reputation. It is absurdly parochial to read some deep significance into this particular failure, among so many, of the biosphere to support human life – or of medical science throughout the ages to cure ageing. The problem of ageing is of the same general type as that of disease. Although it is a complex problem by present-day standards, the complexity is finite and confined to a relatively narrow arena whose basic principles are already fairly well understood. Meanwhile, knowledge in the relevant fields is increasing exponentially.

Sometimes 'immortality' (in this sense) is even regarded as undesirable. For instance, there are arguments from overpopulation; but those are examples of the Malthusian prophetic fallacy: what each additional surviving person would need to survive at present-day standards of living is easily calculated; what knowledge that person would contribute to the solution of the resulting problems is unknowable. There are also arguments about the stultification of society caused by the entrenchment of old people in positions of power; but the traditions of criticism in our society are already well adapted to solving that sort of problem. Even today, it is common in Western countries for powerful politicians or business executives to be removed from office while still in good health.

There is a traditional optimistic story that runs as follows. Our hero is a prisoner who has been sentenced to death by a tyrannical king, but gains a reprieve by promising to teach the king's favourite horse to talk within a year. That night, a fellow prisoner asks what possessed him to make such a bargain. He replies, 'A lot can happen in a year. The horse might die. The king might die. I might die. Or the horse might talk!' The prisoner understands that, while his immediate problems have to do with prison bars and the king and his horse, ultimately the evil he faces is caused by insufficient knowledge. That makes him an optimist. He knows that, if progress is to be made, some of the opportunities and some of the discoveries will be inconceivable in advance. Progress cannot take place at all unless someone is open to, and prepares for, those inconceivable possibilities. The prisoner may or may not discover a way of teaching the horse to talk. But he may discover something else. He may persuade the king to repeal the law that he had broken; he may learn a convincing conjuring trick in which the horse would seem to talk; he may escape; he may think of an achievable task that would please the king even more than making the horse talk. The list is infinite. Even if every such possibility is unlikely, it takes only one of them to be realized for the whole problem to be solved. But if our prisoner is going to escape by creating a new idea, he cannot possibly know that idea today, and therefore he cannot let the assumption that it will never exist condition his planning.

Optimism implies all the other necessary conditions for knowledge to grow, and for knowledge-creating civilizations to last, and hence for the beginning of infinity. We have, as Popper put it, a duty to be optimistic – in general, and about civilization in particular. One can argue that saving civilization will be difficult. That does not mean that there is a low probability of solving the associated problems. When we say that a mathematical problem is hard to solve, we do not mean that it is *unlikely* to be solved. All sorts of factors determine whether mathematicians even address a problem, and with what effort. If an easy problem is not deemed to be interesting or useful, they might leave it unsolved indefinitely, while hard problems are solved all the time.

Usually the hardness of a problem is one of the very factors that cause it to be solved. Thus President John F. Kennedy said in 1962, in a celebrated example of an optimistic approach to the unknown, 'We choose to go to the moon. We choose to go to the moon in this decade and do the other things, not because they are easy, but because they are hard.' Kennedy did not mean that the moon project, being hard, was unlikely to succeed. On the contrary, he believed that it would. What he meant by a hard task was one that depends on facing the unknown. And the intuitive fact to which he was appealing was that although such hardness is always a negative factor when choosing among *means* to pursue an objective, when choosing the objective itself it can be a positive one, because we want to engage with projects that will involve creating new knowledge. And an optimist expects the creation of knowledge to constitute progress – including its unforeseeable consequences.

Thus, Kennedy remarked that the moon project would require a vehicle 'made of new metal alloys, some of which have not yet been invented, capable of standing heat and stresses several times more than have ever been experienced, fitted together with a precision better than the finest watch, carrying all the equipment needed for propulsion, guidance, control, communications, food and survival'. Those were the known problems, which would require as-yet-unknown knowledge. That this was 'on an untried mission, to an unknown celestial body' referred to the unknown problems that made the probabilities, and the outcomes, profoundly unknowable. Yet none of that prevented rational people from forming the expectation that the mission could succeed. This expectation was not a judgement of probability: until far into the project, no one could predict that, because it depended on solutions not yet discovered to problems not yet known. When people were being persuaded to work on the project – and to vote for it, and so on – they were being persuaded that our being confined to one planet was an evil, that exploring the universe was a good, that the Earth's gravitational field was not a barrier but merely a problem, and that overcoming it and all the other problems involved in the project was only a matter of knowing how, and that the nature of the problems made that moment the right one to try to solve them. Probabilities and prophecies were not needed in that argument.

Pessimism has been endemic in almost every society throughout history. It has taken the form of the precautionary principle, and of 'who should rule?' political philosophies and all sorts of other demands for prophecy, and of despair in the power of creativity, and of the misinterpretation of problems as insuperable barriers. Yet there have always been a few individuals who see obstacles as problems, and see problems as soluble. And so, very occasionally, there have been places and moments when there was, briefly, an end to pessimism. As far as I know, no historian has investigated the history of optimism, but my guess is that whenever it has emerged in a civilization there has been a mini-enlightenment: a tradition of criticism resulting in an efflorescence of many of the patterns of human progress with which we are familiar, such as art, literature, philosophy, science, technology and the institutions of an open society. The end of pessimism is potentially a beginning of infinity. Yet I also guess that in every case – with the single, tremendous exception (so far) of our own Enlightenment – this process was soon brought to an end and the reign of pessimism was restored.

The best-known mini-enlightenment was the intellectual and political tradition of criticism in ancient Greece which culminated in the so-called 'Golden Age' of the city-state of Athens in the fifth century BCE. Athens was one of the first democracies, and was home to an astonishing number of people who are regarded to this day as major figures in the history of ideas, such as the philosophers Socrates, Plato and Aristotle, the playwrights Aeschylus, Aristophanes, Euripides and Sophocles, and the historians Herodotus, Thucydides and Xenophon. The Athenian philosophical tradition continued a tradition of criticism dating back to Thales of Miletus over a century earlier and which had included Xenophanes of Colophon (570–480 BCE), one of the first to question anthropocentric theories of the gods. Athens grew wealthy through trade, attracted creative people from all over the known world, became one of the foremost military powers of the age, and built a structure, the Parthenon, which is to this day regarded as one of the great architectural achievements of all time. At the height of the Golden Age, the Athenian leader Pericles tried to explain what made Athens successful. Though he no doubt believed that the city's patron goddess, Athena, was on their side, he evidently did not consider 'the goddess did it' to be a sufficient explanation for the Athenians' success. Instead, he listed specific attributes of Athenian civilization. We do not know exactly how much of what he described was flattery or wishful thinking, but, in assessing the optimism of a civilization, what that civilization aspired to be must be even more important than what it had yet succeeded in becoming.

The first attribute that Pericles cited was Athens' democracy. And he explained why. Not because 'the people should rule', but because it promotes 'wise action'. It involves continual discussion, which is a necessary condition for discovering the right answer, which is in turn a necessary condition for progress:

> Instead of looking upon discussion as a stumbling-block in the way of action, we think it an indispensable preliminary to any wise action at all.
>
> Pericles, 'Funeral Oration', *c.* 431 BCE

He also mentioned *freedom* as a cause of success. A pessimistic civilization considers it immoral to behave in ways that have not been tried many times before, because it is blind to the possibility that the benefits of doing so might offset the risks. So it is intolerant and conformist. But Athens took the opposite view. Pericles also contrasted his city's openness to foreign visitors with the closed, defensive attitude of rival cities: again, he expected that Athens would benefit from contact with new, unforeseeable ideas, even though, as he acknowledged, this policy gave enemy spies access to the city too. He even seems to have regarded the lenient treatment of children as a source of military strength:

> In education, where our rivals from their very cradles by a painful discipline seek after manliness, in Athens we live exactly as we please, and yet are just as ready to encounter every legitimate danger.

A pessimistic civilization prides itself on its children's conformity to the proper patterns of behaviour, and bemoans every real or imagined novelty.

Sparta was, in all the above respects, the opposite of Athens. The epitome of a pessimistic civilization, it was notorious for its citizens' austere 'spartan' lifestyle, for the harshness of its educational system, and for the total militarization of its society. Every male citizen was a full-time soldier, owing absolute obedience to his superiors, who were themselves obliged to follow religious tradition. All other work was done by slaves: Sparta had reduced an entire neighbouring society, the Messenians, to the status of helots (a kind of serf or slave). It had no philosophers, historians, artists, architects, writers – or other knowledge-creating people of any kind apart from the occasional talented general. Thus almost the entire effort of the society was devoted to preserving itself in its existing state – in other words, to preventing improvement. In 404 BCE, twenty-seven years after Pericles' funeral oration, Sparta decisively defeated Athens in war and imposed an authoritarian form of government on it. Although, through the vagaries of international politics, Athens became independent and democratic again soon afterwards, and continued for several generations to produce art, literature and philosophy, it was never again host to rapid, open-ended progress. It became unexceptional. Why? I guess that its optimism was gone.

Another short-lived enlightenment happened in the Italian city-state of Florence in the fourteenth century. This was the time of the early Renaissance, a cultural movement that revived the literature, art and science of ancient Greece and Rome after more than a millennium of intellectual stagnation in Europe. It became an *enlightenment* when the Florentines began to believe that they could improve upon that ancient knowledge. This era of dazzling innovation, known as the Golden Age of Florence, was deliberately fostered by the Medici family, who were in effect the city's rulers – especially Lorenzo de' Medici, known as 'the Magnificent', who was in charge from 1469 to 1492. Unlike Pericles, the Medici were not devotees of democracy: Florence's enlightenment began not in politics but in art, and then philosophy, science and technology, and in those fields it involved the same openness to criticism and desire for innovation both in ideas and in action.

Artists, instead of being restricted to traditional themes and styles, became free to depict what they considered beautiful, and to invent new styles. Encouraged by the Medici, the wealthy of Florence competed with each other in the innovativeness of the artists and scholars whom they sponsored – such as Leonardo da Vinci, Michelangelo and Botticelli. Another denizen of Florence at this time was Niccolò Machiavelli, the first secular political philosopher since antiquity.

The Medici were soon promoting the new philosophy of 'humanism', which valued knowledge above dogma, and virtues such as intellectual independence, curiosity, good taste and friendship over piety and humility. They sent agents all over the known world to obtain copies of ancient books, many of which had not been seen in the West since the fall of the Western Roman Empire. The Medici library made copies which it supplied to scholars in Florence and elsewhere. Florence became a powerhouse of newly revived ideas, new interpretations of ideas, and brand-new ideas.

But that rapid progress lasted for only a generation or so. A charismatic monk, Girolamo Savonarola, began to preach apocalyptic sermons against humanism and every other aspect of the Florentine enlightenment. Urging a return to medieval conformism and self-denial, he proclaimed prophecies of doom if Florence continued on its path. Many citizens were persuaded, and in 1494 Savonarola managed to seize power. He reimposed all the traditional restrictions on art, literature, thought and behaviour. Secular music was banned. Clothing had to be plain. Frequent fasting became effectively compulsory. Homosexuality and prostitution were violently suppressed. The Jews of Florence were expelled. Gangs of ruffians inspired by Savonarola roamed the city searching for taboo artefacts such as mirrors, cosmetics, musical instruments, secular books, and almost anything beautiful. A huge pile of such treasures was ceremonially burned in the so-called 'Bonfire of the Vanities' in the centre of the city. Botticelli is said to have thrown some of his own paintings into the fire. It was the bonfire of optimism.

Eventually Savonarola was himself discarded and burned at the stake. But, although the Medici regained control of Florence, optimism did not. As in Athens, the tradition of art and science continued for a while, and, even a century later, Galileo was sponsored (and then abandoned) by the Medici. But by that time Florence had become just another Renaissance city-state lurching from one crisis to another under the rule of despots. Fortunately, somehow that mini-enlightenment was never quite extinguished. It continued to smoulder in Florence and several other Italian city-states, and finally ignited the Enlightenment itself in northern Europe.

There may have been many enlightenments in history, shorter-lived and shining less brilliantly than those, perhaps in obscure subcultures, families or individuals. For example, the philosopher Roger Bacon (1214–94) is noted for rejecting dogma, advocating observation as a way of discovering the truth (albeit by 'induction'), and making several scientific discoveries. He foresaw the invention of microscopes, telescopes, self-powered vehicles and flying machines – and that mathematics would be a key to future scientific discoveries. He was thus an optimist. But he was not part of any tradition of criticism, and so his optimism died with him.

Bacon studied the works of ancient Greek scientists and of scholars of the 'Islamic Golden Age' – such as Alhazen (965–1039), who made several original discoveries in physics and mathematics. During the Islamic Golden Age (between approximately the eighth and thirteenth centuries), there was a strong tradition of scholarship that valued and drew upon the science and philosophy of European antiquity. Whether there was also a tradition of *criticism* in science and philosophy is currently controversial among historians. But, if there was, it was snuffed out like the others.

It may be that the Enlightenment has 'tried' to happen countless times, perhaps even all the way back to prehistory. If so, those mini-enlightenments put our recent 'lucky escapes' into stark perspective. It may be that there was progress every time – a brief end to stagnation, a brief glimpse of infinity, always ending in tragedy, always snuffed out, usually without trace. Except this once.

The inhabitants of Florence in 1494 or Athens in 404 BCE could be forgiven for concluding that optimism just isn't factually true. For they knew nothing of such things as the reach of explanations or the power of science or even laws of nature as we understand them, let alone the moral and technological progress that was to follow when *the* Enlightenment got under way. At the moment of defeat, it must have seemed at least plausible to the formerly optimistic Athenians that the Spartans might be right, and to the formerly optimistic Florentines that Savonarola might be. Like every other destruction of optimism, whether in a whole civilization or in a single individual, these must have been unspeakable catastrophes for those who had dared to expect progress. But we should feel more than sympathy for those people. We should take it personally. For if any of those earlier experiments in optimism had succeeded, our species would be exploring the stars by now, and you and I would be immortal.

TERMINOLOGY

Blind optimism (recklessness, overconfidence)   Proceeding as if one knew that bad outcomes will not happen.

Blind pessimism (precautionary principle)   Avoiding everything not known to be safe.

The principle of optimism   All evils are caused by insufficient knowledge.

Wealth   The repertoire of physical transformations that one is capable of causing.

MEANINGS OF 'THE BEGINNING OF INFINITY' ENCOUNTERED IN THIS CHAPTER

- Optimism. (And the end of pessimism.)
- Learning how not to fool ourselves.
- Mini-enlightenments like those of Athens and Florence were potential beginnings of infinity.

SUMMARY

Optimism (in the sense that I have advocated) is the theory that all failures – all evils – are due to insufficient knowledge. This is the key to the rational philosophy of the unknowable. It would be contentless if there were fundamental limitations to the creation of knowledge, but there are not. It would be false if there were fields – especially philosophical fields such as morality – in which there were no such thing as objective progress. But truth does exist in all those fields, and progress towards it is made by seeking good explanations. Problems are inevitable, because our knowledge will always be infinitely far from complete. Some problems are hard, but it is a mistake to confuse hard problems with problems unlikely to be solved. Problems are soluble, and each particular evil is a problem that can be solved. An optimistic civilization is open and not afraid to innovate, and is based on traditions of criticism. Its institutions keep improving, and the most important knowledge that they embody is knowledge of how to detect and eliminate errors. There may have been many short-lived enlightenments in history. Ours has been uniquely long-lived.

10

A Dream of Socrates

SOCRATES *is staying at an inn near the Temple of the Oracle at Delphi. Together with his friend* CHAEREPHON, *he has today asked the Oracle who the wisest man in the world is,* * *so that they might go and learn from him. But, to their annoyance, the priestess (who provides the Oracle's voice on behalf of the god Apollo) merely announced, 'No one is wiser than Socrates.' Sleeping now on an uncomfortable bed in a tiny and exorbitantly expensive room,* SOCRATES *hears a deep, melodious voice intoning his name.*

HERMES: Greetings, Socrates.

SOCRATES: [*Draws the blanket over his head.*] Go away. I've already made too many offerings today and you're not going to wring any more out of me. I am too 'wise' for that, hadn't you heard?

HERMES: I seek no offering.

SOCRATES: Then what do you want? [*He turns and sees* HERMES, *who is naked.*] Well, I'm sure that some of my associates camped outside will be glad to –

HERMES: It is not them I seek, but you, O Socrates.

SOCRATES: Then you shall be disappointed, stranger. Now kindly leave me to my hard-earned rest.

HERMES: Very well. [*He makes towards the door.*]

SOCRATES: Wait.

HERMES: [*Turns and raises a quizzical eyebrow.*]

*In the story as told by Plato in his *Apology*, Chaerophon asks the Oracle *whether* there is anyone wiser than Socrates, and is told no. But would he really have wasted this expensive and solemn privilege on a question with only two possible answers, one flattering, the other frustrating, and neither very interesting?

SOCRATES: [*slowly and deliberately*] I am asleep. Dreaming. And you are the god Apollo.

HERMES: What makes you think so?

SOCRATES: These precincts are sacred to you. It is night-time and there is no lamp, yet I see you clearly. This is not possible in real life. So you must be coming to me in a dream.

HERMES: You reason coolly. Are you not afraid?

SOCRATES: Bah! I ask you in return: are you a benevolent or a malevolent god? If benevolent, then what do I have to fear? If malevolent, then I disdain to fear you. We Athenians are a proud people – and protected by our goddess, as you surely know. Twice we defeated the Persian Empire against overwhelming odds,* and now we are defying Sparta. It is our custom to defy anyone who seeks our submission.

HERMES: Even a god?

SOCRATES: A benevolent god would not seek it. On the other hand, it is also our custom to give a hearing to anyone who offers us honest criticism, seeking to persuade us freely to change our minds. For we want to do what is right.

HERMES: Those two customs are two sides of the same valuable coin, Socrates. I give you Athenians great credit for honouring them.

SOCRATES: My city is surely deserving of your favour. But why would an immortal want to converse with such a confused and ignorant person as me? I think I can guess your reason: you have repented of your little joke via the Oracle, haven't you? Indeed, it was rather cruel of you to send us only a mocking answer, considering the distance we have come and the offerings we have made. So please tell me the truth this time, O fount of wisdom: who is really the wisest man in the world?

HERMES: I reveal no facts.

SOCRATES: [*Sighs.*] Then I beg you – I have always wanted to know this: what is the nature of virtue?

HERMES: I reveal no moral truths either.

---

*In this dialogue, Socrates sometimes exaggerates the attributes and achievements of his beloved home city-state, Athens. In this case he is ignoring the contributions of other Greek city-states to the defeats of two invasion attempts by the Persian Empire, both of them before he was born.

SOCRATES: Yet, as a benevolent god, you must have come here to impart *some* sort of knowledge. What sort will you deign to grant me?

HERMES: Knowledge about knowledge, Socrates. *Epistemology.* I have already mentioned some.

SOCRATES: You have? Oh – you said that you honour Athenians for our openness to persuasion. And for our defiance of bullies. But it is well known that those are virtues! Surely telling me what I already know doesn't count as a 'revelation'.

HERMES: Most Athenians would indeed call those virtues. But how many really believe it? How many are willing to criticize a *god* by the standards of reason and justice?

SOCRATES: [*Ponders.*] All who are just, I suppose. For how can anyone be just if he follows a god of whose moral rightness he is not persuaded? And how is it possible to be persuaded of someone's moral rightness without first forming a view about which qualities *are* morally right?

HERMES: Your associates out there on the lawn – are they unjust?

SOCRATES: No.

HERMES: And are they aware of the connections you have just described between reason, morality and the reluctance to defer to gods?

SOCRATES: Perhaps not *sufficiently* aware – yet.

HERMES: So it is not true that every just person knows these things.

SOCRATES: Agreed. Perhaps it is only every *wise* person.

HERMES: Everyone who is at least as wise as you, then. Who else is in that exalted category?

SOCRATES: Is there some high purpose in your continuing to mock me, wise Apollo, by asking me the same question that we asked you today? It seems to me that your joke is wearing thin.

HERMES: Have you, Socrates, never mocked anyone?

SOCRATES: [*with dignity*] If, on occasion, I make fun of someone, it is because I hope he will help me to seek a truth that neither he nor I yet knows. I do not mock from on high, as you do. I want only to goad my fellow mortal into helping me look beyond that which is easy to see.

HERMES: But what in the world *is* easy to see? What things are the *easiest* to see, Socrates?

SOCRATES: [*Shrugs.*] Those that are before our eyes.

HERMES: And what is before your eyes at this moment?

SOCRATES: You are.

HERMES: Are you sure?

SOCRATES: Are you going to start asking me *how I can be sure* of whatever I say? And then, whatever reason I give, are you going to ask how I can be sure of *that*?

HERMES: No. Do you think I have come here to play hackneyed debating tricks?

SOCRATES: Very well: obviously I can't be *sure* of anything. But I don't want to be. I can think of nothing more boring – no offence meant, wise Apollo – than to attain the state of being perfectly secure in one's beliefs, which some people seem to yearn for. I see no use for it – other than to provide a semblance of an argument when one doesn't have a real one. Fortunately that mental state has nothing to do with what I do yearn for, which is to discover the truth of how the world is, and why – and, even more, of how it should be.

HERMES: Congratulations, Socrates, on your epistemological wisdom. The knowledge that you seek – *objective knowledge* – is hard to come by, but attainable. That mental state that you do not seek – *justified belief* – is sought by many people, especially priests and philosophers. But, in truth, beliefs cannot be justified, except in relation to other beliefs, and even then only fallibly. So the quest for their justification can lead only to an infinite regress – each step of which would itself be subject to error.

SOCRATES: Again, I know this.

HERMES: Indeed. And, as you have rightly remarked, it doesn't count as a 'revelation' if I tell you what you already know. Yet – notice that that remark is precisely what people who seek justified belief do not agree with.

SOCRATES: *What?* I'm sorry, but that was too convoluted a comment for my allegedly wise mind to comprehend. Please explain what I am to notice about those people who seek 'justified belief'.

HERMES: Merely this. Suppose they just happen to be aware of the explanation of something. You and I would say that they *know* it. But to them, no matter how good an explanation it is, and no matter how true and important and useful it may be, they still do not consider it to be knowledge. It is only if a god then comes along and reassures them that it is true (or if they imagine such a god or other authority) that they count it as knowledge. So, to them it *does* count as a revelation if the authority tells them what they are already fully aware of.

SOCRATES: I see that. And I see that they are foolish, because, for all they know, the 'authority' [*gestures at* HERMES] may be toying with them. Or trying to teach them some important lesson. Or they may be misunderstanding the authority. Or they may be mistaken in their belief that it is an authority –

HERMES: Yes. So the thing *they* call 'knowledge', namely justified belief, is a chimera. It is *unattainable* to humans except in the form of self-deception; it is *unnecessary* for any good purpose; and it is *undesired* by the wisest among mortals.

SOCRATES: I know.

HERMES: Xenophanes knew it too; but he is no longer among the mortals –

SOCRATES: Is that what you meant when you told the Oracle that no one is wiser than I?

HERMES: [*Ignores the question.*] Hence, also, I wasn't referring to justified belief when I asked whether you are sure that I am before your eyes. I was only questioning how you can claim to be 'seeing clearly' what is before your eyes when you also claim to be asleep!

SOCRATES: Oh! Yes, you have caught me in an error – but surely only a trivial one. Indeed, you may not be literally before my eyes. Perhaps you are at home on Olympus, sending me a mere likeness of yourself. But in that case you are controlling that likeness and I am seeing it, and referring to it as 'you', so I am seeing 'you'.

HERMES: But that is not what I asked. I asked what is here *before your eyes*. In reality.

SOCRATES: All right. Before my eyes, in reality, there is – a small room. Or, if you want a literal reply, what is before my eyes is – eyelids, since I expect that they are shut. Yet I see from your expression that you want even more precision. Very well: before my eyes are the *inside surfaces* of my eyelids.

HERMES: And can you see those? In other words, is it really 'easy to see' what is before your eyes?

SOCRATES: Not at the moment. But that is only because I am dreaming.

HERMES: Is it *only* because you are dreaming? Are you saying that if you were awake you would now be seeing the inside surfaces of your eyelids?

SOCRATES: [*carefully*] If I were awake with my eyes still closed, then yes.

HERMES: What *colour* do you see when you close your eyes?

SOCRATES: In a room as dimly lit as this one – black.

HERMES: Do you think that the inside surfaces of your eyelids are black?

SOCRATES: I suppose not.

HERMES: So would you really be seeing them?

SOCRATES: Not exactly.

HERMES: And if you were to open your eyes, would you be able to see the room?

SOCRATES: Only very vaguely. It is dark.

HERMES: So I ask again: is it true that, if you were awake, you could easily see what was before your eyes?

SOCRATES: All right – not always. But nevertheless, when I am awake, and with my eyes open, *and* in bright light –

HERMES: But not *too* bright, I suppose?

SOCRATES: Yes, yes. If you want to keep quibbling, I must accept that when one is dazzled by the sun one may see even less well than in the dark. Likewise one may see one's own face behind a mirror where there is in reality only empty space. One may sometimes see a mirage, or be fooled by a pile of crumpled clothes that happens to resemble a mythical creature –

HERMES: Or one may be fooled by dreaming of one . . .

SOCRATES: [*Smiles.*] Quite so. And, conversely, whether sleeping or waking, we often *fail* to see things that *are* there in reality.

HERMES: You have no idea how many such things there are . . .

SOCRATES: No doubt. But still, when one is *not* dreaming, and conditions are *good* for seeing –

HERMES: And how can you tell whether 'conditions are good' for seeing?

SOCRATES: Ah! Now you are trying to catch me in a circularity. You want me to say that one can tell that conditions are good for seeing when one can easily see what is there.

HERMES: I want you *not* to say so.

SOCRATES: It seems to me that you have been asking questions about *me* – what is in front of me, what I can easily see, whether I am sure, and so on. But I seek fundamental truths, of which I estimate that not a single one is predominantly about me. So let me stress again: I am *not* sure what is in front of my eyes – ever – with my eyes open or closed, asleep or awake. Nor can I be sure what is *probably* in front of my eyes, for how could I estimate the probability that I am dreaming when I think I am awake? Or that my whole previous life has been but a dream in which it has pleased one of you immortals to imprison me?

HERMES: Indeed.

SOCRATES: I might even be a victim of a mundane deception, such as those of conjurers. We know that a conjurer is deceiving us because he shows us something that cannot be – and then asks for money! But if he were to forgo his fee and show me something that *can* be but is not, how could I ever know? Perhaps this entire vision of you is not a dream after all but some cunning conjurer's trick. On the other hand, perhaps you really are here in person and I am awake after all. None of this can I ever be *sure* is so, or not so. I can, however, conceive of *knowing* some of it.

HERMES: Precisely. And is the same true of your *moral* knowledge? In regard to what is right and wrong, could you be mistaken, or misled, by the equivalent of mirages or tricks?

SOCRATES: That seems harder to imagine. For in regard to moral knowledge I need my senses very little: it is mainly just my own thoughts. I *reason* about what is right and wrong, or what makes a person virtuous or wicked. I can be mistaken, of course, in these mental deliberations, but not so easily *deceived* by outside tricks or illusions, for they affect only our senses and not our reason.

HERMES: How, then, do you account for the fact that you Athenians are constantly squabbling among yourselves about what qualities constitute virtue or vice, and what actions are right or wrong?

SOCRATES: Why is that puzzling? We disagree because it is easy to be mistaken. Yet, despite that, we also *agree* about many such issues.

From this I speculate that, where we have so far failed to agree, it is not because anything is actively deceiving us, but simply because some issues are hard to reason about – just as there are many truths in geometry that even Pythagoras did not know but which future geometers may discover. As that other 'wise mortal' Xenophanes wrote:

> The gods did not reveal, from the beginning,
> All things to us; but in the course of time,
> Through seeking we may learn and know things better.*

That is what we Athenians have done in regard to moral knowledge. Through seeking we have learned, and agreed upon, the easy things. And in future, by the same means – namely by refusing to hold any of our ideas immune from criticism – we may learn some matters not so light.

HERMES: There is much truth in what you say. So, take it a little further: if it is so hard to be systematically deceived on moral issues, how is it that the Spartans disagree with you about some of those issues on which nearly all Athenians agree – the ones that you have just said are the *easy* ones?

SOCRATES: Because the Spartans learn many mistaken beliefs and values in early childhood.

HERMES: Whereas Athenians begin their flawless education at what age?

SOCRATES: Again, you catch me in an error. Yes of course we too teach our values to our young, and those must include our most serious misconceptions as well as our deepest wisdom. Yet our values include being open to suggestions, tolerant of dissent, and critical of both dissent *and* received opinion. So I suppose that the real difference between the Spartans and us is that their moral education enjoins them to hold their most important ideas immune from criticism. *Not* to be open to suggestions. *Not* to criticize certain ideas such as their traditions or their conceptions of the gods; *not* to seek the truth, because they claim that they already have it.

Hence they do not believe that 'in the course of time they may

*Popper's translation in *The World of Parmenides* (1998).

learn and know things better.' They agree among themselves because their laws and customs enforce conformity. *We* agree among ourselves (to the extent that we do) because, through our tradition of endless critical debate, we have discovered some genuine knowledge. Since there is only one truth of any given matter, as we discover ideas closer to the truth our ideas become closer to each other's, so we agree more. People who converge upon the truth converge with each other.

HERMES: Indeed.

SOCRATES: Moreover, since the Spartans never seek improvement, it is not surprising that they never find it. We, in contrast, have sought it – by constantly criticizing and debating and trying to correct our ideas and behaviour. And thereby we are well placed to learn more in the future.

HERMES: It follows, then, that it is *wrong* of the Spartans to educate their children to hold their city's ideas, laws and customs immune from criticism.

SOCRATES: I thought you weren't going to reveal moral truths!

HERMES: I can't help it if it follows logically from epistemology. But, anyway, you already know this one.

SOCRATES: Yes, I do. And I see what you are getting at. You are showing me that there *are* such things as mirages and tricks in regard to moral knowledge. Some of them are embedded in the Spartans' traditional moral choices. Their whole way of life misleads and traps them – because one of their mistaken beliefs is that they must take no steps to prevent their way of life from misleading and trapping them!

HERMES: Yes.

SOCRATES: Are there such traps embedded in *our* way of life? [*Frowns.*] Of course, I think there aren't – but I would think that, wouldn't I? As Xenophanes also wrote, it's all too easy to attribute universal truth to mere local appearances:

> The Ethiops say that their gods are flat-nosed and black
> While the Thracians say that theirs have blue eyes and red hair.
> Yet if cattle or horses or lions had hands and could draw
> And could sculpture like men, then the horses would draw their gods
> Like horses, and cattle like cattle . . .

HERMES: So now you are imagining some Spartan Socrates who considers *their* ways virtuous and yours decadent –

SOCRATES: And who considers *us* to be stuck in a trap, since we shall never willingly 'correct' ourselves by adopting Spartan ways. Yes.

HERMES: But does this Spartan Socrates, if he exists, worry that the Athenian Socrates may be right, and he wrong? Was there a Spartan Xenophanes who suspected that the gods might not be as the Greeks think they are?

SOCRATES: Most certainly not!

HERMES: So, since one of their 'ways' is to preserve all their ways unchanged, then if he *were* right, and you wrong –

SOCRATES: Then the Spartans must also have been right ever since they embarked on their present way of life. The gods must have revealed the perfect way of life to them at the outset. So – did you?

HERMES: [*Raises his eyebrows.*]

SOCRATES: Of course you didn't. Now I see that the difference between our ways and theirs is not merely a matter of perspective, nor just a matter of degree.* Let me restate it:

If the Spartan Socrates is right that Athens is trapped in falsehoods but Sparta is not, then Sparta, being unchanging, must already be perfect, and hence right about everything else too. Yet in fact they know almost nothing. One thing that they *clearly* don't know is how to persuade other cities that Sparta is perfect, even cities that have a policy of listening to arguments and criticism . . .

HERMES: Well, logically it could be that the 'perfect way of life' involves having few accomplishments and being wrong about most things. But, yes, you are glimpsing something important here –

SOCRATES: Whereas if I am right that Athens is *not* in such a trap, that implies nothing about whether we are right or wrong about any other matter. Indeed, our very idea that improvement is possible implies that there *must* be errors and inadequacies in our current ideas.

I thank you, generous Apollo, for this 'glimpse' into that important difference.

HERMES: Yet there is even more of a difference than you think. Bear

---

*I shall say more about the difference between those two kinds of society – which I call *static* and *dynamic* societies – in Chapter 15.

in mind that the Spartans and Athenians alike are but fallible men and are subject to misconceptions and errors in all their thinking –

SOCRATES: Wait! We are fallible in *all* our thinking? Is there literally no idea that we may safely hold immune from criticism?

HERMES: Like what?

SOCRATES: [*Ponders for a while. Then:*] What about the truths of arithmetic, like two plus two equals four? Or the fact that Delphi exists? What about the geometrical fact that the angles of a triangle sum to two right angles?

HERMES: Revealing no facts, I cannot confirm that all three of those propositions are even true! But more important is this: how did you come to choose those particular propositions as candidates for immunity from criticism? Why Delphi and not Athens? Why two plus two and not three plus four? Why not the theorem of Pythagoras? Was it because you decided that the propositions you chose would best make your point because they were the most obviously, unambiguously true of all the propositions you considered using?

SOCRATES: Yes.

HERMES: But then how did you determine how obviously and unambiguously true each of those candidate propositions was, compared with the others? Did you not criticize them? Did you not quickly attempt to think of ways or reasons that they might conceivably be false?

SOCRATES: Yes, I did. I see. Had I held them immune from criticism, I would have had no way of arriving at that conclusion.

HERMES: So you are, after all, a thoroughgoing fallibilist – though you mistakenly believed you were not.

SOCRATES: I merely doubted it.

HERMES: You doubted *and* criticized fallibilism itself, as a true fallibilist should.

SOCRATES: That is so. Moreover, had I not criticized it, I could not have come to understand why it is true. My doubt *improved* my knowledge of an important truth – as knowledge held immune from criticism never can be improved!

HERMES: This too you already knew. For it is why you always encourage everyone to criticize even that which seems most obvious to you –

SOCRATES: And why I set an example by doing it to them!

HERMES: Perhaps. Now consider: what would happen if the fallible Athenian voters made a mistake and enacted a law that was very unwise and unjust –

SOCRATES: Which, alas, they often do –

HERMES: Imagine a specific case, for the sake of argument. Suppose that they were somehow firmly persuaded that *thieving* is a high virtue from which many practical benefits flow, and that they abolished all laws forbidding it. What would happen?

SOCRATES: Everyone would start thieving. Very soon those who were best at thieving (and at living among thieves) would become the wealthiest citizens. But most people would no longer be secure in their property (even most thieves), and all the farmers and artisans and traders would soon find it impossible to continue to produce anything worth stealing. So disaster and starvation would follow, while the promised benefits would not, and they would all realize that they had been mistaken.

HERMES: Would they? Let me remind you again of the fallibility of human nature, Socrates. Given that they were firmly persuaded that thievery was beneficial, wouldn't their first reaction to those setbacks be that there was *not enough* thievery going on? Wouldn't they enact laws to encourage it still further?

SOCRATES: Alas, yes – at first. Yet, no matter how firmly they were persuaded, these setbacks would be *problems* in their lives, which they would want to solve. A few among them would eventually begin to suspect that increased thievery might not be the solution after all. So they would think about it more. They would have been convinced of the benefits of thievery by some explanation or other. Now they would try to explain why the supposed solution didn't seem to be working. Eventually they would find an explanation that seemed better. So gradually they would persuade others of that – and so on until a majority again opposed thievery.

HERMES: Aha! So salvation would come about through persuasion.

SOCRATES: If you like. Thought, explanation and persuasion. And now they would understand better *why* thievery is harmful, through their new explanations.*

---

*Which some would mistakenly think were 'derived from experience'.

HERMES: By the way, the little story we have just imagined is exactly how Athens really does look, from my point of view.

SOCRATES: [*somewhat resentfully*] How you must laugh at us!

HERMES: Not at all, Athenian. As I said, I honour you. Now, let us consider what would happen if, instead of legalizing thievery, their error had been to *ban debate*. And to ban philosophy and politics and elections and that whole constellation of activities, and to consider them shameful.

SOCRATES: I see. That would have the effect of banning *persuasion*. And hence it would block off that path to salvation that we have discussed. This is a rare and deadly sort of error: it prevents itself from being undone.

HERMES: Or at least it makes salvation immensely more difficult, yes. This is what *Sparta* looks like, to me.

SOCRATES: I see. And to me too, now that you point it out. In the past I have often pondered the many differences between our two cities, for I must confess that there was – and still is – much that I admire about the Spartans. But I had never realized before now that those differences are all superficial. Beneath their evident virtues and vices, beneath even the fact that they are bitter enemies of Athens, Sparta is the victim – and the servant – of a profound evil. This is a momentous revelation, noble Apollo, better than a thousand declarations of the Oracle, and I cannot adequately express my gratitude.

HERMES: [*Nods in acknowledgement.*]

SOCRATES: I also see why you urge me always to bear human fallibility in mind. In fact, since you mentioned that *some* moral truths follow logically from epistemological considerations, I am now wondering whether they *all* do. Could it be that the moral imperative *not to destroy the means of correcting mistakes* is the only moral imperative? That all other moral truths follow from it?

HERMES: [*Is silent.*]

SOCRATES: As you wish. Now, in regard to Athens, and to what you were saying about epistemology: if our prospects for discovering new knowledge are so good, why were you stressing the unreliability of the senses?

HERMES: I was correcting your description of the quest for knowledge as striving to 'see beyond what is easy to see'.

SOCRATES: I meant that metaphorically: 'see' in the sense of 'understand'.

HERMES: Yes. Nevertheless, you have conceded that even those things that you thought were the easiest to see *literally* are in fact not easy to see at all without prior knowledge about them. In fact *nothing* is easy to see without prior knowledge. All knowledge of the world is hard to come by. Moreover –

SOCRATES: Moreover, it follows that we do not come by it through *seeing*. It does not flow into us through our senses.

HERMES: Exactly.

SOCRATES: Yet you say that objective knowledge is attainable. So, if it does not come to us through the senses, where does it come from?

HERMES: Suppose I were to tell you that all knowledge comes from *persuasion*.

SOCRATES: Persuasion again! Well, I would reply – with all due respect – that that makes no sense. Whoever persuades me of something must first have discovered it himself, so in such a case the relevant issue is where *his* knowledge came from –

HERMES: Quite right, unless –

SOCRATES: And, in any case, when I learn something through persuasion, it *is* coming to me via my senses.

HERMES: No, there you are mistaken. It only seems that way to you.

SOCRATES: *What?*

HERMES: Well, you are learning things from me now, aren't you? Are they coming to you through your senses?

SOCRATES: Yes, of course they are. Oh – no they're not. But that is only because you, a supernatural being, are bypassing my senses and sending me knowledge in a dream.

HERMES: Am I?

SOCRATES: I thought you said you're not here to play debating tricks! Are you denying your own existence now? When sophists do that, I usually take them at their word and stop arguing with them.

HERMES: A policy that again bespeaks your wisdom, Socrates. But I have not denied my existence. I was only questioning *what difference it makes* whether I am real or not. Would it make you change your mind about anything that you have learned about epistemology during this conversation?

SOCRATES: Perhaps not . . .

HERMES: *Perhaps* not? Come now, Socrates, you were boasting earlier that you and your fellow citizens are always open to persuasion.

SOCRATES: Yes, I see.

HERMES: Now, if I *am* only a figment of your imagination, then who has persuaded you?

SOCRATES: Presumably I myself – unless this dream is coming neither from you nor from within myself, but from some other source . . .

HERMES: But did you not say that you are open to persuasion *by anyone*? If dreams emanate from an unknown source, what difference should that make? If they are persuasive, are you not honour-bound as an Athenian to accept them?

SOCRATES: It seems that I am. But what if a dream were to emanate from a malevolent source?

HERMES: That makes no fundamental difference either. Suppose that the source purports to tell you a fact. Then, if you suspect that the source is malevolent, you will try to understand what evil it is trying to perpetrate by telling you the alleged fact. But then, depending on your explanation, you may well decide to believe it anyway –

SOCRATES: I see. For instance, if an enemy announces that he is planning to kill me, I may well believe him despite his malevolence.

HERMES: Yes. Or you may not. And if your closest friend purports to tell you a fact, you may likewise wonder whether *he* has been misled by a malevolent third party – or is simply mistaken for any of countless reasons. Thus situations can easily arise in which you disbelieve your closest friend and believe your worst enemy. What matters in all cases is the explanation you create, within your own mind, for the facts, and for the observations and advice in question.

But the case here is simpler. As I said, I reveal no facts. I'm only making arguments.

SOCRATES: I see. I have no need to trust the source if the argument itself is persuasive. And no way of using *any* source unless I also have a persuasive argument.

Wait a moment – I've just realized something. You 'reveal no facts'. But the god Apollo *does* reveal facts, hundreds of them every day, through the Oracle. Aha, I understand now. You are not Apollo, but a different god.

HERMES: [*Is silent.*]

SOCRATES: You're evidently a god of knowledge . . . but several gods have an interest in knowledge. Athena herself does – but I can tell that you are not she.

HERMES: No you can't.

SOCRATES: Yes I can. I don't mean from your appearance. I mean I can infer it from the detached way you speak of Athens. So – I think you are Hermes. God of knowledge, and of messages, and of information flow –

HERMES: A fine thought. But, by the way, what makes you think that Apollo *reveals facts* through the Oracle?

SOCRATES: Oh!

HERMES: We have agreed that by 'reveal' we mean telling the supplicant something that he doesn't yet know . . .

SOCRATES: Are *all* its replies just jokes and tricks?

HERMES: [*Is silent.*]

SOCRATES: As you wish, fleet Hermes. Then let me try to understand your argument about knowledge. I asked where knowledge comes from, and you directed my attention to this very dream. You asked whether it would make any difference to how I regard the knowledge I am learning from you if it turns out not to have been supernaturally inspired after all. And I had to agree that it would not. So am I to conclude that . . . all knowledge originates from the same source as dreams? Which is within ourselves?

HERMES: Of course it does. Do you remember what Xenophanes wrote just after he said that objective knowledge is attainable by humans?

SOCRATES: Yes. The passage continues:

> But as for certain truth, no man has known it,
> Nor will he know it; neither of the gods,
> Nor yet of all things of which I speak.
> And even if by chance he were to utter
> The perfect truth, he would himself not know it –

So there he's saying that, although objective knowledge is attainable, justified belief ('certain truth') is not.

HERMES: Yes, we've covered all that. But your answer is in the next line.

SOCRATES: 'For all is but a woven web of guesses.' Guesses!

HERMES: Yes. Conjectures.

SOCRATES: But wait! What about when knowledge *does not* come from guesswork – as when a god sends me a dream? What about when I simply hear ideas from other people? *They* may have guessed them, but I then obtain them merely by listening.

HERMES: You do not. In all those cases, you still have to guess in order to acquire the knowledge.

SOCRATES: I do?

HERMES: Of course. Have you yourself not often been misunderstood, even by people trying hard to understand you?

SOCRATES: Yes.

HERMES: Have you, in turn, not often misunderstood what someone means, even when he is trying to tell you as clearly as he can?

SOCRATES: Indeed I have. Not least during this conversation!

HERMES: Well, this is not an attribute of philosophical ideas only, but of all ideas. Remember when you all got lost on your way here from the ship? And why?

SOCRATES: It was because – as we realized with hindsight – we completely misunderstood the directions given to us by the captain.

HERMES: So, when you got the wrong idea of what he meant, despite having listened attentively to every word he said, *where did that wrong idea come from*? Not from him, presumably . . .

SOCRATES: I see. It must come from within ourselves. It must be a guess. Though, until this moment, it had never even remotely occurred to me that I had been guessing.

HERMES: So why would you expect that anything different happens when you do understand someone correctly?

SOCRATES: I see. When we hear something being said, we *guess* what it means, without realizing what we are doing. That is beginning to make sense to me.

Except – guesswork isn't knowledge!

HERMES: Indeed, most guesses are not new knowledge. Although guesswork is the *origin* of all knowledge, it is also a source of error, and therefore what happens to an idea *after* it has been guessed is crucial.

SOCRATES: So – let me combine that insight with what I know of criticism. A guess might come from a dream, or it might just be a wild speculation or random combination of ideas, or anything. But then we do not just accept it blindly or because we imagine it is 'authorized', or because we *want* it to be true. Instead we criticize it and try to discover its flaws.

HERMES: Yes. That is what you *should* do, at any rate.

SOCRATES: Then we try to remedy those flaws by altering the idea, or dropping it in favour of others – and the alterations and other ideas are themselves guesses. And are themselves criticized. Only when we fail in these attempts either to reject or to improve an idea do we provisionally accept it.

HERMES: That can work. Unfortunately, people do not always do what can work.

SOCRATES: Thank you, Hermes. It is exciting to learn of this single process through which all knowledge originates, whether it is our knowledge of a sea captain's directions to Delphi, or knowledge of right and wrong that we have carefully refined for years, or theorems of arithmetic or geometry – or epistemology revealed to us by a god –

HERMES: It all comes from within, from conjecture and criticism.

SOCRATES: Wait! It comes from within, *even if revealed by a god*?

HERMES: And is just as fallible as ever. Yes. Your argument covers that case just like any other.

SOCRATES: Marvellous! But now – what about objects that we just *experience* in the natural world. We reach out and touch an object, and hence experience it *out there*. Surely that is a different kind of knowledge, a kind which – fallible or not – really does come from without, at least in the sense that our own experience is out there, at the location of the object.*

HERMES: You loved the idea that all those other different kinds of knowledge originate in the same way, and are improved in the same way. Why is 'direct' sensory experience an exception? What if it just *seems* radically different?

---

*The ancient Greeks were not very clear about where sensory experiences are located. Even in the case of vision, many in Socrates' time believed that the eye *emits* something like light, and that the sensation of seeing an object consists of some sort of interaction between the object and that light.

SOCRATES: But surely you are now asking me to believe in a sort of all-encompassing conjuring trick, resembling the fanciful notion that the whole of life is really a dream. For it would mean that the sensation of touching an object does not happen where we experience it happening, namely in the hand that touches, but in the mind – which I believe is located somewhere in the brain. So all my sensations of touch are located inside my skull, where in reality nothing can touch while I still live. And whenever I think I am seeing a vast, brilliantly illuminated landscape, all that I am really experiencing is likewise located entirely inside my skull, where in reality it is constantly dark!

HERMES: Is that so absurd? Where do you think all the sights and sounds of *this dream* are located?

SOCRATES: I accept that *they* are indeed in my mind. But that is my point: most dreams portray things that are simply not there in the external reality. To portray things that *are* there is surely impossible without some input that does not come from the mind but from those things themselves.

HERMES: Well reasoned, Socrates. But is that input needed in the *source* of your dream, or only in your ongoing criticism of it?

SOCRATES: You mean that we first guess what is there, and then – what? – we test our guesses against the input from our senses?

HERMES: Yes.

SOCRATES: I see. And then we hone our guesses, and then fashion the best ones into a sort of waking dream of reality.*

HERMES: Yes. A waking dream that *corresponds* to reality. But there is more. It is a dream of which you then gain *control*. You do that by controlling the corresponding aspects of the external reality.

SOCRATES: [*Gasps.*] It is a wonderfully unified theory, and consistent, as far as I can tell. But am I really to accept that I myself – the thinking being that I call 'I' – has no direct knowledge of the physical world at all, but can only receive arcane hints of it through flickers and shadows that happen to impinge on my eyes and other senses? And that what I *experience* as reality is never more than a

---

*Our experience of the world is indeed a form of virtual-reality rendering which happens wholly inside the brain.

waking dream, composed of conjectures originating from within myself?

HERMES: Do you have an alternative explanation?

SOCRATES: No! And the more I contemplate this one, the more delighted I become. (A sensation of which I should beware! Yet I am also persuaded.) Everyone knows that man is the paragon of animals. But if this epistemology you tell me is true, then we are infinitely more marvellous creatures than that. Here we sit, for ever imprisoned in the dark, almost-sealed cave of our skull, *guessing*. We weave stories of an outside world – *worlds*, actually: a physical world, a moral world, a world of abstract geometrical shapes, and so on – but we are not satisfied with merely weaving, nor with mere stories. We want true explanations. So we seek explanations that remain robust when we test them against those flickers and shadows, and against each other, and against criteria of logic and reasonableness and everything else we can think of. And when we can change them no more, we have understood some *objective truth*. And, as if that were not enough, what we understand we then control. It is like magic, only real. We are like gods!

HERMES: Well, *sometimes* you discover *some* objective truth, and exert *some* control as a result. But often, when you think you have achieved any of that, you haven't.

SOCRATES: Yes, yes. But having discovered some truths, can we not make better guesses and further criticisms and tests, and so understand more and control more, as Xenophanes says?

HERMES: Yes.

SOCRATES: So we *are* like gods!

HERMES: Somewhat. And yes, to answer your next question, you can indeed become ever more like gods in ever more ways, *if you choose to*. (Though you will always remain fallible.)

SOCRATES: Why on earth would we not choose to? Oh, I see: Sparta and suchlike . . .

HERMES: Yes. But also because some may argue that *fallible gods* are not a good thing –

SOCRATES: All right. But, *if* we choose to, are you saying that there is no upper bound to how much we can eventually understand, and control, and achieve?

HERMES: Funny you should ask that. Generations from now, a book will be written which will provide a compelling –

[*At that moment there is a knocking at the door.* SOCRATES *glances towards the sound, and then back to where* HERMES *had been, but the god has vanished.*]

CHAEREPHON: [*through the door*] Sorry to wake you, old chap, but I hear that unless we vacate these rooms before the house slaves arrive to clean them, they're liable to charge us for another day.

SOCRATES: [*Emerges, and motions* CHAEREPHON'S SLAVE *into the room to pack* SOCRATES' *modest travelling bag.*] Chaerephon – our trip hasn't been wasted after all! I met Hermes.

CHAEREPHON: What?

SOCRATES: Yes, the god. In a dream, or maybe in person. Or maybe I just dreamed I met him. But it doesn't matter, because, as he pointed out, it makes no difference.

CHAEREPHON: [*Confused.*] What? Why not?

SOCRATES: Because I learned a whole new branch of philosophy – and more!

[*A group of Socrates'* COMPANIONS *is approaching. Sprinting eagerly ahead of the rest is the teenage poet Aristocles, whom his friends call* PLATO *('the Broad') because of his wrestler's build.*]

PLATO: Socrates! Good morning! Thank you again a thousandfold for letting me come on this pilgrimage! [*Launches straight into philosophy without waiting for a reply.*] But I was thinking last night: does it really count as a *revelation* if the Oracle tells us only what we already know? We already knew that there's no one wiser than you, so I thought: shouldn't we go back and demand a free question? But then I thought –

CHAEREPHON: Aristocles, Socrates has –

PLATO: No, wait! Don't tell me the answer. Let me tell you my best guess first. So I thought: yes, we already knew he's the wisest. And that he's modest. But we didn't know quite *how* modest. So that's what the god revealed to us! That Socrates is so modest that he'd contradict even a god saying he's wise.

COMPANIONS: [*Laugh.*]

PLATO: And another thing: *we* knew of Socrates' excellence, but now Apollo has revealed it to *the whole world.*

CHAEREPHON: [*under his breath*] Then I wish 'the whole world' had chipped in for the fee.

PLATO: What was that? Did I get it right?

[SOCRATES *draws breath to answer, but* PLATO *again continues.*] Oh, and Socrates, may I call you 'Master'?

SOCRATES: No.

PLATO: Yes, yes, of course. Sorry. It's just that I've been hanging out with some Spartan kids at the gymnasium, and they talk like that all the time. 'My master says this. My master says that. My master does not permit . . .' and so on and so on. It got so that I became a bit envious that I don't have a master myself, so –

COMPANION NO. 1: Eww, Plato!

PLATO: Yeah, but –

CHAEREPHON: [*catching up*] *Spartan* kids? Aristocles, that is most improper. We are at war!

PLATO: Not here in Delphi we're not. They'd *never* violate the sacred truce of the Oracle. They're very devout, you know. Nice kids, despite their funny accents. We spoke a lot about wrestling – in between actual wrestling, that is. We were up all night, wrestling by candle-light. I've never done that before. They're really good! Though they do sometimes cheat as well. [*Smiles indulgently in recollection.*] But, even so, I wasn't going to let our city be humiliated. I won a few bouts for Athens, you'll be glad to know. That was intense! They taught me some great moves. I can't wait to try them out back home. For some reason none of them are much into poetry, though.

SOCRATES: They don't honour poets in Sparta. Not living ones, anyway.

PLATO: Oh! Pity. I dashed off a poem in commemoration of our wrestling competition. Or rather, between the lines, it's really about why Athens is better than Sparta. It's a mathematical argument . . . Anyway, I've just sent a slave over to their compound to recite it to them, but if they don't honour poets perhaps they won't appreciate it. Oh well. It goes like this –

CHAEREPHON: Aristocles – last night Socrates was visited by the god Hermes!

PLATO: Wow! Why didn't you call us, Socrates? That would have trumped even wrestling with Spartans!

SOCRATES: I couldn't call anyone because it happened in a dream – or something. I'm not even sure that it was really the god. But, as he pointed out to me, it doesn't matter.

PLATO: Why not? Oh, I guess that, once the experience is over, all that matters is what you learned from it. So, what did he want? I bet he wanted to poach you away from the cult of Apollo. Don't do it, Socrates! Apollo is much better. Not that there's anything *wrong* with Hermes, but he has no Oracle. And he's not as cool –

CHAEREPHON: [*shocked*] Show some respect, Aristocles – to Socrates *and* to the gods!

SOCRATES: He *is* showing respect, Chaerephon, in his own way.

PLATO: [*mystified*] Of course I respect them, Chaerephon. And you know I'd literally worship Socrates if he'd let me. Oh, and I respect you too, old man. Greatly. I beg you to forgive me if I have offended you: I know I get too enthusiastic sometimes. [*Pauses briefly.*] But, Socrates – what did you ask the god and what did he reply?

SOCRATES: It wasn't quite like that. He came to reveal to me a new branch of philosophy: *epistemology* – knowledge about knowledge, which also has implications for morality and other fields. Much of it I already knew, or partially knew in various special cases. But he gave me a god's-eye overview, which was breathtaking. Interestingly, he mainly did this by asking *me* questions, and inviting *me* to think about certain things. It seems an effective technique – I may try it sometime.

PLATO: Tell us everything, Socrates! Start with the most interesting thing he asked, and your reply.

SOCRATES: Well – one thing he asked me to do was to imagine a 'Spartan Socrates'.

PLATO: A Spartan *what*? Oh! I see! *That* must be whom the Oracle meant. How sneaky Apollo is! It's the *Spartan* Socrates who's the wisest man in the world – though only by the breadth of a hair, I'll bet! But, being Spartan, he's probably the greatest warrior as well. Awesome! Of course I know you were a great warrior in your day too, Socrates. But still – a *Spartan* Socrates! So are we going to Sparta to see him right away? Please!

CHAEREPHON:⎫ Aristocles – *the war!*
SOCRATES:　⎭ Sorry to disappoint you, Aristocles, but it was a

purely intellectual exercise. There is no 'Spartan Socrates'. In fact I know of no Spartan philosophers at all. In a way, that is what much of my conversation with Hermes was about.

PLATO: Please tell us more.

[*While saying this,* PLATO *gestures to his own* SLAVE, *who, well trained, tosses him a wax-covered writing tablet from a stack that he is carrying.* PLATO *catches it in one hand and pulls out a stylus.*]

SOCRATES: At one stage, Hermes made me aware of the fundamental distinction between the Athenian approach to life and the Spartan. It is that –

PLATO: Wait! Let's all guess! This sounds fascinating.

I'll start – because this is basically what my poem was about. Well, the Spartan half of the riddle is easy: Sparta glories in *war*. And she values all the associated virtues such as courage, endurance and so on.

[*The other* COMPANIONS *of Socrates murmur their assent.*]

We, on the other hand – well, we value *everything*, don't we! Everything good, that is.

COMPANION NO. 1: Everything good? That seems a bit circular, Plato, unless you're going to define 'good' in some way that's independent of 'what we Athenians value'. I think I can put it more elegantly: *fighting*, versus *having something to fight for*.

COMPANION NO. 2: Nice. But that's basically 'War versus Philosophy', isn't it?

PLATO: [*taking mock offence*] And *poetry*.

COMPANION NO. 3: Could it be that Athens, whose patron deity is female, represents the creative spirit in the world, while Sparta favours Ares, the god of bloodlust and slaughter, whom Athena defeated and humbled –

PLATO: No, no, they're actually not that keen on Ares. They prefer Artemis. And, strangely enough, they also revere Athena. Did you know that?

CHAEREPHON: Speaking as an Athenian who is older than all of you and who has seen plenty of war, may I just say that it seems to me that Athens, despite all its glorious martial achievements, would be just as happy to lead a quiet life and be friends with all the Greeks, and not least with the Spartans. But unfortunately the Spartans like nothing better than to annoy us whenever they possibly can. Though

I must admit that in that respect they are not especially worse than anyone else. Including our allies!

SOCRATES: Those are very interesting conjectures, all of which I think do capture aspects of the differences between the cities. And yet I suspect – and I may of course be mistaken –

PLATO: A Spartan Socrates wouldn't be *modest*. Is that the difference?

SOCRATES: No. (By the way, I think that if anything, he *would* be.)

I suspect that we have all been labouring under a misconception about Sparta. Could it be that the Spartans do not seek war, as such, at all? At least, not since they conquered their neighbours, centuries ago, and made them helots. Perhaps, since then, they have acquired an entirely different concern that is of overriding importance to them; and perhaps they fight *only* when that concern is under threat.

COMPANION NO. 2: What is it? Keeping the helots down?

SOCRATES: No, that would be only a means, not the end in itself. I think that the god told me what their overarching concern is. And he also told me what ours is – though alas *we* also fight for all sorts of other reasons, of which we often repent.

Those two overarching concerns are these: we Athenians are concerned above all with *improvement*; the Spartans seek only – *stasis*. Two opposite objectives. If you think about it, I believe you'll soon agree that *this* is the single source of all the myriad differences between the two cities.

PLATO: I never thought of it that way before, but I think I do agree. Let me try out the theory. Here's one difference between the cities: Sparta has no philosophers. That's because the job of a philosopher is to understand things better, which is a form of change, so they don't want it. Another difference: they don't honour living poets, only dead ones. Why? Because dead poets don't write anything new, but live ones do. A third difference: their education system is insanely harsh; ours is famously lax. Why? Because they don't want their kids to dare to question anything, so that they won't ever think of changing anything. How am I doing?

SOCRATES: You are quick on the uptake as usual, Aristocles. However –

CHAEREPHON: Socrates, I think I know plenty of Athenians who do not seek improvement! We have many politicians who think they're perfect. And many sophists who think they know everything.

SOCRATES: But what, specifically, do those politicians believe to be perfect? Their own grandiose plans for how to *improve* the city. Similarly, each sophist believes that everyone should adopt his ideas, which he sees as an *improvement* over everything that has been believed before. The laws and customs of Athens are set up to accommodate all these many rival ideas of perfection (as well as more modest proposals for improvement), to subject them to criticism, to winnow out from them what may be the few tiny seeds of truth, and to test out those that seem the most promising. Thus those myriad individuals who can conceive of no improvement of themselves nevertheless add up to a *city* that relentlessly seeks nothing else for itself, day and night.

CHAEREPHON: Yes, I see.

SOCRATES: In Sparta there are no such politicians, and no such sophists. And no gadflies such as me, because any Spartan who did doubt or disapprove of the way things have always been done would keep it to himself. What few new ideas they have are intended to sustain the city more securely in its current state. As for war, I know that there are Spartans who glory in war, and would love to conquer and enslave the whole world, just as they once set out to conquer their neighbours. Yet the institutions of their city, and the deep assumptions that are built into the minds of even the hotheads, embody a visceral fear of any such step into the unknown. Perhaps it is significant that the statue of Ares that stands outside Sparta represents him *chained*, so that he will always be there to protect the city. Is that not the same as preventing the god of violence from breaking discipline? From being loosed upon the world to cause random mayhem, with its terrifying risk of change?

CHAEREPHON: Perhaps it is. In any case, I understand now, Socrates, how a city can have 'overarching concerns' that are not shared by all its citizens. However, I'm afraid I still don't see how your theory accounts for the *enmity* between our cities. First of all, I cannot recall the Spartans ever objecting to our propensity to improve ourselves. Instead, they cite all sorts of specific grievances about how we are allegedly violating treaties, undermining their allies, plotting to build an empire on the mainland and so on. Second – not that I want to criticize the god, of course! –

SOCRATES: It is not impious to criticize the gods, Chaerephon, but rational. Hermes thinks so too, for what it's worth . . .

PLATO: [*Scribbles, 'It is not impious to criticize gods.'*]

CHAEREPHON: Well, even if the god is right about those two 'over-arching concerns' of stasis and improvement, each city holds its respective concern *only for itself*. It has no ambition to impose it on anyone else. So, although Athens chooses to race forwards while Sparta chooses to tie itself down, and although these choices may logically be 'opposite', how can they possibly be a source of *enmity*?

SOCRATES: My guess is this. The very existence of Athens, however peaceful, is a deadly threat to Sparta's stasis. And therefore, in the long run, the condition for the continued stasis of Sparta (which means its continued *existence*, as they see it) is the destruction of progress in Athens (which from our perspective would constitute the destruction of Athens).

CHAEREPHON: I still do not see specifically what the threat is.

SOCRATES: Well, suppose that in future both cities were to continue to succeed with their overarching concerns. The Spartans would stay exactly as they are now. But we Athenians are already the envy of other Greeks with our wealth and diverse achievements. What will happen when we improve further, and begin to outshine *everyone in the world* at *everything*? Spartans seldom travel or interact with foreigners, but they cannot keep themselves entirely in ignorance of developments elsewhere. Even going to war gives them some inkling of what life is like in other cities that are wealthier, and freer, than they. One day, some Spartan youths visiting Delphi will find that it is the Athenians who have the better 'moves' and the greater skill. And what if, in a generation or two, Athenian warriors have developed some better 'moves' *on the battlefield*?

PLATO: But, Socrates, even if this is true, the Spartans are unaware of it! So how can they fear it?

SOCRATES: They need no prescience. Do you think that a Spartan messenger, on reaching Athens, does not gasp in admiration like everyone else when he sees what stands on our Acropolis?* And, however much he may mutter (perhaps justly) about our hubris and irresponsibility, do you think that he does not reflect, on his way home, that his city can never and will never attract that sort of admiration from anyone? Do you think that the Spartan elders are not at this very moment worrying about the growing reputation of *democracy* in many cities, including some of their allies?

By the way, we ourselves should be at least as wary of democracy as I think the Spartans are of bloodlust and battle rage, for it is intrinsically as dangerous. We could not do without our democracy any more than the Spartans could do without their military training. And, just as they have moderated the destructiveness of bloodlust through their traditions of discipline and caution, we have moderated the destructiveness of democracy through our traditions of virtue, tolerance and liberty. We are utterly dependent on those traditions to keep our monster under control and on our side, just as the Spartans are dependent on *their* traditions to keep *their* monster from devouring them along with everyone else in sight. We might do well to put up a statue of *democracy chained*, to symbolize the fundamental safeguard of our city.

PLATO: [*Scribbles, 'Democracy is a monster, dangerous if not chained.'*]

SOCRATES: The Spartans – and many others who do not understand us – must also be wondering every day how we Athenians can possibly be holding our own against them at the one thing in the world at which they are the best, namely warfare. This despite the fact that at the same time we are excelling more than ever at philosophy and poetry and drama and mathematics and architecture and all those other fields of human endeavour that the Spartans seldom if ever bother with.

PLATO: [*Scribbles, 'Spartans are world's best at warfare but suck at everything else.'*]

SOCRATES: They need not know the reason if they can see the fact. But the reason is: we can improve because we are constantly striving to; they hardly ever improve, because they are trying *not* to! That is the Achilles' heel of Sparta.

PLATO: [*Scribbles, 'Sparta's Achilles' heel is that they don't improve.'*]

*Namely the Parthenon.

So all they need is *philosophers*. With philosophers, they'd be invincible!

SOCRATES: [*Chuckles.*] In a sense, that is the case, Aristocles. But –

PLATO: [*Scribbles, 'Socrates says that, with philosophers, Sparta would be invincible.'*]

CHAEREPHON: [*Worried.*] Then should we really be discussing this here at a public inn? What if someone overhears and tells them the secret?

PLATO: [*Scribbles, 'Note to self: Don't tell them!'*]

SOCRATES: Don't worry, old friend. If the Spartans in general were capable of understanding that 'secret', they'd have implemented it long ago – and there'd be no war between our cities. If some individual Spartan tried to advocate new philosophical ideas, he would soon find himself on trial for heresy or any number of other crimes.

PLATO: Unless . . .

SOCRATES: Unless what?

PLATO: Unless the one who had taken up philosophy was a king.

SOCRATES: Trust you to find the logical loophole, Aristocles. Theoretically you're right, but in Sparta, even the kings are not allowed to change anything important. If one were to try, he would be deposed by the ephors.

PLATO: Well, they have two kings, five ephors and twenty-eight senators. So mathematics tells us that if only fifteen senators, three ephors and one king were to take up philosophy –

SOCRATES: [*Laughs.*] Yes, Aristocles. I concede. If the rulers of Sparta were to take up our style of philosophy, and were then seriously to embark upon criticizing and reforming their traditions –

PLATO: [*Slightly distracted, scribbles, 'Theorem: a king who's a philosopher is the same as a philosopher who's a king. So, what if a philosopher became king?'*] Or perhaps it's more likely that *one* benevolent king would have seized power –

SOCRATES: Whatever. *If* they succeeded in such reforms, then their city might indeed evolve into something truly great. But don't hold your breath.

PLATO: [*Scribbles, 'Socrates says a city with a philosopher king would be truly great.'*] I won't hold my breath. But, in the long run, how shall we teach philosophy to kings, Socrates? [*Scribbles, 'Is the role of philosophers to educate kings?'*]

SOCRATES: I'm not sure that philosophy should be the first step in the education of a leader. One must have something to philosophize *about*. He should know history, and literature, and arithmetic – and, perhaps above all, he should be familiar with the deepest knowledge we have, namely geometry.

PLATO: [*Scribbles, 'Let no one unversed in geometry enter here!'*]

CHAEREPHON: Well, *I* judge a city by how it treats its *philosophers*.

SOCRATES: [*Smiles.*] An excellent criterion, Chaerephon, with which I had better not quibble! By the way, Aristocles, I am not in the least modest. And, to prove it, I can tell you that Hermes persuaded me that I *am* wise after all – at least in one respect that he especially values, namely that I am aware that *justified belief* is impossible and useless and undesirable.

PLATO: [*Scribbles, 'Socrates is the wisest man in the world because he is the only one who* knows *he has no knowledge, because genuine knowledge is impossible!'*] Wait! Justified belief is impossible? Really? Are you sure?

SOCRATES: [*Laughs loudly, while the* OTHERS *look on, puzzled.*] Sorry, but it's a somewhat perverse question, Aristocles.

PLATO: Oh! I see.

[*Smiles ruefully, as do the* OTHERS *when they realize that Plato has just asked for a justification of the belief that one cannot justify beliefs.*]

SOCRATES: No, I am not sure of anything. I never have been. But the god explained to me why that must be so, starting with the fallibility of the human mind and the unreliability of sensory experience.

PLATO: [*Scribbles, 'It's only knowledge of the material world that's impossible, useless and undesirable.'*]

SOCRATES: He gave me a marvellous perspective on how we perceive the world. Each of your eyes is like a dark little cave, one on whose rear wall some stray shadows fall from outside. You spend your whole life at the back of that cave, able to see nothing but that rear wall, so you cannot see reality directly at all.

PLATO: [*Scribbles, 'It is as if we were prisoners, chained inside a cave and permitted to look only at the rear wall. We can never know the*

reality outside because we see only fleeting, distorted shadows of it.']
   [Note: Socrates is slightly improving on Hermes, and Plato has
   been increasingly misinterpreting Socrates.]

SOCRATES: He then went on to explain to me that objective knowledge
   is indeed possible: it comes from within! It begins as conjecture, and
   is then *corrected* by repeated cycles of criticism, including comparison
   with the evidence on our 'wall'.

PLATO: [*Scribbles, 'The only true knowledge is that which comes from
   within. (How? Remembered from a previous life?)'*]

SOCRATES: In this way, we frail and fallible humans can come to know
   objective reality – provided we use philosophically sound methods
   as I have described (which most people do not).

PLATO: [*Scribbles, 'We can come to know the true world beyond the
   illusory world of experience. But only by pursuing the kingly art of
   philosophy.'*]

CHAEREPHON: Socrates, I think it *was* the god speaking to you, for I
   strongly feel that I have glimpsed a divine truth through you today.
   It will take me a long time to reorganize my ideas to take account
   of this new epistemology that he revealed to you. It seems a
   tremendously far-reaching, and important, subject.

SOCRATES: Indeed. I have some reorganizing to do myself.

PLATO: Socrates, you really ought to write all this down – together
   with all your other wisdom – for the benefit of the whole world, and
   posterity.

SOCRATES: No need, Aristocles. Posterity is right here, listening.
   Posterity is all of *you*, my friends. What is the point of writing down
   things that are going to be endlessly tinkered with and improved?
   Rather than make a permanent record of all my misconceptions as
   they are at a particular instant, I would rather offer them to others
   in two-way debate. That way I benefit from criticism and may even
   make improvements myself. Whatever is valuable will survive such
   debates and be passed on without any effort from me. Whatever is
   not valuable would only make me look a fool to future generations.

PLATO: If you say so, Master.

Since Socrates left us no writings, historians of ideas can only guess at
what he really thought and taught, using the indirect evidence of his

portrayal by Plato and a few others who were there at the time and whose accounts have survived. This is known as the 'Socratic problem', and is the source of much controversy. One common view is that the young Plato conveyed Socrates' philosophy fairly faithfully, but that later he used the character of Socrates more as a vehicle for conveying his own views; that he did not even intend his dialogues to represent the real Socrates, but used them only as convenient ways of expressing arguments that have a to-and-fro form.

Perhaps I had better stress – in case it is not already obvious – that I am doing the same. I do not intend the above dialogue accurately to represent the philosophical opinions of the historical Socrates and Plato. I have set it at that moment in history, with those participants, because Socrates and his circle were among the foremost contributors to the 'Golden Age of Athens', which should have become a beginning of infinity but did not. And also because one thing that we do know about the ancient Greeks is that the philosophical *problems* they considered important have dominated Western philosophy ever since: How is knowledge obtained? How can we distinguish between true and false, right and wrong, reason and unreason? Which sorts of knowledge (moral, empirical, theological, mathematical, justified . . .) are possible, and which are mere chimeras? And so on. And therefore, although the theory of knowledge presented in the dialogue is largely that of the twentieth-century philosopher Karl Popper, together with some addenda of my own, I guess that Socrates would have understood and liked it. In some universes that were very like ours at the time, he thought of it himself.

I do want to make one indirect comment on the Socratic problem, though: we habitually underestimate the difficulty of communication – just as Socrates does at the end of the dialogue, when he assumes that each party to a debate necessarily knows what the other is saying, and Plato increasingly gets the wrong end of the stick. In reality, the communication of new ideas – even mundane ones like directions – depends on guesswork on the part of both the recipient and the communicator, and is inherently fallible. Hence there is no reason to expect that the young Plato, just because he was intelligent and highly educated, and by all accounts a near-worshipper of Socrates, made the fewest mistakes in conveying Socrates' theories. On the contrary, the default assumption should be that misunderstandings are ubiquitous and that neither intelligence nor the intention to be accurate is any guarantee against them. It could easily be that the young Plato misunderstood everything that Socrates said to him, and that the older Plato gradually succeeded in understanding it, and is therefore the more reliable guide. Or it could be that Plato slipped ever further into misinterpretation, and into positive errors of his own. Evidence, argument and explanation are needed to distinguish between these and many other possibilities. It is a difficult task for historians. Objective knowledge, though attainable, is hard to attain.

All this holds as much for knowledge written down as for knowledge spoken in person. So there would still be a 'Socratic problem' even if Socrates had written books. Indeed, there is such a problem in regard to the prolific Plato, and sometimes even in regard to living philosophers. What does the philosopher mean by such and such a term or assertion? What problem is the assertion intended to solve, and how? These are not themselves philosophical problems. They are problems in the history of philosophy. Yet nearly all philosophers, especially academic ones, have devoted a great deal of their attention to them. Courses in philosophy place great weight on reading original texts, and commentaries on them, in order to understand the theories that were in the minds of various great philosophers.

This focus on history is odd, and is in marked contrast to all other academic disciplines (except perhaps history itself). For example, in all the physics courses that I took at university, both as an undergraduate and as a graduate student, I cannot recall a single instance where any original papers or books by the great physicists of old were studied or were even on the reading list. Only when a course touched upon very recent discoveries did we ever read the work of their discoverers. So we learned Einstein's theory of relativity without ever hearing from Einstein; we knew Maxwell, Boltzmann, Schrödinger, Heisenberg and so on only as names. We read their *theories* in textbooks whose authors were physicists (not historians of physics) who themselves may well never have read the works of those pioneers.

Why? The immediate reason is that the original sources of scientific theories are almost never good sources. How could they be? All subsequent expositions are intended to be improvements on them, and some succeed, and improvements are cumulative. And there is a deeper reason. The originators of a fundamental new theory initially share many of the misconceptions of previous theories. They need to develop an understanding of how and why those theories are flawed, and how the new theory explains everything that they explained. But most people who subsequently learn the new theory have quite different concerns. Often they just want to take the theory for granted and use it to make predictions, or to understand some complex phenomenon in combination with other theories. Or they may want to understand nuances of it that have nothing to do with why it is superior to the old theories. Or they may want to improve it. But what they no longer care about is tracking down and definitively meeting every last objection that would naturally be made by someone thinking in terms of older, superseded theories. There is rarely any reason for scientists to address the obsolete problem-situations that motivated the great scientists of the past.

Historians of science, in contrast, must do precisely that – and they encounter much the same difficulties as the historians of philosophy who address the Socratic problem. Why, then, do scientists not encounter these difficulties when learning scientific theories? What is it that allows such theories to be communicated through chains of intermediaries with such apparent ease? What has happened to the 'difficulty of communication' that I stressed above?

The first, seemingly paradoxical, half of the answer is that, when they learn a theory, scientists are not interested in what the theory's originator, or anyone else along the chain of communication, believed. When physicists read a textbook on the theory of relativity, their immediate objective is to learn the *theory*, and not the opinions of Einstein or of the textbook's author. If that seems strange, imagine, for the sake of argument, that a historian were to discover that Einstein wrote his papers only as a joke, or at gunpoint, and was actually a lifelong believer in Kepler's laws. This would be a bizarre and important discovery about the *history* of physics, and all the textbooks about that would have to be rewritten. But our knowledge of physics itself would be unaffected, and physics textbooks would not need any change at all.

The second half of the answer is that the reason why the scientists are trying to learn the theory, and also why they have such disregard for faithfulness to the original, is that they want to know how the world is. Crucially, this is the same objective that the originator of the theory had. If it is a good theory – if it is a superb theory, as the fundamental theories of physics nowadays are – then it is exceedingly hard to vary while still remaining a viable explanation. So the learners, through criticism of their initial guesses and with the help of their books, teachers and colleagues, seeking a viable explanation, will arrive at the same theory as the originator. *That* is how the theory manages to be passed faithfully from generation to generation, despite no one caring about its faithfulness one way or the other.

Slowly, and with many setbacks, the same is becoming true in non-scientific fields. The way to converge with each other is to converge upon the truth.

I I

The Multiverse

The idea of a 'doppelgänger' (a 'double' of a person) is a frequent theme of science fiction. For instance, the classic television series *Star Trek* featured several types of doppelgänger story involving malfunctions of the 'transporter', the starship's teleportation device, normally used for short-range space travel. Since teleporting something is conceptually similar to making a copy of it at a different location, one can imagine various ways in which the process could go wrong and somehow end up with two instances of each passenger – the original and the copy.

Stories vary in how similar the doppelgängers are to their originals. To share literally all their attributes, they would have to be at exactly the same location as well as looking alike. But what would that mean? Trying to make atoms coincide leads to some problematic physics – for instance, two coinciding nuclei are liable to combine to form atoms of heavier chemical elements. And if two identical human bodies were to coincide even approximately, they would explode simply because water at double its normal density exerts a pressure of hundreds of thousands of atmospheres. In fiction one could imagine different laws of physics to avoid those problems; but, even then, if the doppelgängers continued to coincide with their originals throughout the story, it would not really be about doppelgängers. Sooner or later they have to be different. Sometimes they are the good and evil 'sides' of the same person; sometimes they start with identical minds but become increasingly different through having different experiences.

Sometimes a doppelgänger is not *copied* from an original, but exists from the outset in a 'parallel universe'. In some stories there is a 'rift' between universes through which one can communicate or even travel to meet one's doppelgänger. In others, the universes remain mutually imperceptible, in which case the interest of the story (or, rather, two stories) is in how events are affected by the differences between them. For instance, the movie *Sliding Doors* interleaves two variants of a love story, following the fortunes of two instances of the same couple in two universes which initially differ only in one small detail. In a related genre, known as 'alternative history', one of the two stories need not be told explicitly because it is a part of our own history and is assumed to be known to the audience. For example, the novel *Fatherland*, by Robert Harris, is about a universe in which Germany won the Second World War; Robert Silverberg's *Roma Eterna* is about one in which the Roman Empire did not fall.

In another class of stories, the transporter's malfunction accidentally exiles the passengers to a 'phantom zone' where they are imperceptible to everyone in the ordinary world, but can see and hear them (and each other). So they have the distressing experience of yelling and gesticulating in vain to their shipmates, who are oblivious and walk right through them.

In some stories it is only *copies* of the travellers that are sent to a phantom zone, unbeknown to the originals. Such a story may end with the exiles discovering that they can, after all, have some effect on the ordinary world. They use that effect to signal their existence, and are rescued through a reversal of the process that exiled them. Depending on the fictional science that has been supposed, they then may begin new lives as separate people, or they may merge with their originals. The latter option violates the principle of the conservation of mass, among other laws of physics. But, again, this is fiction.

Nevertheless, there is a certain category of rather pedantic science fiction enthusiasts, myself included, who prefer the fictional science to make sense – to consist of reasonably good explanations. Imagining worlds with different laws of physics is one thing; imagining worlds that do not make sense in their own terms is quite another. For instance, we want to know how it can be that the exiles can see and hear the ordinary world but not touch it. This attitude of ours was nicely parodied in an episode of the television series *The Simpsons*, in which fans of a fantasy-adventure series question its star:

STAR: Next question.

FAN: Yes, over here. [*Clears throat.*] In episode BF12, you were battling
barbarians while riding a wingèd Appaloosa, yet in the very next scene,
my dear, you're clearly atop a wingèd Arabian. Please to explain it.

STAR: Ah, yeah, well, whenever you notice something like that, a wizard
did it.

FAN: I see, all right, yes, but in episode AG4 –

STAR: [*firmly*] Wizard.

FAN: Aw, for glayvin* out loud!

Because that is a parody, the fan is complaining not about the story
itself, but only that there is a *continuity error*: two horses were used
at different times to play the role of a single fictional horse. Nevertheless, there are such things as flawed stories. Consider, for instance, a
story about a quest to discover whether winged horses are real, in
which the characters pursue that quest on winged horses. Though
logically consistent, such a story would not make sense in its own
terms, as an explanation. One could embed it in a context that would
make sense of it – for instance, it could be part of an allegory about
how people often fail to see the meaning of what is right there in front
of them. But in that case any merit in the story would still depend on
how the characters' apparently nonsensical behaviour was *explicable*
in terms of that allegory. Compare that with the explanation that 'a
wizard did it.' Since a wizard could equally well have been said to
conjure *any* events, in *any* story, it is a bad explanation; and that is
why the fan is exasperated by it.

In some stories the plot is not important: the story is really about
something else. But a good plot always rests, implicitly or explicitly,
on good explanations of how and why events happen, given its fictional
premises. In that case, even if those premises are about wizards, the
story is not really about the supernatural: it is about imaginary laws
of physics and imaginary societies, as well as real problems and true
ideas. As I shall explain in Chapter 14, not only do all good science-
fiction plots resemble scientific explanation in this way, in the broadest
sense all good art does.

In that spirit, then, consider the fictional doppelgängers in the

---

*'Glayvin' is a term of indeterminate meaning, coined by *The Simpsons*.

phantom zone. What enables them to *see* the ordinary world? Since they are structurally identical to their originals, their eyes work by absorbing light and detecting the resulting chemical changes, just as real eyes do. But if they absorb some of the light coming from the ordinary world, then they must cast shadows at the places where that light would otherwise have arrived. Also, if the exiles in the phantom zone can see each other, what light are they seeing with? The phantom zone's own light? If so, where does it come from?

On the other hand, if the exiles can see *without* absorbing light, then they must be differently constituted from their originals, at the microscopic level. And in that case we no longer have an explanation of why they outwardly resemble their originals: the 'accidental-copying' idea will no longer do: where did the transporter get the knowledge required to build things that look and behave like human bodies, but function internally in a different way? It would be a case of spontaneous generation.

Similarly, is there air in the phantom zone? If the exiles breathe air, it can't be the ship's air, because they would be heard speaking or even breathing. But nor can it be a copy of the small amount of air that was in the transporter with them, because they are free to move around the ship. So there must be a whole shipful of phantom-zone air. But then what is preventing it from expanding out into space?

It seems that almost everything that happens in the story not only conflicts with the real laws of physics (which is unexceptionable in fiction), but raises problems within the fictional explanation. If the doppelgängers can walk through people, why do they not fall through the floor? In reality, a floor supports people by bending slightly. But if it were to bend in the story, it would also vibrate with their steps and set off sound waves which people in the ordinary world could hear. So there must be a separate floor and walls as well as an entire spaceship hull in the phantom zone. Even the space outside cannot be ordinary space, because if one could get back into ordinary space by leaving the ship, then the exiles could return by that route. But if there is an entire phantom-zone space out there – a parallel universe – how could a mere transporter malfunction have created *that*?

We should not be surprised that good fictional science is hard to invent: it is a variant of real science, and real scientific knowledge is very hard to vary. Thus few if any of the storylines that I have outlined make sense as they stand. But I want to continue with one of my own, making sure that it (eventually) does make sense.

A writer of real science fiction faces two conflicting incentives. One is, as with all fiction, to allow the reader to engage with the story, and the easiest way to do that is to draw on themes that are already familiar. But that is an anthropocentric incentive. For instance, it pushes authors to imagine ways around the absolute speed limit that the laws of physics impose on travel and communication (namely the speed of light). But when authors do that, they relegate *distance* to the role that it has in stories about our home planet: star systems play the same role that remote islands or the Wild West did in the fiction of earlier eras. Similarly, the temptation in parallel-universe stories is to allow communication or travel between universes. But then the story is really about a single universe: once the barrier between the universes is easily penetrable, it becomes no more than an exotic version of the oceans that separate continents. A story that succumbs entirely to this anthropocentric incentive is not really science fiction but ordinary fiction in disguise.

The opposing incentive is to explore the strongest possible version of a fictional-science premise, and its strangest possible implications – which pushes in the anti-anthropocentric direction. This may make the story harder to engage with, but it allows for a much broader range of scientific speculations. In the story that I shall tell here, I shall use a succession of such speculations, increasingly distant from the familiar, as means of explaining the world according to quantum theory.

Quantum theory is the deepest explanation known to science. It violates many of the assumptions of common sense, and of all previous science – including some that no one suspected were being made at all until quantum theory came along and contradicted them. And yet this seemingly alien territory is the reality of which we and everything we experience are part. There is no other. So, in setting a story there, perhaps what I lose in terms of the familiar ingredients of drama I shall gain in terms of opportunity to explain something that is more astounding than any fiction, yet is the purest and most basic fact we know about the physical world.

I had better warn the reader that the account that I shall give –

known as the 'many-universes interpretation' of quantum theory (rather inadequately, since there is much more to it than 'universes') – remains at the time of writing a decidedly minority view among physicists. In the next chapter I shall speculate why that is so despite the fact that many well-studied phenomena have no other known explanation. For the moment, suffice it to say that the very idea of *science as explanation*, in the sense that I am advocating in this book (namely an account of what is really out there), is itself still a minority view even among theoretical physicists.

Let me begin with perhaps the simplest possible 'parallel-universe' speculation: a 'phantom zone' has existed all along (ever since its own Big Bang). Until our story begins, it has been an exact doppelgänger of the entire universe, atom for atom and event for event.

All the flaws that I mentioned in the phantom-zone stories derive from the asymmetry that things in the ordinary world affect things in the phantom zone but not vice versa. So let me eliminate those flaws by imagining, for the moment, that the universes are completely imperceptible to each other. Since we are heading towards real physics, let me also retain the speed-of-light limit on communication, and let the laws of physics be universal and symmetrical (i.e. they make no distinction between the universes). Moreover, they are deterministic: nothing random ever happens, which is why the universes have remained alike – so far. So how can they ever become different? That is a key question in the theory of the multiverse, which I shall answer below.

All these basic properties of my fictional world can be thought of as conditions on the flow of information: one cannot send a message to the other universe; nor can one change anything in one's own universe sooner than light could reach that thing. Nor can one bring new information – even random information – into the world: everything that happens is determined by laws of physics from what has gone before. However, one can, of course, bring new *knowledge* into the world. Knowledge consists of explanations, and none of those conditions prevents the creation of new explanations. All this is true of the real world too.

We can temporarily think of the two universes as being literally parallel. Suppress the third dimension of space and think of a universe as being two-dimensional, like an infinitely flat television. Then place a second such television parallel to it, showing exactly the same pictures (symbolizing the objects in the two universes). Now forget the material of which the televisions are made. Only the pictures exist. This is to stress that a universe is not a receptacle containing physical objects: it *is* those objects. In real physics, even space is a physical object, capable of warping and affecting matter and being affected by it.

So now we have two perfectly parallel, identical universes, each including an instance of our starship, its crew and its transporter, and of the whole of space. Because of the symmetry between them, it is now misleading to call one of them 'the ordinary universe' and the other 'the phantom zone'. So I shall just call them 'universes'. The two of them together (which comprise the whole of physical reality in the story so far) are the *multiverse*. Similarly, it is misleading to speak of the 'original' object and its 'doppelgänger': they are simply the two *instances* of the object.

If our science-fiction speculation were to stop there, the two universes would have to remain identical for ever. There is nothing logically impossible about that. Yet it would make our story fatally flawed both as fiction and as scientific speculation – and for the same reason: it is a story of two universes, but only one *history*. That is to say, there is only one script about what is really there in both universes. Considered as fiction, therefore, it is really a single-universe story in a pointless disguise. Considered as scientific speculation, it describes a world that would not be explicable to its inhabitants. For how could they ever argue that their history takes place in two universes and not three or thirty? Why not two today and thirty tomorrow? Moreover, since their world has only one history, all their good explanations about nature would be about that history. That single history would be what they meant by their 'world' or 'universe'. Nothing of the underlying two-ness of their reality would be accessible to them, nor would it make any more sense to them as an explanation than would three-ness or thirty-ness – yet they would be factually mistaken.

A remark about explanation: Although the story so far would be a bad explanation from the inhabitants' point of view, it is not necessarily bad from ours. Imagining inexplicable worlds can help us to understand the nature of explicability. I have already imagined some inexplicable worlds for that very reason in previous chapters, and I shall imagine more in this chapter. But, in the end, I want to tell of an explicable world, and it will be ours.

A remark about terminology: The *world* is the whole of physical reality. In classical (pre-quantum) physics, the world was thought to consist of one *universe* – something like a whole three-dimensional space for the whole of time, and all its contents. According to quantum physics, as I shall explain, the world is a much larger and more complicated object, a *multiverse*, which includes many such universes (among other things). And a *history* is a sequence of events happening to objects and possibly their identical counterparts. So, in my story so far, the world is a multiverse that consists of two universes but has only a single history.

So our two universes must not stay identical. Something like a transporter malfunction will have to make them different. Yet, as I said, that may seem to have been ruled out by those restrictions on information flow. The laws of physics in the fictional multiverse are deterministic and symmetrical. So what can the transporter possibly do that would make the two universes differ? It may seem that whatever one instance of it does to one universe, its doppelgänger must be doing to the other, so the universes can only remain the same.

Surprisingly, that is not so. It *is* consistent for two identical entities to become different under deterministic and symmetrical laws. But, for that to happen, they must initially be more than just exact images of each other: they must be *fungible* (the g is pronounced as in 'plunger'), by which I mean identical in literally every way except that there are two of them. The concept of fungibility is going to appear repeatedly in my story. The term is borrowed from legal terminology, where it refers to the legal fiction that *deems* certain entities to be identical for purposes such as paying debts. For example, dollar bills are fungible in law, which means that, unless otherwise agreed, borrowing a dollar does not require one to return the specific banknote that one borrowed. Barrels of oil (of a given grade) are fungible too. Horses are not: borrowing someone's horse means that one has to return that specific horse; even its identical twin will not do. But the physical fungibility I am referring to here is not about deeming. It means *being* identical, and that is a very different and counter-intuitive property. Leibniz, in his doctrine of 'the identity of indiscernibles', went so far as to rule out its existence on principle. But he was mistaken. Even aside from the physics of the multiverse, we now know that photons, and under some conditions even atoms, can be fungible. This is achieved in lasers and in devices called 'atomic lasers' respectively. The latter emit bursts of extremely cold, fungible atoms. For how this is possible without causing transmutation, explosions and so on, see below.

You will not find the concept of fungibility discussed or even mentioned in many textbooks or research papers on quantum theory, even the small minority that endorse the many-universes interpretation. Nevertheless, it is everywhere just beneath the conceptual surface, and I believe that making it explicit helps to explain quantum phenomena without fudging. As will become clear, it is an even weirder attribute than Leibniz guessed – much weirder than multiple universes for instance, which are, after all, just common sense, repeated. It allows radically new types of *motion* and *information flow*, different from anything that was imagined before quantum physics, and hence a radically different structure of the physical world.

It so happens that, in some situations, money is not only legally fungible but physically too; and, being so familiar, it provides a good model for thinking about fungibility. For example, if the balance in your (electronic) bank account is one dollar, and the bank adds a second dollar as a loyalty bonus and later withdraws a dollar in charges, there is no meaning to whether the dollar they withdrew is the one that was there originally or the one that they had added – or is composed of a little of each. It is not merely that *we cannot know* whether it was the same dollar, or have decided not to care: because of the physics of the situation there really is no such thing as taking the original dollar, nor such a thing as taking the one added subsequently.

Dollars in bank accounts are what may be called 'configurational' entities: they are states or configurations of objects, not what we usually think of as physical objects in their own right. Your bank balance resides in the *state* of a certain information-storage device. In a sense you own that state (it is illegal for anyone to alter it without your consent), but you do not own the device itself or any part of it. So in that sense a dollar is an abstraction. Indeed, it is a piece of *abstract knowledge*. As I discussed in Chapter 4, knowledge, once embodied in physical form in a suitable environment, causes itself to remain so. And thus, when a physical dollar wears out and is destroyed by the mint, the abstract dollar causes the mint to transfer it into electronic form, or into a new instance in paper form. It is an abstract replicator – though, unusually for a replicator, it causes itself *not* to proliferate, but rather to be copied into ledgers and into backups of computer memories.

Another example of fungible configurational entities in classical physics is amounts of energy: if you pedal your bicycle until you have built up a kinetic energy of ten kilojoules, and then brake until half that energy has been dissipated as heat, there is no meaning to whether the energy dissipated was the first five kilojoules that you had added or the second, or any combination. But it is meaningful that *half* the energy that was there has been dissipated. It turns out that, in quantum physics, elementary particles are configurational entities too. The vacuum, which we perceive as empty at everyday scales and even at atomic scales, is not really emptiness, but a richly structured entity known as a 'quantum field'. Elementary particles are higher-energy configurations of this entity: 'excitations of the vacuum'. So, for instance, the photons in a laser are configurations of the vacuum inside its 'cavity'. When two or more such excitations with identical attributes (such as energy and spin) are present in the cavity, there is no such thing as which one was there first, nor which one will be the next to leave. There is only such a thing as the attributes of any one of them, and how many of them there are.

If the two universes of our fictional multiverse are initially fungible, our transporter malfunction can make them acquire different attributes in the same way that a bank's computer can withdraw one of two fungible dollars and not the other from an account containing two dollars. The laws of physics could, for instance, say that, when the transporter malfunctions, then *in one of the universes and not the other* there will be a small voltage surge in the transported objects. The laws, being symmetrical, could not possibly specify *which* universe the surge will take place in. But, precisely because the universes are initially fungible, they do not have to.

It is a rather counter-intuitive fact that if objects are merely identical (in the sense of being exact copies), and obey deterministic laws that make no distinction between them, then they can never become different; but *fungible* objects, which on the face of it are even more alike, can. This is the first of those weird properties of fungibility that Leibniz never thought of, and which I consider to be at the heart of the phenomena of quantum physics.

Here is another. Suppose that your account contains a hundred dollars and you have instructed your bank to transfer one dollar from this account to the tax authority on a specified date in the future. So the bank's computer now contains a deterministic rule to that effect. Suppose that you have done this because the dollar already belongs to the tax authority. (Say it had mistakenly sent you a tax refund, and has given you a deadline to repay it.) Since the dollars in the account are fungible, there is no such thing as *which one* belongs to the tax authority and which belong to you. So we now have a situation in which a collection of entities, though fungible, do not all have the same owner! Everyday language struggles to describe this situation: each dollar in the account shares literally all its attributes with the others, yet it is not the case that all of them have the same owner. So, could we say that in this situation they have no owner? That would be misleading, because evidently the tax authority does own one of them and you do own the rest. Could one say that they all have two owners? Perhaps, but only because that is a vague term. Certainly there is no point in saying that one cent of each of the dollars is owned by the tax authority, because that simply runs into the problem that the cents in the account are all fungible too. But, in any case, notice that the problem raised by this 'diversity within fungibility' is one of language only. It is a problem of how to describe some aspects of the situation in words. No one finds the situation itself paradoxical: the computer has been instructed to execute definite rules, and there will never be any ambiguity about what will happen as a result.

Diversity within fungibility is a widespread phenomenon in the multiverse, as I shall explain. One big difference from the case of fungible money is that in the latter case we never have to wonder about – or predict – what it would be like to *be* a dollar. That is to say, what it would be like to be fungible, and then to become differentiated. Many applications of quantum theory require us to do exactly that.

But first: I suggested temporarily visualizing our two universes as being next to each other in space – just as some science-fiction stories refer to doppelgänger universes as being 'in other dimensions'. But now we have to abandon that image and make them coincide: whatever that 'extra dimension' was supposed to denote, it would make them non-fungible.* It is not that they coincide *in* anything, such as an external space: they are not in space. An instance of space is part of each of them. That they 'coincide' means only that they are not separate in any way.

It is hard to imagine perfectly identical things coinciding. For instance, as soon as you imagine just one of them, your imagination has already violated their fungibility. But, although imagination may baulk, reason does not.

Now our story can begin to have a non-trivial plot. For example, the voltage surge that happens in one of the two universes when the transporter malfunctions could cause some of the neurons in a passenger's brain to misfire in that universe. As a result, in that universe, that passenger spills a cup of coffee on another passenger. As a result, they have a shared experience which they do not have in the other universe, and this leads to romance – just as in *Sliding Doors*.

The voltage surges need not be 'malfunctions' of the transporter. They could be a regular effect of the way it works. We accept much larger unpredictable jolts during others forms of travel such as flying or bronco-riding. Let us imagine that a tiny surge is produced in one of the universes whenever the transporter is operated in both, but that it is too small to be noticeable unless measured with a sensitive voltmeter, or unless it nudges something that happens to be on the brink of changing but would recede from the brink if not nudged.

In principle, a phenomenon could appear unpredictable to observers for one or more of three reasons. The first is that it is affected by some fundamentally random (indeterministic) variable. I have excluded that possibility from our story because there are no such variables in real physics. The second, which is at least partly responsible for most everyday unpredictability, is that the factors affecting the phenomenon,

---

*Identical entities that were at different locations *in an otherwise empty space* would not be fungible, but some philosophers have argued that they would be 'indiscernible' in Leibniz's sense. If so, then this is yet another respect in which fungibility is worse than Leibniz imagined.

though deterministic, are either unknown or too complex to take account of. (This is especially so when they involve the creation of knowledge, as I discussed in Chapter 9.) The third – which had never been imagined before quantum theory – is that two or more initially fungible instances of the observer become different. That is what those transporter-induced jolts bring about, and it makes their outcomes strictly unpredictable despite being described by deterministic laws of physics.

These remarks about unpredictable phenomena could be expressed without ever referring explicitly to fungibility. And indeed that is what multiverse researchers usually do. Nevertheless, as I have said, I believe that fungibility is essential to the explanation of quantum randomness and most other quantum phenomena.

All three of these radically different causes of unpredictability could in principle feel exactly the same to observers. But, in an explicable world, there must be a way of finding out which of them (or which combination of them) is the actual source of any apparent randomness in nature. How could one find out that it is fungibility and parallel universes that are responsible for a given phenomenon?

In fiction, there is always the temptation to introduce inter-universe communication for this purpose, making the universes no longer 'parallel'. As I have said, that would really make it a single-universe story – but we might try to disguise that fact by saying that such communication is *difficult*. For example, it might be that there is a way of adjusting the transporter in either universe so that it produces a voltage surge in the other. Then one could use it to transmit a message there. But we could imagine that this is very expensive, or dangerous, so that the ship's regulations limit its use. 'Personal communication' with one's own doppelgänger is especially prohibited. Nevertheless, one crew member illicitly ignores this prohibition during the night watch, and is startled to receive a message 'HAVE MARRIED SONAK.' We know, but the character does not, that this marriage is a knock-on effect of the coffee-spilling incident which was itself a knock-on effect of the voltage surge in the other universe. Then the transmission ends and no more such messages are received. We know – but again the character does not – that this is because the illicit use of the equipment has been detected in the other universe and stricter safeguards have been implemented. The story could then explore what might happen when the crew member acts upon that startling message.

How *should* one react to the news that one's doppelgänger has married? Should one seek out the spouse's doppelgänger in one's own universe – whom one has never even met personally, let alone formed a romantic relationship with? Or whom, in the time-honoured tradition of love stories, one finds annoying. It can't do any harm. Or can it?

Ideas originating in the other universe are at least as fallible as those in ours; and if they are difficult to obtain, that makes error-correction harder. Knowledge-creation depends on error-correction. So perhaps the message would have continued 'ALREADY REGRETTING IT'. Or perhaps Sonak had just turned up in the transporter room in the other universe, making it impossible to send that warning. Or perhaps the couple are happy at the moment, but will shortly have a disastrous break-up resulting in divorce. In all those cases, that inter-universe communication, far from being helpful, could cause a doubling of the number of disastrous marriage decisions made by the two instances of that crew member.

More generally, the news that your doppelgänger seems happy having made a particular decision in the other universe does not imply that you will be happy if you make the 'corresponding' decision. Once there are differences between the universes (and without such differences news from the other universe is not news), there is no good reason to expect the outcome of a decision to be unaffected by them. In one universe, you met because of an accidental shared experience; in the other, because you have illegally used the ship's equipment. Can that affect the happiness of a marriage? Perhaps not, but you can only know that if you have a good explanatory theory of which factors affect the outcomes of marriages and which do not. And if you have such a theory, then perhaps you have no need to be skulking in transporter rooms.

Still more generally, the benefit of inter-universe communication would be, in effect, that it permits new forms of information processing. In the fictional case I have described, since the two universes have been identical until quite recently, communicating with one's other-universe counterpart achieves the same effect as running a computer simulation of an alternative version of a period of one's own life, without having to know all the relevant physical variables explicitly. This computation is infeasible in any other way, and could be helpful in testing explanatory theories of how various factors affect outcomes. Nevertheless, it is no substitute for thinking of those theories in the first place.

Therefore, if such communication is a scarce resource, a more efficient way of using it might be to exchange the theories themselves: if your doppelgänger solves a problem and tells you the solution, then you can see for yourself that it is a good explanation even if you have no way of knowing how your doppelgänger arrived at it.

Another efficient use of inter-universe communication might be to share the work of a lengthy computation. For instance, the story might be that some crew members have been poisoned and will die within hours unless the antidote is administered. To find the antidote requires computer simulations of the effects of many variants of a drug. So the two instances of the ship's computer can each search half the list of variants, thus running through the full list in half the time. When the cure is found in one universe, its number in the list can be transmitted to the other universe, the result can be checked there, and the crew in both universes are saved. Again, evidence that there is computer power accessible in this way through the transporter would be evidence that there really was a computer out there, performing different calculations from one's own. Reflecting on the details (about what the doppelgängers breathe and so on) would then let the inhabitants know that the other universe as a whole was a real place with similar structure and complexity to their own. So their world would be explicable.

Since there is no inter-universe communication in real quantum physics, we shall not allow it in our story, and so that specific route to explicability is not open. The history in which our crew members are married and the one in which they still hardly know each other cannot *communicate* with each other or *observe* each other. Nevertheless, as we shall see, there are circumstances in which histories can still *affect* each other in ways that do not amount to communication, and the need to explain those effects provides the main argument that our own multiverse is real.

After the universes in our story begin to differ inside one starship, everything else in the world exists in pairs of identical instances. We must continue to imagine those pairs as being fungible. This is necessary because the universes are not 'receptacles' – there is nothing to them apart from the objects that they contain. If they did have an independent reality, then each of the objects in such a pair would have a property of being in one particular universe and not the other, which would make them non-fungible.

Typically, the region in which the universes are different will then grow. For instance, when the couple decide to marry, they send messages to their home planets announcing this. When the messages arrive, the two instances of each of those planets become different. Previously only the two instances of the starship were different, bur soon, even before anyone broadcasts it intentionally, some of the information will have leaked out. For instance, people in the starship are moving differently in the two universes as a result of the marriage decision, so light bounces off them differently and some of it leaves the starship through portholes, making the two universes slightly different wherever it goes. The same is true of heat radiation (infra-red light), which leaves the starship through every point on the hull. Thus, starting with the voltage happening in only one universe, a *wave of differentiation* between the universes spreads in all directions through space. Since information travelling in either universe cannot exceed the speed of light, nor can the wave of differentiation. And since, at its leading edge, it mostly travels at or near that speed, differences in the head start that some directions have over others will become an ever smaller proportion of the total distance travelled, and so the further the wave travels the more nearly spherical it becomes. So I shall call it a '*sphere* of differentiation'.

Even inside the sphere of differentiation, there are comparatively few differences between the universes: the stars still shine, the planets still have the same continents. Even the people who hear of the wedding, and behave differently as a result, retain most of the same data in their brains and other information-storage devices, and they still breathe the same type of air, eat the same types of food, and so on.

However, although it may seem intuitively reasonable that news of the marriage leaves most things unchanged, there is a different common-sense intuition that seems to prove that it must change everything, if only slightly. Consider what happens when the news reaches a planet – say, in the form of pulse of photons from a communication laser.

Even before any human consequences, there is the physical impact of those photons, which one might expect to impart momentum to every atom exposed to the beam – which will be every atom in something like that half of the surface of the planet which is facing the beam. Those atoms would then vibrate a little differently, affecting the atoms below through interatomic forces. As each atom affected others, the effect would spread rapidly through the planet. Soon, every atom in the planet would have been affected – though most of them by unimaginably tiny amounts. Nevertheless, however small such an effect was, it would be enough to break the fungibility between each atom and its other-universe counterpart. Hence it would seem that nothing would be left fungible after the wave of differentiation had passed.

These two opposite intuitions reflect the ancient dichotomy between the discrete and the continuous. The above argument – that everything in the sphere of differentiation must become different – depends on the reality of *extremely small physical changes* – changes that would be many orders of magnitude too small to be measurable. The existence of such changes follows inexorably from the explanations of *classical* physics, because in classical physics most fundamental quantities (such as energy) are continuously variable. The opposing intuition comes from thinking about the world in terms of information processing, and hence in terms of discrete variables such as the contents of people's memories. Quantum theory adjudicates this conflict in favour of the discrete. For a typical physical quantity, there is a *smallest possible change* that it can undergo in a given situation. For instance, there is a smallest possible amount of energy that can be transferred from radiation to any particular atom. The atom cannot absorb any less than that amount, which is called a 'quantum' of energy. Since this was the first distinctive feature of quantum physics to be discovered, it gave its name to the field. Let us incorporate it into our fictional physics as well.

Hence it is not the case that all the atoms on the surface of the planet are changed by the arrival of the radio message. In reality, the typical response of a large physical object to very small influences is that most of its atoms remain strictly unchanged, while, to obey the conservation laws, a few exhibit a discrete, relatively large change of one quantum.

The discreteness of variables raises questions about motion and change. Does it mean that changes happen instantaneously? They do not – which raises the further question: what is the world like halfway through that change? Also if a few atoms are strongly affected by some influence, and the rest are unaffected, what determines which are the ones to be affected? The answer has to do with fungibility, as the reader may guess, and as I shall explain below.

The effects of a wave of differentiation usually diminish rapidly with distance – simply because physical effects in general do. The sun, from even a hundredth of a light year away, looks like a cold, bright dot in the sky. It barely affects anything. At a thousand light years, nor does a supernova. Even the most violent of quasar jets, when viewed from a neighbouring galaxy, would be little more than an abstract painting in the sky. There is only one known phenomenon which, if it ever occurred, would have effects that did not fall off with distance, and that is the creation of a certain type of knowledge, namely a beginning of infinity. Indeed, knowledge can aim itself at a target, travel vast distances having scarcely any effect, and then utterly transform the destination.

In our story, too, if we wanted the transporter malfunction to have a significant physical effect at astronomical distances, it would have to be via knowledge. All those torrents of photons streaming out of the starship and carrying, intentionally or unintentionally, information about a wedding will have a noticeable effect on the distant planet only if someone there cares about the possibility of such information enough to set up scientific instruments that could detect it.

Now, as I have explained, our imaginary laws of physics which say that a voltage surge happens 'in one universe but not the other' cannot be deterministic unless the universes are fungible. So, what happens when the transporter is used again, after the universes are no longer fungible? Imagine a second starship, of the same type as the first and far away. What happens if the second starship runs its transporter immediately after the first one did?

One logically possible answer would be that *nothing* happens – in other words, the laws of physics would say that, once the two universes are different, all transporters just work normally and never produce a voltage surge again. However, that would also provide a way of communicating faster than light, albeit unreliably and only once. You set up a voltmeter in the transporter room and run the transporter. If the voltage surges, you know that the other starship, however far away, has not yet run its transporter (because, if it had, that would have put a permanent end to such surges everywhere). The laws governing the real multiverse do not allow information to flow in that way. If we want our fictional laws of physics to be universal from the inhabitants' point of view, the second transporter must do exactly what the first one did. It must cause a voltage surge in one universe and not in the other.

But in that case something must determine *which* universe the second surge will happen in. 'In one universe but not the other' is no longer a deterministic specification. Also, a surge must not happen if the transporter is run *only* in the other universe. That would constitute inter-universe communication. It must depend on both instances of the transporter being run simultaneously. Even that could allow some inter-universe communication, as follows. In the universe where a surge has once happened, run the transporter at a prearranged time and observe the voltmeter. If no surge happens, then the transporter in the other universe is switched off. So we are at an impasse. It is remarkable how much subtlety there can be in the apparently straightforward, binary distinction between 'same' and different' – or between 'affected' and 'unaffected'. In the real quantum theory, too, the prohibitions on inter-universe communication and faster-than-light communication are closely connected.

There is a way – I think it is the only way – to meet simultaneously the requirements that our fictional laws of physics be universal and deterministic, and forbid faster-than-light and inter-universe communication: *more universes*. Imagine an uncountably infinite number of them, initially all fungible. The transporter causes previously fungible ones to become different, as before; but now the relevant law of physics says, 'The voltage surges in *half the universes* in which the transporter is used.' So, if the two starships both run their transporters, then, after the two spheres of differentiation have overlapped, there will be universes of four different kinds: those in which a surge happened only in the first starship, only in the second, in neither, and in both. In other words, in the overlap region there are four different histories, each taking place in one quarter of the universes.

Our fictional theory has not provided enough structure in its multiverse to give a meaning to 'half the universes', but the real quantum theory does. As I explained in Chapter 8, the method that a theory provides for giving a meaning to proportions and averages for infinite sets is called a *measure*. A familiar example is that classical physics assigns *lengths* to infinite sets of points arranged in a line. Let us suppose that our theory provides a measure for universes.

Now we are allowed storylines such as the following. In the universes in which the couple married, they spend their honeymoon on a human-colonized planet that the starship is visiting. As they are teleporting back up, the voltage surge in *half* those universes causes someone's electronic notepad to play a voice message suggesting that one of the newlyweds has already been unfaithful. This sets off a chain of events that ends in divorce. So now our original collection of fungible universes contains three different histories: in one, comprising *half* the original set of universes, the couple in question are still single; in the second, comprising a *quarter* of the original set, they are married; and in the third, comprising the remaining quarter, they are divorced.

Thus the three histories do not occupy equal proportions of the multiverse. There are twice as many universes in which the couple never married as there are universes in which they divorced.

Now suppose that scientists on the starship know about the multiverse and understand the physics of the transporter. (Though note that we have not yet given them a way of discovering those things.) Then they know that, when they run the transporter, an infinite number of fungible instances of themselves, all sharing the same history, are doing so at the same time. They know that a voltage surge will occur in half the universes in that history, which means that it will split into two histories of equal measure. Hence they know that, if they use a voltmeter capable of detecting the surge, half of the instances of themselves are going to find that it has recorded one, and the other half are not. But they also know that it is meaningless to ask (not merely impossible to know) *which* event they will experience. Consequently they can make two closely related predictions. One is that, despite the perfect determinism of everything that is happening, *nothing* can reliably predict for them whether the voltmeter will detect a surge.

The other prediction is simply that the voltmeter will record a surge with probability one-half. Thus the outcomes of such experiments are *subjectively random* (from the perspective of any observer) even though everything that is happening is completely determined objectively. This is also the origin of quantum-mechanical randomness and probability in real physics: it is due to the measure that the theory provides for the multiverse, which is in turn due to what kinds of physical processes the theory allows and forbids.

Notice that when a random outcome (in this sense) is about to happen, it is a situation of diversity within fungibility: the diversity is in the variable 'what outcome they are *going* to see'. The logic of the situation is the same as in cases like that of the bank account I discussed above, except that this time the fungible entities are people. They are fungible, yet half of them are going to see the surge and the other half not.

In practice they could test this prediction by doing the experiment many times. Every formula purporting to predict the sequence of outcomes will eventually fail: that tests the unpredictability. And in the overwhelming majority of universes (and histories) the surge will happen approximately half the time: that tests the predicted value of the probability. Only a tiny proportion of the instances of the observers will see anything different.

Our story continues. In one of the histories, the newspapers on the astronauts' home planets report the engagement. They fill many column-inches with reports about the accident that brought the astronauts together and so on. In the other history, where there is no astronaut-engagement news, one newspaper fills the same space on the page with a short story. It happens to be about a romance on a starship. Some of the sentences in that story are identical to sentences in the news items in the other history. The same words, printed in the same column in the same newspaper, are fungible between the two histories; but they are fiction in one history and fact in the other. So here the fact/fiction attribute has diversity within fungibility.

The number of distinct histories will now increase rapidly. Whenever the transporter is used, it takes only microseconds for the sphere of differentiation to engulf the whole starship, so, if it is typically used ten times per day, the number of distinct histories inside the whole starship will double about ten times a day. Within a month there will be more distinct histories than there are atoms in our visible universe. Most of them will be extremely similar to many others, because in only a small proportion will the precise timing and magnitude of the voltage surge be just right to precipitate a noticeable, *Sliding Doors*-type change. Nevertheless, the number of histories continues to increase exponentially, and soon there are so many variations on events that several significant changes have been caused *somewhere* in the multiversal diversity of the starship. So the total number of such histories increases exponentially too, even though they continue to constitute only a small proportion of all histories that are present.

Soon after that, in an even smaller but still exponentially growing number of histories, uncanny chains of 'accidents' and 'unlikely coincidences' will have come to dominate events. I put those terms in quotation marks because those events are not in the least accidental. They have all happened inevitably, according to deterministic laws of physics. All of them were caused by the transporter.

Here is another situation where, if we are not careful, common sense makes false assumptions about the physical world, and can make descriptions of situations sound paradoxical even though the situations themselves are quite straightforward. Dawkins gives an example in his book *Unweaving the Rainbow*, analysing the claim that a television psychic was making accurate predictions:

> There are about 100,000 five-minute periods in a year. The probability that any given watch, say mine, will stop in a designated five-minute period is about 1 in 100,000. Low odds, but there are 10 million people watching the [television psychic's] show. If only half of them are wearing watches, we could expect about 25 of those watches to stop in any given minute. If only a quarter of these ring in to the studio, that is 6 calls, more than enough to dumbfound a naive audience. Especially when you add in the calls from people whose watches stopped the day before, people whose watches didn't stop but whose grandfather clocks did, people who died of heart attacks and their bereaved relatives phoned in to say that their 'ticker' gave out, and so on.

As this example shows, the fact that certain circumstances can *explain* other events without being in any way involved in *causing* them is very familiar despite being counter-intuitive. The 'naive' audience's mistake is a form of parochialism: they observe a phenomenon – people phoning in because their watches stopped – but they are failing to understand it as part of a wider phenomenon, most of which they do not observe. Though the unobserved parts of that wider phenomenon have in no way affected what we, the viewers, observe, they are essential to its explanation. Similarly, common sense and classical physics contain the parochial error that only one history exists. This error, built into our language and conceptual framework, makes it sound odd to say that an event can be in one sense extremely unlikely and in another certain to happen. But there is nothing odd about it in reality.

We are now seeing the interior of the spaceship as an overwhelmingly complex jumble of superposed objects. Most locations on board are packed with people, some of them on very unusual errands, and all unable to perceive each other. The spaceship itself is on many slightly different courses, due to slightly different behaviours of the crew. Of course we are 'seeing' this only in our mind's eye. Our fictional laws of physics ensure that no observer in the multiverse itself would see anything like that. Consequently, on closer inspection (in our mind's eye), we also see that there is great order and regularity in that apparent chaos. For instance, although there is a flurry of human figures in the Captain's chair, we see that most of them are the Captain; and although there is a flurry of human figures in the Navigator's chair, we see that few of them are the Captain. Regularities of that kind are ultimately due to the fact that all the universes, despite their differences, obey the same laws of physics (including their initial conditions).

We also see that any particular instance of the Captain only ever interacts with one instance of the Navigator, and one instance of the First Officer; and those instances of the Navigator and First Officer are precisely the ones that interact with each other. These regularities are due to the fact that the histories are nearly autonomous: what happens in each of them depends almost entirely on previous events in that history alone – with transporter-induced voltage surges being the only exceptions. In the story so far, this autonomy of the histories is rather a trivial fact, since we began by making the *universes* autonomous. But it is going to be worth becoming even more pedantic for a moment: what exactly is the difference between the instance of you that I can interact with and the ones that are imperceptible to me? The latter are 'in other universes' – but, remember, universes consist only of the objects in them, so that amounts only to saying I can see the ones that I can see. The upshot is that our laws of physics must also say that every object carries within it information about which instances of it could interact with which instances of other objects (except when the instances are fungible, when there is no such thing as 'which'). Quantum theory describes such information. It is known as *entanglement* information.*

So far in the story we have set up a vast, complex world which looks very unfamiliar in our mind's eye, but to the overwhelming majority of the inhabitants looks almost exactly like the single universe of our everyday experience and of classical physics, plus some apparently random jiggling whenever the transporter operates. A tiny minority of the histories have been significantly affected by very 'unlikely' events, but even in those the information *flow* – what affects what – is still very tame and familiar. For instance, a version of the ship's log that contains records of bizarre coincidences will be perceptible to people who remember those coincidences, but not to other instances of those people.

Thus the information in the fictional multiverse flows along a branching tree, whose branches – histories – have different thicknesses (measures) and never rejoin once they have separated. Each behaves exactly as if the others did not exist. If that were the whole story, that multiverse's imaginary laws of physics would still be fatally flawed as explanations in the same way that they have been all along: there would be no difference between their predictions and those of much more straightforward laws saying that there is only one universe – one history – in which the transporter *randomly* introduces a change in the objects that it teleports. Under those laws, instead of branching into two autonomous histories on such occasions, the single history randomly does or does not undergo such a change. Thus the entire stupendously complicated multiverse that we have imagined – with its multiplicity of entities including people walking through each other and its bizarre occurrences and its entanglement information – would collapse into nothing, like the galaxy in Chapter 2 that became an emulsion flaw. The multiverse explanation of the same events would be a bad

---

*That this information is carried entirely locally in objects is currently somewhat controversial. For a detailed technical discussion see the paper 'Information Flow in Entangled Quantum Systems' by myself and Patrick Hayden (*Proceedings of the Royal Society* A456 (2000)).

explanation, and so the world would be inexplicable to the inhabitants if it were true.

It may seem that, by imposing all those conditions on information flow, we have gone to a lot of trouble to achieve that very attribute – to hide, from the inhabitants, the Byzantine intricacies of their world. In the words of Lewis Carroll's White Knight in *Through the Looking Glass*, it is as if we were

> . . . thinking of a plan
> To dye one's whiskers green,
> And always use so large a fan
> That they could not be seen.

Now it is time to start removing the fan.

In quantum physics, information flow in the multiverse is not as tame as in that branching tree of histories I have described. That is because of one further quantum phenomenon: under certain circumstances, the laws of motion allow histories to rejoin (becoming fungible again). This is the time-reverse of the splitting (differentiation of history into two or more histories) that I have already described, so a natural way to implement it in our fictional multiverse is for the transporter to be capable of undoing its own history-splitting.

If we represent the original splitting like this

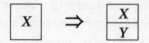

where X is the normal voltage and Y is the anomalous one introduced by the transporter, then the rejoining of histories can be represented as

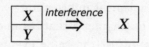

In an interference phenomenon, differentiated histories rejoin.

This phenomenon is known as *interference*: the presence of the Y-history *interferes* with what the transporter usually does to an X-history. Instead, the X and Y histories merge. This is rather like the doppelgängers merging with their originals in some phantom-zone stories, except that here we do not need to repeal the principle of the conservation of mass or any other conservation law: the total measure of all the histories remains constant.

Interference is the phenomenon that can provide the inhabitants of the multiverse with evidence of the existence of multiple histories in their world without allowing the histories to communicate. For example, suppose that they run the transporter twice in quick succession (I shall explain in a moment what 'quick' means):

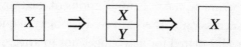

An interference experiment

If they did this repeatedly (with, say, different copies of the transporter on each occasion), they could soon infer that the intermediate result *could not* be just randomly X or Y, because if it were then the final outcome would sometimes be Y (because of ⊠→$\frac{X}{Y}$), while in fact it is always X. Thus the inhabitants would no longer be able to explain away what they see by assuming that only one, randomly chosen, value of the voltage is real at the intermediate stage.

Although such an experiment would provide evidence that multiple histories not only exist but affect each other strongly (in the sense that they behave differently according to whether the other is present or absent), it does not involve inter-history *communication* (sending a message of one's choice to the other history).

In our story, just as we did not allow splitting to happen in a way that would allow communication faster than light, so we must ensure the same for interference. The simplest way is to require that the rejoining take place only if no wave of differentiation has happened. That is to say, the transporter can undo the voltage surge only if this has not yet caused any differential effects on anything else. When a wave of differentiation, set off by two different values X and Y of some variable, has left an object, the object is *entangled* with all the differentially affected objects.

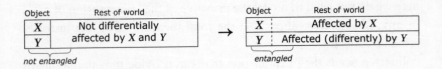

Entanglement

So our rule, in short, is that interference can happen only in objects that are unentangled with the rest of the world. This is why, in the interference experiment, the two applications of the transporter have to be 'in quick succession'. (Alternatively, the object in question has to be sufficiently well isolated for its voltages not to affect its surroundings.) So we can represent a generic interference experiment symbolically as follows:

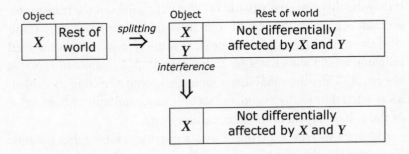

If an object is unentangled, it can be made to undergo interference by something acting on it alone.

(The arrows '$\Rightarrow$' and '$\Downarrow$' represent the action of the transporter.) Once the object is entangled with the rest of the world in regard to the values X and Y, no operation on the object alone can create interference between those values. Instead, the histories are merely split further, in the usual way:

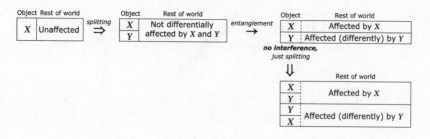

In entangled objects, further splitting happens instead of interference.

When two or more values of a physical variable have differently affected something in the rest of the world, knock-on effects typically continue indefinitely, as I have described, with a wave of differentiation entangling more and more objects. If the differential effects can all be undone, then interference between those original values becomes possible again; but the laws of quantum mechanics dictate that undoing them requires fine control of *all* the affected objects, and that rapidly becomes infeasible. The process of its becoming infeasible is known as *decoherence*. In most situations, decoherence is very rapid, which is why splitting typically predominates over interference, and why interference – though ubiquitous on microscopic scales – is quite hard to demonstrate unambiguously in the laboratory.

Nevertheless, it can be done, and quantum interference phenomena constitute our main evidence of the existence of the multiverse, and of what its laws are. A real-life analogue of the above experiment is standard in quantum optics laboratories. Instead of experimenting on voltmeters (whose many interactions with their environment quickly cause decoherence), one uses individual photons, and the variable being acted upon is not voltage but which of two possible paths the photon is on. Instead of the transporter, one uses a simple device called a semi-silvered mirror (represented by the grey sloping bars in the diagrams below). When a photon strikes such a mirror, it bounces off in half the universes, and passes straight through in the other half, as shown on next page:

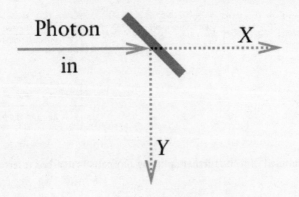

Semi-silvered mirror

The attributes of travelling in the X or Y directions behave analogously to the two voltages X and Y in our fictitious multiverse. So passing through the semi-silvered mirror is the analogue of the transformation $\boxed{X} \rightarrow \boxed{\frac{X}{Y}}$ above. And when the two instances of a single photon, travelling in directions X and Y, strike the second semi-silvered mirror at the same time, they undergo the transformation $\boxed{\frac{X}{Y}} \rightarrow \boxed{X}$, which means that both instances emerge in the direction X: the two histories rejoin. To demonstrate this, one can use a set-up known as a 'Mach–Zehnder interferometer', which performs those two transformations (splitting and interference) in quick succession:

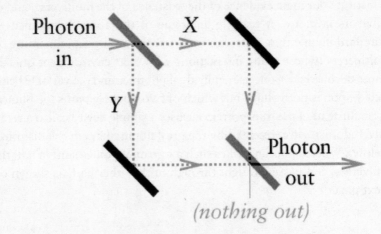

Mach–Zehnder interferometer

The two ordinary mirrors (the black sloping bars) are merely there to steer the photon from the first to the second semi-silvered mirror.

If a photon is introduced travelling rightwards (X) *after* the first mirror instead of before as shown, then it appears to emerge randomly, rightwards or downwards, from the last mirror (because then, $\boxed{X} \rightarrow \boxed{\frac{X}{Y}}$ happens there). The same is true of a photon introduced travelling downwards (Y) after the first mirror. But a photon introduced as shown in the diagram invariably emerges rightwards, never downwards. By doing the experiment repeatedly with and without detectors on the paths, one can verify that only one photon is ever present per history, because only one of those detectors is ever observed to fire during such an experiment. Then, the fact that the intermediate histories X and Y *both* contribute to the deterministic final outcome X makes it inescapable that both are happening at the intermediate time.

In the real multiverse, there is no need for the transporter or any other special apparatus to cause histories to differentiate and to rejoin. Under the laws of quantum physics, elementary particles are undergoing such processes of their own accord, all the time. Moreover, histories may split into more than two – often into many trillions – each characterized by a slightly different direction of motion or difference in other physical variables of the elementary particle concerned. Also, in general the resulting histories have unequal measures. So let us now dispense with the transporter in the fictional multiverse too.

The rate of growth in the number of distinct histories is quite mind-boggling – even though, thanks to interference, there is now a certain amount of spontaneous rejoining as well. Because of this rejoining, the flow of information in the real multiverse is not divided into strictly autonomous subflows – branching, autonomous histories. Although there is still no communication between histories (in the sense of message-sending), they are intimately affecting each other, because the effect of interference on a history depends on what other histories are present.

Not only is the multiverse no longer perfectly partitioned into histories, individual particles are not perfectly partitioned into instances. For example, consider the following interference phenomenon, where X and Y now represent different values of the position of a single particle:

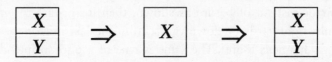

How instances of a particle lose their identity during interference. Has the instance of the particle at X stayed at X or moved to Y? Has the instance of the particle at Y returned to Y or moved to X?

Because these two groups of instances of the particle, initially at different positions, have gone through a moment of being fungible, there is no such thing as which of them has ended up at which final position. This sort of interference is going on all the time, even for a single particle in a region of otherwise empty space. So there is in general no such thing as the 'same' instance of a particle at different times.

Even within the same history, particles in general do not retain their identities over time. For example, during a collision between two atoms, the histories of the event split into something like this

and something like this

So, for each particle individually, the event is rather like a collision with a semi-silvered mirror. Each atom plays the role of the mirror for the other atom. But the multiversal view of both particles looks like this

where at the end of the collision some of the instances of each atom have become fungible with what was originally a different atom.

For the same reason, there is no such thing as the *speed* of one instance of the particle at a given location. Speed is defined as distance travelled divided by time taken, but that is not meaningful in situations where there is no such thing as a particular instance of the particle over time. Instead, a collection of fungible instances of a particle in general have several speeds – meaning that in general they will do different things an instant later. (This is another instance of 'diversity within fungibility'.)

Not only can a fungible collection with the same position have different speeds, a fungible group with the same speed can have different positions. Furthermore, it follows from the laws of quantum physics that, for *any* fungible collection of instances of a physical object, some of their attributes must be diverse. This is known as the 'Heisenberg uncertainty principle', after the physicist Werner Heisenberg, who deduced the earliest version from quantum theory.

Hence, for instance, an individual electron always has a range of different locations *and* a range of different speeds and directions of motion. As a result, its typical behaviour is to spread out gradually in space. Its quantum-mechanical law of motion resembles the law governing the spread of an ink blot – so if it is initially located in a very small region it spreads out rapidly, and the larger it gets the more slowly it spreads. The entanglement information that it carries ensures that no two instances of it can ever contribute to the same history. (Or, more precisely, at times and places where there *are* histories, it exists in instances which can never collide.) If a particle's range of speeds is centred not on zero but on some other value, then the whole of the 'ink blot' moves, with its centre obeying approximately the laws of motion in classical physics. In quantum physics this is how motion, in general, works.

This explains how particles in the same history can be fungible too, in something like an atomic laser. Two 'ink-blot' particles, each of which is a multiversal object, can coincide perfectly in space, and their entanglement information can be such that no two of their instances are ever at the same point in the same history.

Now, put a proton into the middle of that gradually spreading cloud of instances of a single electron. The proton has a positive charge, which attracts the negatively charged electron. As a result, the cloud stops spreading when its size is such that its tendency to spread outwards due to its uncertainty-principle diversity is exactly balanced by its attraction to the proton. The resulting structure is called an atom of hydrogen.

Historically, this explanation of what atoms are was one of the first triumphs of quantum theory, for atoms could not exist at all according to classical physics. An atom consists of a positively charged nucleus surrounded by negatively charged electrons. But positive and negative charges attract each other and, if unrestrained, accelerate towards each other, emitting energy in the form of electromagnetic radiation as they go. So it used to be a mystery why the electrons do not 'fall' on to the nucleus in a flash of radiation. Neither the nucleus nor the electrons individually have more than one ten-thousandth of the diameter of the atom, so what keeps them so far apart? And what makes atoms stable at that size? In non-technical accounts, the structure of atoms is sometimes explained by analogy with the solar system: one imagines electrons in orbit around the nucleus like planets around the sun. But that does not match the reality. For one thing, gravitationally bound objects do slowly spiral in, emitting gravitational radiation (the process has been observed for binary neutron stars), and the corresponding electromagnetic process in an atom would be over in a fraction of a second. For another, the existence of solid matter, which consists of atoms packed closely together, is evidence that atoms cannot easily penetrate each other, yet solar systems certainly could. Furthermore, it turns out that, in the hydrogen atom, the electron in its lowest-energy state is not orbiting at all but, as I said, just sitting there like an ink blot – its uncertainty-principle tendency to spread exactly balanced by the electrostatic force. In this way, the phenomena of interference and diversity within fungibility are integral to the structure and stability of all static objects, including all solid bodies, just as they are integral to all motion.

The term 'uncertainty principle' is misleading. Let me stress that it has nothing to do with uncertainty or any other distressing psychological sensations that the pioneers of quantum physics might have felt. When an electron has more than one speed or more than one position, that has nothing to do with anyone being uncertain what the speed is, any more than anyone is 'uncertain' which dollar in their bank account belongs to the tax authority. The diversity of attributes in both cases is a physical fact, independent of what anyone knows or feels.

Nor, by the way, is the uncertainty principle a 'principle', for that suggests an independent postulate that could logically be dropped or replaced to obtain a different theory. In fact one could no more drop it from quantum theory than one could omit eclipses from astronomy. There is no 'principle of eclipses': their existence can be deduced from theories of much greater generality, such as those of the solar system's geometry and dynamics. Similarly, the uncertainty principle is deduced from the principles of quantum theory.

Thanks to the strong internal interference that it is continuously undergoing, a typical electron is an irreducibly multiversal object, and not a collection of parallel-universe or parallel-histories objects. That is to say, it has multiple positions and multiple speeds without being divisible into autonomous sub-entities each of which has one speed and one position. Even different electrons do not have completely separate identities. So the reality is an electron *field* throughout the whole of space, and disturbances spread through this field as waves, at the speed of light or below. This is what gave rise to the often-quoted misconception among the pioneers of quantum theory that electrons (and likewise all other particles) are 'particles and waves at the same time'. There is a field (or 'waves') in the multiverse for every individual particle that we observe in a particular universe.

Although quantum theory is expressed in mathematical language, I have now given an account in English of the main features of the reality that it describes. So at this point the fictional multiverse that I have been describing is more or less the real one. But there is one thing left to tidy up. My 'succession of speculations' was based on universes, and on instances of objects, and then on corrections to those ideas in order to describe the multiverse. But the real multiverse is not 'based on' anything, nor is it a correction to anything. Universes, histories, particles and their instances are not referred to by quantum theory at all – any more than are planets, and human beings and their lives and loves. Those are all approximate, emergent phenomena in the multiverse.

A history is part of the multiverse in the same sense that a geological stratum is part of the Earth's crust. One history is distinguished from the others by the values of physical variables, just as a stratum is distinguished from others by its chemical composition and by the types of fossils found in it and so on. A stratum and a history are both channels of information flow. They preserve information because, although their contents change over time, they are approximately *autonomous* – that is to say, the changes in a particular stratum or history depend almost entirely on conditions inside it and not elsewhere. It is because of that autonomy that a fossil found today can be used as evidence of what was present when that stratum was formed. Similarly, it is why, within a history, using classical physics, one can successfully predict some aspects of the future of that history from its past.

A stratum, like a history, has no separate existence over and above the objects in it: it *consists* of them. Nor does a stratum have well-defined edges. Also, there are regions of the Earth – for instance, near volcanoes – where strata have merged (though I think there are no geological processes that split and remerge strata in the way that histories split and remerge). There are regions of the Earth – such as the core – where there have never been strata. And there are regions – such as the atmosphere – where strata do form but their contents interact and mix on much shorter timescales than in the crust. Similarly, there are regions of the multiverse that contain short-lived histories, and others that do not even approximately contain histories.

However, there is one big difference between the ways in which strata and histories emerge from their respective underlying phenomena. Although not every atom in the Earth's crust can be unambiguously assigned to a particular stratum, most of the atoms that form a stratum can. In contrast, every atom in an everyday object is a multiversal object, not partitioned into nearly autonomous instances and nearly autonomous histories, yet everyday objects such as starships and betrothed couples, which are made of such particles, are partitioned very accurately into nearly autonomous histories with exactly one instance, one position, one speed of each object in each history.

That is because of the suppression of interference by entanglement. As I explained, interference almost always happens either very soon after splitting or not at all. That is why the larger and more complex an object or process is, the less its gross behaviour is affected by interference. At that 'coarse-grained' level of emergence, events in the multiverse consist of autonomous histories, with each coarse-grained history consisting of a swathe of many histories differing only in microscopic details but affecting each other through interference. Spheres of differentiation tend to grow at nearly the speed of light, so, on the scale of everyday life and above, those coarse-grained histories can justly be called 'universes' in the ordinary sense of the word. Each of them somewhat resembles the universe of classical physics. And they can usefully be called 'parallel' because they are nearly autonomous. To the inhabitants, each looks very like a single-universe world.

Microscopic events which are accidentally amplified to that coarse-grained level (like the voltage surge in our story) are rare in any one coarse-grained history, but common in the multiverse as a whole. For example, consider a single cosmic-ray particle travelling in the direction of Earth from deep space. That particle must be travelling in a range of slightly different directions, because the uncertainty principle implies that in the multiverse it must spread sideways like an ink blot as it travels. By the time it arrives, this ink blot may well be wider than the whole Earth – so most of it misses and the rest strikes everywhere on the exposed surface. Remember, this is just a single particle, which may consist of fungible instances. The next thing that happens is that they cease to be fungible, splitting through their interaction with atoms at their points of arrival into a finite but huge number of instances, each of which is the origin of a separate history.

In each such history, there is an autonomous instance of the cosmic-ray particle, which will dissipate its energy in creating a 'cosmic-ray shower' of electrically charged particles. Thus, in different histories, such a shower will occur at different locations. In some, that shower will provide a conducting path down which a lightning bolt will travel. Every atom on the surface of the Earth will be struck by such lightning in *some* history. In other histories, one of those cosmic-ray particles will strike a human cell, damaging some already damaged DNA in such a way as to make the cell cancerous. Some non-negligible proportion of all cancers are caused in this way. As a result, there exist histories in which any given person, alive in our history at any time, is killed soon afterwards by cancer. There exist other histories in which the course of a battle, or a war, is changed by such an event, or by a lightning bolt at exactly the right place and time, or by any of countless other unlikely, 'random' events. This makes it highly plausible that there exist histories in which events have played out more or less as in alternative-history stories such as *Fatherland* and *Roma Eterna* – or in which events in your own life played out very differently, for better or worse.

A great deal of fiction is therefore close to a fact somewhere in the multiverse. But not all fiction. For instance, there are no histories in which my stories of the transporter malfunction are true, because they require different laws of physics. Nor are there histories in which the fundamental constants of nature such as the speed of light or the charge on an electron are different. There is, however, a sense in which different laws of physics *appear* to be true for a period in some histories, because of a sequence of 'unlikely accidents'. (There may also be universes in which there are different laws of physics, as required in anthropic explanations of fine-tuning. But as yet there is no viable theory of such a multiverse.)

Imagine a single photon from a starship's communication laser, heading towards Earth. Like the cosmic ray, it arrives all over the surface, in different histories. In each history, only one atom will absorb the photon and the rest will initially be completely unaffected. A receiver for such communications would then detect the relatively large, discrete change undergone by such an atom. An important consequence for the construction of measuring devices (including eyes) is that no matter how far away the source is, the kick given to an atom by an arriving photon is always the same: it is just that the weaker the signal is, the fewer kicks there are. If this were not so – for instance, if classical physics were true – weak signals would be much more easily swamped by random local noise. This is the same as the advantage of digital over analogue information processing that I discussed in Chapter 6.

Some of my own research in physics has been concerned with the theory of *quantum computers*. These are computers in which the information-carrying variables have been protected by a variety of means from becoming entangled with their surroundings. This allows a new mode of computation in which the flow of information is not confined to a single history. In one type of quantum computation, enormous numbers of different computations, taking place simultaneously, can affect each other and hence contribute to the output of a computation. This is known as *quantum parallelism*.

In a typical quantum computation, individual bits of information are represented in physical objects known as 'qubits' – quantum bits – of which there is a large variety of physical implementations but always with two essential features. First, each qubit has a variable that can take one of two discrete values, and, second, special measures are taken to protect the qubits from entanglement – such as cooling them to temperatures close to absolute zero. A typical algorithm using quantum parallelism begins by causing the information-carrying variables in some of the qubits to acquire both their values simultaneously. Consequently, regarding those qubits as a register representing (say) a number, the number of separate instances of the register as a whole is exponentially large: two to the power of the number of qubits. Then, for a period, classical computations are performed, during which waves of differentiation spread to some of the other qubits – but no further, because of the special measures that prevent this. Hence, information is processed separately in each of that vast number of autonomous histories. Finally, an interference process involving all the affected qubits combines the information in those histories into a single history. Because of the intervening computation, which has processed the information, the final state is not the same as the initial one, as in the simple interference experiment I discussed above, namely $\boxed{x} \rightarrow \boxed{\frac{x}{y}} \rightarrow \boxed{x}$, but is some function of it, like this:

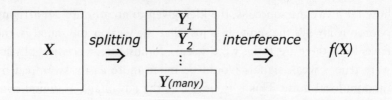

A typical quantum computation. $Y_1 \ldots Y_{many}$ are intermediate results that depend on the input X. All of them are needed to compute the output $f(X)$ efficiently.

Just as the starship crew members could achieve the effect of large amounts of computation by sharing information with their doppelgängers computing the same function on different inputs, so an algorithm that makes use of quantum parallelism does the same. But, while the fictional effect is limited only by starship regulations that we may invent to suit the plot, quantum computers are limited by the laws of physics that govern quantum interference. Only certain types of parallel computation can be performed with the help of the multiverse in this way. They are the ones for which the mathematics of quantum interference happens to be just right for combining into a single history the information that is needed for the final result.

In such computations, a quantum computer with only a few hundred qubits could perform far more computations in parallel than there are atoms in the visible universe. At the time of writing, quantum computers with about ten qubits have been constructed. 'Scaling' the technology to larger numbers is a tremendous challenge for quantum technology, but it is gradually being met.

I mentioned above that, when a large object is affected by a small influence, the usual outcome is that the large object is strictly unaffected. I can now explain why. For example, in the Mach–Zehnder interferometer, shown earlier, two instances of a single photon travel on two different paths. On the way, they strike two different mirrors. Interference will happen only if the photon does not become entangled with the mirrors – but it *will* become entangled if either mirror retains the slightest record that it has been struck (for that would be a differential effect of the instances on the two different paths). Even a single quantum of change in the amplitude of the mirror's vibration on its supports, for instance, would be enough to prevent the interference (the subsequent merging of the photon's two instances).

When one of the instances of the photon bounces off either mirror, its momentum changes, and hence by the principle of the conservation of momentum (which holds universally in quantum physics, just as in classical physics), the mirror's momentum must change by an equal and opposite amount. Hence it seems that, in each history, one mirror but not the other must be left vibrating with slightly more or less energy after the photon has struck it. That energy change would be a record of which path the photon took, and hence the mirrors would be entangled with the photon.

Fortunately, that is not what happens. Remember that, at a sufficiently fine level of detail, what we crudely see as a single history of the mirror, resting passively or vibrating gently on its supports, is actually a vast number of histories with instances of all its atoms continually splitting and rejoining. In particular, the total energy of the mirror takes a vast number of possible values around the average, 'classical' one. Now, what happens when a photon strikes the mirror, changing that total energy by one quantum?

Oversimplifying for a moment, imagine just five of those countless instances of the mirror, with each instance having a different vibrational energy ranging from two quanta below the average to two quanta above it. Each instance of the photon strikes one instance of the mirror and imparts one additional quantum of energy to it. So, after that impact, the average energy of the instances of the mirror will have increased by one quantum, and there will now be instances with energies ranging from one quantum below the old average to three above. But since, at this fine level of detail, there is no autonomous history associated with any of those values of the energy, it is not meaningful to ask whether an instance of the mirror with a particular energy after the impact is *the same* one that previously had that energy. The objective physical fact is only that, of the five instances of the mirror, four have energies that were present before, and one does not. Hence, only that one – whose energy is three quanta higher than the previous average – carries any record of the impact of the photon. And that means that in only one-fifth of the universes in which the photon struck has the wave of differentiation spread to the mirror, and only in those will subsequent interference between instances of that photon that have or have not hit the mirror be suppressed.

With realistic numbers, that is more like one in a trillion trillion – which means that there is only a probability of one in a trillion trillion that interference will be suppressed. This is considerably lower than the probability that the experiment will give inaccurate results due to imperfect measuring instruments, or that it will be spoiled by a lightning strike.

Now let us look at the arrival of that single quantum of energy, to see how that discrete change can possibly happen without any discontinuity. Consider the simplest possible case: an atom absorbs a photon, including all its energy. This energy transfer does not take place instantaneously. (Forget anything that you may have read about 'quantum jumps': they are a myth.) There are many ways in which it can happen but the simplest is this. At the beginning of the process, the atom is in (say) its 'ground state', in which its electrons have the least possible energy allowed by quantum theory. That means that all its instances (within the relevant coarse-grained history) have that energy. Assume that they are also fungible. At the end of the process, all those instances are still fungible, but now they are in the 'excited state', which has one additional quantum of energy. What is the atom like halfway through the process? *Its instances are still fungible*, but now half of them are in the ground state and half in the excited state. It is as if a continuously variable amount of money changed ownership gradually from one discrete owner to another.

This mechanism is ubiquitous in quantum physics, and is the general means by which transitions between discrete states happen in a continuous way. In classical physics, a 'tiny effect' always means a tiny change in some measurable quantities. In quantum physics, physical variables are typically discrete and so cannot undergo tiny changes. Instead, a 'tiny effect' means a tiny change in the *proportions* that have the various discrete attributes.

This also raises the issue of whether time itself is a continuous variable. In this discussion I am assuming that it is. However, the quantum mechanics of time is not yet fully understood, and will not be until we have a quantum theory of gravity (the unification of quantum theory with the general theory of relativity), so it may turn out that things are not as simple as that. One thing we can be fairly sure of, though, is that, in that theory, *different times are a special case of different universes*. In other words, time is an entanglement phenomenon, which places all equal clock readings (of correctly prepared clocks – or of any objects usable as clocks) into the same history. This was first understood by the physicists Don Page and William Wooters, in 1983.

In this full version of the quantum multiverse, how is our science-fiction story to continue? Almost all the attention that the quantum theory has attracted, from physicists, philosophers and science-fiction authors alike, has focused on its parallel-universes aspect. That is ironic, because it is in the parallel-universe approximation that the world most resembles that of classical physics, yet that is the very aspect of quantum theory that many people seem to find viscerally unacceptable.

Fiction can explore the possibilities opened up by parallel universes. For instance, since our story is a romance, the characters may well wonder about their counterparts in other histories. The story could compare their speculations with what we 'know' happened in the other histories. The character whose spouse's unfaithfulness was revealed by a 'random' event might wonder whether that event provided a lucky escape from what was a doomed marriage anyway. Are they still married in the history in which the unfaithfulness was not subsequently revealed? Are they still happy? Can it be true happiness if it is 'based on a lie'? As we see them speculating on these matters, we see the 'still married' history and know the (fictional) fact of the matter.

They might also speculate about less parochial issues. The story could say that their sun is part of a cluster of dozens of stars, all within a sphere of a few light-weeks' radius. This has puzzled their scientists for decades, since the composition of the stars shows that they originated from far and wide and became gravitationally bound through a series of very unlikely coincidences. In most universes, these scientists calculate, life cannot evolve in such dense star clusters, because there are too many collisions. So in most universes that contain humans there are no fleets of starships visiting inhabited star systems one after another. They have been trying to discover a mechanism by which the proximity of nearby stars might somehow precipitate the formation of intelligent life, but they have failed. Should they consider it just an astronomically unlikely coincidence? But they do not like leaving things unexplained. *Something* must have selected them, they conclude. It did. Those people are not just a story. They are real, living, thinking human beings, wondering at this very moment where they came from. But they will never find out. In that one respect, they are unlucky: they were indeed selected by coincidence. Another way of putting that is that they were selected by the very story that I am now telling about them. All fiction that does not violate the laws of physics is fact.

Some fiction in which the laws of physics *appear* to be violated is also fact, somewhere in the multiverse. This involves a subtle issue about how the multiverse is structured – how histories emerge. A history is approximately autonomous. If I boil some water in a kettle and make tea, I am in a history in which I switched on the kettle and the water became gradually hotter because of the energy being poured into it by the kettle, causing bubbles to form and so on, and eventually hot tea forms. That is a history because one can give explanations and make predictions about it without ever mentioning either that there are other histories in the multiverse where I chose to make coffee instead or that the microscopic motion of the water molecules is slightly affected by parts of the multiverse that are outside that history. It is irrelevant to that explanation that a small measure of that history differentiates itself during that process and does other things. In some tiny sliver of it, the kettle transforms itself into a top hat, and the water into a rabbit which then hops away, and I get neither tea nor coffee but am very surprised. That is a history too, *after* that transformation. But there is no way of correctly explaining what was happening during it, or predicting the probabilities, without referring to other parts of the multiverse – enormously larger parts (i.e. with larger measures) – in which there was no rabbit. So that history began at the transformation, and its causal connection with what happened before that cannot be expressed in history terms but only in multiverse terms.

In simple cases like that, there is a ready-made approximative language in which we can minimize mention of the rest of the multiverse: the language of random events. This allows us to acknowledge that most of the high-level objects concerned still behaved autonomously except for being affected by something outside themselves – as when I am affected by the rabbit. This constitutes some continuity between a history and a previous history from which it split, and we can refer to the former as a 'history that has been affected by random events'. However, this is never literally what has happened: the part of that 'history' prior to the 'random event' is fungible with the rest of the broader history and therefore has no separate identity from it: it is not separately explicable.

But the broader of those two histories still is. That is to say, the rabbit history is fundamentally different from the tea history, in that the latter remains very accurately autonomous throughout the period. In the rabbit history I end up with memories that are identical to what they would be in a history in which water became a rabbit. But those are misleading memories. There was no such history; the history containing those memories began only after the rabbit had formed. For that matter, there are also places in the multiverse – of far larger measure than that one – in which *only* my brain was affected, producing exactly those memories. In effect, I had a hallucination, caused by random motion of the atoms in my brain. Some philosophers make a big issue of that sort of thing, claiming that it casts doubt on the scientific status of quantum theory, but of course they are empiricists. In reality, misleading observations, misleading memories and false interpretations are common even in the mainstreams of history. We have to work hard to avoid fooling ourselves with them.

So it is not quite true that, for instance, there are histories in which magic appears to work. There are only histories in which magic appears to *have* worked, but will never work again. There are histories in which I appear to have walked through a wall, because all the atoms of my body happened to resume their original courses after being deflected by atoms in the wall. But those histories began at the wall: the true explanation of what happened involves many other instances of me and it – or we can roughly explain it in terms of random events of very low probability. It is a bit like winning a lottery: the winner cannot properly explain what has just happened without invoking the existence of many losers. In the multiverse, the losers are other instances of oneself.

The 'history' approximation breaks down completely only when histories not only split but merge – that is to say, in interference phenomena. For example, there are certain molecules that exist in two or more structures at once (a 'structure' being an arrangement of atoms, held together by chemical bonds). Chemists call this phenomenon 'resonance' between the two structures, but the molecule is not alternating between them: it has them simultaneously. There is no way of explaining the chemical properties of such molecules in terms of a single structure, because when a 'resonant' molecule participates in a chemical reaction with other molecules, there is quantum interference.

In science fiction, we have a mandate to speculate, even to levels of implausibility that would make for quite bad explanations in real science. But the best explanation of ourselves in real science is that we – sentient beings in this gigantic, unfamiliar structure in which material things have no continuity, in which even something as basic as motion or change is different from anything in our experience – are embedded in multiversal objects. Whenever we observe anything – a scientific instrument or a galaxy or a human being – what we are actually seeing is a single-universe perspective on a larger object that extends some way into other universes. In some of those universes, the object looks exactly as it does to us, in others it looks different, or is absent altogether. What an observer sees as a married couple is actually just a sliver of a vast entity that includes many fungible instances of such a couple, together with other instances of them who are divorced, and others who have never married.

We are channels of information flow. So are histories, and so are all relatively autonomous objects within histories; but we sentient beings are extremely unusual channels, along which (sometimes) *knowledge grows*. This can have dramatic effects, not only within a history (where it can, for instance, have effects that do not diminish with distance), but also across the multiverse. Since the growth of knowledge is a process of error-correction, and since there are many more ways of being wrong than right, knowledge-creating entities rapidly become more alike in different histories than other entities. As far as is known, knowledge-creating processes are unique in both these respects: all other effects diminish with distance in space, and become increasingly different across the multiverse, in the long run.

But that is only as far as is known. Here is an opportunity for some wild speculations that could inform a science-fiction story. What if there is something other than *information flow* that can cause coherent, emergent phenomena in the multiverse? What if knowledge, or something other than knowledge, could emerge from that, and begin to have purposes of its own, and to conform the multiverse to those purposes, as we do? Could we communicate with it? Presumably not in the usual sense of the term, because that would be information flow; but perhaps the story could propose some novel analogue of communication which, like quantum inference, did not involve sending messages. Would we be trapped in a war of mutual extermination with such an entity? Or is it possible that we could nevertheless have something in common with it? Let us shun parochial resolutions of the issue – such as a discovery that what bridges the barrier is *love*, or *trust*. But let us remember that, just as we are at the top rank of significance in the great scheme of things, anything else that could create explanations would be too. And there is always room at the top.

TERMINOLOGY

Fungible   Identical in every respect.

The world   The whole of physical reality.

Multiverse   The world, according to quantum theory.

Universe   Universes are quasi-autonomous regions of the multiverse.

History   A set of fungible universes, over time. One can also speak of the history of parts of a universe.

Parallel universes   A somewhat misleading way of referring to the multiverse. Misleading because the universes are not perfectly 'parallel' (autonomous), and because the multiverse has much more structure – especially fungibility, entanglement and the measures of histories.

Instances   In parts of the multiverse that contain universes, each multiversal object consists approximately of 'instances', some identical, some not, one in each of the universes.

Quantum   The smallest possible change in a discrete physical variable.

Entanglement   Information in each multiversal object that determines which parts (instances) of it can affect which parts of other multiversal objects.

Decoherence   The process of its becoming infeasible to undo the effect of a wave of differentiation between universes.

Quantum interference   Phenomena caused by non-fungible instances of a multiversal object becoming fungible.

Uncertainty principle   The (badly misnamed) implication of quantum theory that, for any fungible collection of instances of a physical object, some of their attributes must be diverse.

Quantum computation Computation in which the flow of information is not confined to a single history.

SUMMARY

The physical world is a multiverse, and its structure is determined by how information flows in it. In many regions of the multiverse, information flows in quasi-autonomous streams called histories, one of which we call our 'universe'. Universes approximately obey the laws of classical (pre-quantum) physics. But we know of the rest of the multiverse, and can test the laws of quantum physics, because of the phenomenon of quantum interference. Thus a universe is not an exact but an emergent feature of the multiverse. One of the most unfamiliar and counter-intuitive things about the multiverse is fungibility. The laws of motion of the multiverse are deterministic, and apparent randomness is due to initially fungible instances of objects becoming different. In quantum physics, variables are typically discrete, and how they change from one value to another is a multiversal process involving interference and fungibility.

12

A Physicist's History of Bad Philosophy

With Some Comments on Bad Science

> By the way, what I have just outlined is what I call a 'physicist's history of physics', which is never correct . . .
>
> Richard Feynman, *QED:*
> *The Strange Theory of Light and Matter* (1985)

READER: So, I am an emergent, quasi-autonomous flow of information in the multiverse.

DAVID: You are.

READER: And I exist in multiple instances, some of them different from each other, some not. And those are the *least* weird things about the world according to quantum theory.

DAVID: Yes.

READER: But your argument is that we have no option but to accept the theory's implications, because it is the only known explanation of many phenomena and has survived all known experimental tests.

DAVID: What other option would you *like* to have?

READER: I'm just summarizing.

DAVID: Then yes: quantum theory does have universal reach. But if all you want to explain is how we know that there are other universes, you don't have to go via the full theory. You need look no further than what a Mach–Zehnder interferometer does to a single photon: the path that was not taken affects the one that was. Or, if you want the same thing writ large, just think of a quantum computer: its output will depend on intermediate results being computed in vast numbers of *different* histories of the *same* few atoms.

305

READER: But that's just a few *atoms* existing in multiple instances. Not people.

DAVID: Are you claiming to be made of something other than atoms?

READER: Ah, I see.

DAVID: Also, imagine a vast cloud of instances of a single photon, some of which are stopped by a barrier. Are they absorbed by the barrier that we see, or is each absorbed by a different, quasi-autonomous barrier at the same location?

READER: Does it make a difference?

DAVID: Yes. If they were all absorbed by the barrier we see, it would vaporize.

READER: So it would.

DAVID: And we can ask – as I did in the story of the starship and the twilight zone – what is holding up those barriers? It must be other instances of the floor. And of the planet. And then we can consider the experimenters who set all this up and who observe the results, and so on.

READER: So that trickle of photons through the interferometer really does provide a window on a vast multiplicity of universes.

DAVID: Yes. It's another example of reach – just a small portion of the reach of quantum theory. The explanation of those experiments in isolation isn't as hard to vary as the full theory. But in regard to the existence of other universes it's incontrovertible all the same.

READER: And that's all there is to it?

DAVID: Yes.

READER: But then why is it that only a small minority of quantum physicists agree?

DAVID: Bad philosophy.

READER: What's that?

Quantum theory was discovered independently by two physicists who reached it from different directions: Werner Heisenberg and Erwin Schrödinger. The latter gave his name to the *Schrödinger equation*, which is a way of expressing the quantum-mechanical laws of motion.

Both versions of the theory were formulated between 1925 and 1927, and both explained motion, especially within atoms, in new and astonishingly counter-intuitive ways. Heisenberg's theory said that the physical variables of a particle do not have numerical values. Instead, they are *matrices*: large arrays of numbers which are related in complicated, probabilistic ways to the outcomes of observations of those variables. With hindsight, we now know that that multiplicity of information exists because a variable has different values for different instances of the object in the multiverse. But, at the time, neither Heisenberg nor anyone else believed that his matrix-valued quantities literally described what Einstein called 'elements of reality'.

The Schrödinger equation, when applied to the case of an individual particle, described a wave moving through space. But Schrödinger soon realized that for two or more particles it did not. It did not represent a wave with multiple crests, nor could it be resolved into two or more waves; mathematically, it was a single wave in a higher-dimensional space. With hindsight, we now know that such waves describe what proportion of the instances of each particle are in each region of space, and also the entanglement information among the particles.

Although Schrödinger's and Heisenberg's theories seemed to describe very dissimilar worlds, neither of which was easy to relate to existing conceptions of reality, it was soon discovered that, if a certain simple rule of thumb was added to each theory, they would always make identical *predictions*. Moreover, these predictions turned out to be very successful.

With hindsight, we can state the rule of thumb like this: whenever a measurement is made, all the histories but one cease to exist. The surviving one is chosen at random, with the probability of each possible outcome being equal to the total measure of all the histories in which that outcome occurs.

At that point, disaster struck. Instead of trying to improve and integrate those two powerful but slightly flawed explanatory theories, and to explain why the rule of thumb worked, most of the theoretical-physics community retreated rapidly and with remarkable docility into instrumentalism. If the predictions work, they reasoned, why worry about the explanation? So they tried to regard quantum theory as being *nothing but* a set of rules of thumb for predicting the observed outcomes of experiments, saying nothing (else) about reality. This move is still popular today, and is known to its critics (and even to some of its proponents) as the 'shut-up-and-calculate interpretation of quantum theory'.

This meant ignoring such awkward facts as (1) the rule of thumb was grossly inconsistent with both theories; hence it could be used only in situations where quantum effects were too small to be noticed. Those happened to include the moment of measurement (because of entanglement with the measuring instrument, and consequent decoherence, as we now know). And (2) it was not even *self*-consistent when applied to the hypothetical case of an observer performing a quantum measurement on another observer. And (3) both versions of quantum theory were clearly describing *some* sort of physical process that *brought about* the outcomes of experiments. Physicists, both through professionalism and through natural curiosity, could hardly help wondering about that process. But many of them tried not to. Most of them went on to train their students not to. This counteracted the scientific tradition of criticism in regard to quantum theory.

Let me define 'bad philosophy' as philosophy that is not merely false, but actively prevents the growth of other knowledge. In this case, instrumentalism was acting to prevent the explanations in Schrödinger's and Heisenberg's theories from being improved or elaborated or unified.

The physicist Niels Bohr (another of the pioneers of quantum theory) then developed an 'interpretation' of the theory which later became known as the 'Copenhagen interpretation'. It said that quantum theory, including the rule of thumb, was a complete description of reality. Bohr excused the various contradictions and gaps by using a combination of instrumentalism and studied ambiguity. He denied the 'possibility of speaking of phenomena as existing objectively' – but said that only the outcomes of observations should count as phenomena. He also said that, although observation has no access to 'the real essence of phenomena', it does reveal relationships between them, and that, in addition, quantum theory blurs the distinction between observer and observed. As for what would happen if one observer performed a quantum-level observation on another, he avoided the issue – which became known as the 'paradox of Wigner's friend', after the physicist Eugene Wigner.

In regard to the unobserved processes between observations, where both Schrödinger's and Heisenberg's theories seemed to be describing a multiplicity of histories happening at once, Bohr proposed a new fundamental principle of nature, the 'principle of complementarity'. It said that accounts of phenomena could be stated only in 'classical language' – meaning language that assigned single values to physical variables at any one time – but classical language could be used only in regard to some variables, including those that had just been measured. One was not permitted to ask what values the other variables had. Thus, for instance, in response to the question 'Which path did the photon take?' in the Mach–Zehnder interferometer, the reply would be that there is no such thing as which path when the path is not observed. In response to the question 'Then how does the photon know which way to turn at the final mirror, since this depends on what happened on both paths?', the reply would be an equivocation called 'particle–wave duality': the photon is both an extended (non-zero volume) and a localized (zero-volume) object at the same time, and one can choose to observe either attribute but not both. Often this is expressed in the saying 'It is both a wave and a particle simultaneously.' Ironically, there is a sense in which those words are precisely true: in that experiment the entire multiversal photon is indeed an extended object (wave), while instances of it (particles, in histories) are localized. Unfortunately, that is not what is meant in the Copenhagen interpretation. There the idea is that quantum physics defies the very foundations of reason: particles have mutually exclusive attributes, period. And it dismisses criticisms of the idea as invalid because they constitute attempts to use 'classical language' outside its proper domain (namely describing outcomes of measurements).

Later, Heisenberg called the values about which one was not permitted to ask *potentialities*, of which only one would become actual when a measurement was completed. How can potentialities that do not happen affect actual outcomes? That was left vague. What caused the transition between 'potential' and 'actual'? The implication of Bohr's anthropocentric language – which was made explicit in most subsequent presentations of the Copenhagen interpretation – was that the transition is caused by human consciousness. Thus consciousness was said to be acting at a fundamental level in physics.

For decades, various versions of all that were taught as fact – vagueness, anthropocentrism, instrumentalism and all – in university physics courses. Few physicists claimed to understand it. None did, and so students' questions were met with such nonsense as 'If you think you've understood quantum mechanics then you don't.' Inconsistency was defended as 'complementarity' or 'duality'; parochialism was hailed as philosophical sophistication. Thus the theory claimed to stand outside the jurisdiction of normal (i.e. all) modes of criticism – a hallmark of bad philosophy.

Its combination of vagueness, immunity from criticism, and the prestige and perceived authority of fundamental physics opened the door to countless systems of pseudo-science and quackery supposedly based on quantum theory. Its disparagement of plain criticism and reason as being 'classical', and therefore illegitimate, has given endless comfort to those who want to defy reason and embrace any number of irrational modes of thought. Thus quantum theory – the deepest discovery of the physical sciences – has acquired a reputation for endorsing practically every mystical and occult doctrine ever proposed.

Not every physicist accepted the Copenhagen interpretation or its descendants. Einstein never did. The physicist David Bohm struggled to construct an alternative that was compatible with realism, and produced a rather complicated theory which I regard as the multiverse theory in heavy disguise – though he was strongly opposed to thinking of it in that way. And in Dublin in 1952 Schrödinger gave a lecture in which at one point he jocularly warned his audience that what he was about to say might 'seem lunatic'. It was that, when his equation seems to be describing several different histories, they are 'not alternatives but all really happen simultaneously'. This is the earliest known reference to the multiverse.

Here was an eminent physicist joking that he might be considered mad. Why? For claiming that his own equation – the very one for which he had won the Nobel prize – might be *true*.

Schrödinger never published that lecture, and seems never to have taken the idea further. Five years later, and independently, the physicist Hugh Everett published a comprehensive theory of the multiverse, now known as the *Everett interpretation* of quantum theory. Yet it took several more decades before Everett's work was even noticed by more than a handful of physicists. Even now that it has become well known, it is endorsed by only a small minority. I have often been asked to explain this unusual phenomenon. Unfortunately I know of no entirely satisfactory explanation. But, to understand why it is perhaps not quite as bizarre and isolated an event as it may appear, one has to consider the broader context of bad philosophy.

Error is the normal state of our knowledge, and is no disgrace. There is nothing bad about *false* philosophy. Problems are inevitable, but they can be solved by imaginative, critical thought that seeks good explanations. That is good philosophy, and good science, both of which have always existed in some measure. For instance, children have always learned language by making, criticizing and testing conjectures about the connection between words and reality. They could not possibly learn it in any other way, as I shall explain in Chapter 16.

Bad philosophy has always existed too. For instance, children have always been told, 'Because I say so.' Although that is not always intended as a philosophical position, it is worth analysing it as one, for in four simple words it contains remarkably many themes of false *and* bad philosophy. First, it is a perfect example of bad explanation: it could be used to 'explain' anything. Second, one way it achieves that status is by addressing only the form of the question and not the substance: it is about who said something, not what they said. That is the opposite of truth-seeking. Third, it reinterprets a request for *true explanation* (why should something-or-other be as it is?) as a request for *justification* (what entitles you to assert that it is so?), which is the justified-true-belief chimera. Fourth, it confuses the nonexistent *authority for ideas* with *human* authority (power) – a much-travelled path in bad political philosophy. And, fifth, it claims by this means to stand outside the jurisdiction of normal criticism.

Bad philosophy before the Enlightenment was typically of the because-I-say-so variety. When the Enlightenment liberated philosophy and science, they both began to make progress, and increasingly there was good philosophy. But, paradoxically, *bad* philosophy became *worse*.

I have said that empiricism initially played a positive role in the history of ideas by providing a defence against traditional authorities and dogma, and by attributing a central role – albeit the wrong one – to experiment in science. At first, the fact that empiricism is an impossible account of how science works did almost no harm, because no one took it literally. Whatever scientists may have *said* about where their discoveries came from, they eagerly addressed interesting problems, conjectured good explanations, tested them, and only lastly claimed to have induced the explanations from experiment. The bottom line was that they succeeded: they made progress. Nothing prevented that harmless (self-)deception, and nothing was inferred from it.

Gradually, though, empiricism did begin to be taken literally, and so began to have increasingly harmful effects. For instance, the doctrine of *positivism*, developed during the nineteenth century, tried to eliminate from scientific theories everything that had not been 'derived from observation'. Now, since nothing is ever derived from observation, what the positivists tried to eliminate depended entirely on their own whims and intuitions. Occasionally these were even good. For instance, the physicist Ernst Mach (father of Ludwig Mach of the Mach–Zehnder interferometer), who was also a positivist philosopher, influenced Einstein, spurring him to eliminate untested assumptions from physics – including Newton's assumption that time flows at the same rate for all observers. That happened to be an excellent idea. But Mach's positivism also caused him to oppose the resulting theory of relativity, essentially because it claimed that spacetime really exists even though it cannot be 'directly' observed. Mach also resolutely denied the existence of atoms, because they were too small to observe. We laugh at this silliness now – when we have microscopes that can see atoms – but the role of philosophy should have been to laugh at it *then*.

Instead, when the physicist Ludwig Boltzmann used atomic theory to unify thermodynamics and mechanics, he was so vilified by Mach and other positivists that he was driven to despair, which may have contributed to his suicide just before the tide turned and most branches of physics shook off Mach's influence. From then on there was nothing to discourage atomic physics from thriving. Fortunately also, Einstein soon rejected positivism and became a forthright defender of realism. That was why he never accepted the Copenhagen interpretation. I wonder: if Einstein had continued to take positivism seriously, could he ever have thought of the general theory of relativity, in which spacetime not only exists but is a dynamic, unseen entity bucking and twisting under the influence of massive objects? Or would spacetime theory have come to a juddering halt like quantum theory did?

Unfortunately, most philosophies of science since Mach's have been even worse (Popper's being an important exception). During the twentieth century, anti-realism became almost universal among philosophers, and common among scientists. Some denied that the physical world exists at all, and most felt obliged to admit that, even if it does, science has no access to it. For example, in 'Reflections on my Critics' the philosopher Thomas Kuhn wrote:

> There is [a step] which many philosophers of science wish to take and which I refuse. They wish, that is, to compare [scientific] theories as representations of nature, as statements about 'what is really out there'.
>
> Imre Lakatos and Alan Musgrave, eds., *Criticism and the Growth of Knowledge* (1979)

Positivism degenerated into *logical positivism*, which held that statements not verifiable by observation are not only worthless but meaningless. This doctrine threatened to sweep away not only explanatory scientific knowledge but the whole of philosophy. In particular: logical positivism itself is a philosophical theory, and it cannot be verified by observation; hence it asserts its own meaninglessness (as well as that of all other philosophy).

The logical positivists tried to rescue their theory from that implication (for instance by calling it 'logical', as distinct from philosophical), but in vain. Then Wittgenstein embraced the implication and declared all philosophy, including his own, to be meaningless. He advocated remaining silent about philosophical problems, and, although he never attempted to live up to that aspiration, he was hailed by many as one of the greatest geniuses of the twentieth century.

One might have thought that this would be the nadir of philosophical thinking but unfortunately there were greater depths to plumb. During the second half of the twentieth century, mainstream philosophy lost contact with, and interest in, trying to understand science as it was actually being done, or how it should be done. Following Wittgenstein, the predominant school of philosophy for a while was 'linguistic philosophy', whose defining tenet was that what seem to be philosophical problems are actually just puzzles about how words are used in everyday life, and that philosophers can meaningfully study only that.

Next, in a related trend that originated in the European Enlightenment but spread all over the West, many philosophers moved away from trying to understand *anything*. They actively attacked the idea not only of explanation and reality, but of truth, and of reason. Merely to criticize such attacks for being self-contradictory like logical positivism – which they were – is to give them far too much credence. For at least the logical positivists and Wittgenstein were interested in making a *distinction* between what does and does not make sense – albeit that they advocated a hopelessly wrong one.

One currently influential philosophical movement goes under various names such as postmodernism, deconstructionism and structuralism, depending on historical details that are unimportant here. It claims that because all ideas, including scientific theories, are conjectural and impossible to justify, they are essentially arbitrary: they are no more than stories, known in this context as 'narratives'. Mixing extreme cultural relativism with other forms of anti-realism, it regards objective truth and falsity, as well as reality and knowledge of reality, as mere conventional forms of words that stand for an idea's being endorsed by a designated group of people such as an elite or consensus, or by a fashion or other arbitrary authority. And it regards science and the Enlightenment as no more than one such fashion, and the objective knowledge claimed by science as an arrogant cultural conceit.

Perhaps inevitably, these charges are true of postmodernism itself: it is a narrative that resists rational criticism or improvement, precisely because it rejects all criticism as mere narrative. Creating a successful postmodernist theory is indeed purely a matter of meeting the criteria of the postmodernist community – which have evolved to be complex, exclusive and authority-based. Nothing like that is true of rational ways of thinking: creating a good explanation is hard not because of what anyone has decided, but because there is an objective reality that does not meet *anyone's* prior expectations, including those of authorities. The creators of bad explanations such as myths are indeed just making things up. But the method of seeking good explanations creates an engagement with reality, not only in science, but in good philosophy too – which is why it works, and why it is the antithesis of concocting stories to meet made-up criteria.

Although there have been signs of improvement since the late twentieth century, one legacy of empiricism that continues to cause confusion, and has opened the door to a great deal of bad philosophy, is the idea that it is possible to split a scientific theory into its predictive rules of thumb on the one hand and its assertions about reality (sometimes known as its 'interpretation') on the other. This does not make sense, because – as with conjuring tricks – without an explanation it is impossible to recognize the circumstances under which a rule of thumb is supposed to apply. And it especially does not make sense in fundamental physics, because the predicted outcome of an observation is itself an unobserved physical process.

Many sciences have so far avoided this split, including most branches of physics – though relativity may have had a narrow escape, as I mentioned. Hence in, say, palaeontology, we do not speak of the existence of dinosaurs millions of years ago as being 'an interpretation of our best theory of fossils': we claim that it is *the explanation* of fossils. And, in any case, the theory of evolution is not primarily about fossils or even dinosaurs, but about their genes, of which not even fossils exist. We claim that there really were dinosaurs, and that they had genes whose chemistry we know, even though there is an infinity of possible rival 'interpretations' of the same data which make all the same predictions and yet say that neither the dinosaurs nor their genes were ever there.

One of them is the 'interpretation' that dinosaurs are only a manner of speaking about certain sensations that palaeontologists have when they gaze at fossils. The sensations are real, but the dinosaurs were not. Or, if they were, we can never know of them. The latter is one of many tangles that one gets into via the justified-true-belief theory of knowledge – for in reality here we are, knowing of them. Then there is the 'interpretation' that the fossils themselves come into existence only when they are extracted from the rock in a manner chosen by the palaeontologist and experienced in a way that can be communicated to other palaeontologists. In that case, fossils are certainly no older than the human species. And they are evidence not of dinosaurs, but only of those acts of observation. Or one can say that dinosaurs *are* real, but not as animals, only as a set of relationships between different people's experiences of fossils. One can then infer that there is no sharp distinction between dinosaurs and palaeontologists, and that 'classical language', though unavoidable, cannot express the ineffable relationship between them. None of those 'interpretations' is *empirically* distinguishable from the rational explanation of fossils. But they are ruled out for being bad explanations: all of them are general-purpose means of denying anything. One can even use them to deny that Schrödinger's equation is true.

Since explanationless prediction is actually impossible, the methodology of excluding explanation from a science is just a way of holding one's explanations immune from criticism. Let me give an example from a distant field: psychology.

I have mentioned *behaviourism*, which is instrumentalism applied to psychology. It became the prevailing interpretation in that field for several decades, and, although it is now largely repudiated, research in psychology continues to downplay explanation in favour of stimulus-response rules of thumb. Thus, for instance, it is considered good science to conduct behaviouristic experiments to measure the extent to which a human psychological state such as, say, loneliness or happiness is genetically coded (like eye colour) or not (such as date of birth). Now, there are some fundamental problems with such a study from an explanatory point of view. First, how can we measure whether different people's ratings of their own psychological state are commensurable? That is to say, some proportion of the people claiming to have happiness level 8 might be quite unhappy but also so pessimistic that they cannot imagine anything much better. And some of the people who claim only level 3 might in fact be happier than most, but have succumbed to a craze that promises extreme future happiness to those who can learn to chant in a certain way. And, second, if we were to find that people with a particular gene tend to rate themselves happier than people without it, how can we tell whether the gene is coding for happiness? Perhaps it is coding for less reluctance to *quantify* one's happiness. Perhaps the gene in question does not affect the brain at all, but only how a person looks, and perhaps better-looking people are happier on average because they are treated better by others. There is an infinity of possible explanations. But the study is not seeking explanations.

It would make no difference if the experimenters tried to eliminate the subjective self-assessment and instead observed happy and unhappy *behaviour* (such as facial expressions, or how often a person whistles a happy tune). The connection with happiness would still involve comparing subjective interpretations which there is no way of calibrating to a common standard; but in addition there would be an extra level of interpretation: some people believe that behaving in 'happy' ways is a remedy for unhappiness, so, for those people, such behaviours might be a proxy for *un*happiness.

For these reasons, no behavioural study can detect whether happiness is inborn or not. Science simply cannot resolve that issue until we have explanatory theories about what objective attributes people are referring to when they speak of their happiness, and also about what physical chain of events connects genes to those attributes.

So how does explanation-free science address the issue? First, one explains that one is not measuring happiness directly, but only a proxy such as the behaviour of marking checkboxes on a scale called 'happiness'. All scientific measurements use chains of proxies. But, as I explained in Chapters 2 and 3, each link in the chain is an additional source of error, and we can avoid fooling ourselves only by criticizing the theory of each link – which is impossible unless an explanatory theory links the proxies to the quantities of interest. That is why, in genuine science, one can claim to have measured a quantity only when one has an explanatory theory of how and why the measurement procedure should reveal its value, and with what accuracy.

There are circumstances under which there *is* a good explanation linking the measurable proxy such as marking checkboxes with a quantity of interest, and in such cases there need be nothing unscientific about the study. For example, political opinion surveys may ask whether respondents are 'happy' with a given politician facing re-election, under the theory that this gives information about which checkbox the respondents will choose in the election itself. That theory is then tested at the election. There is no analogue of such a test in the case of happiness: there is no independent way of measuring it. Another example of bona-fide science would be a clinical trial to test a drug purported to alleviate (particular identifiable types of) unhappiness. In that case, the objective of the study is, again, to determine whether the drug causes *behaviour* such as saying that one is happier (without also experiencing adverse side effects). If a drug passes that test, the issue of whether it really makes the patients happier, or merely alters their personality to have lower standards or something of that sort, is inaccessible to science until such time as there is a testable explanatory theory of what happiness is.

In explanationless science, one may acknowledge that actual happiness and the proxy one is measuring are not necessarily equal. But one nevertheless calls the proxy 'happiness' and moves on. One chooses a large number of people, ostensibly at random (though in real life one is restricted to small minorities such as university students, in a particular country, seeking additional income), and one excludes those who have detectable extrinsic reasons for happiness or unhappiness (such as recent lottery wins or bereavement). So one's subjects are just 'typical people' – though in fact one cannot tell whether they are statistically representative without an explanatory theory. Next, one defines the 'heritability' of a trait as its degree of statistical correlation with how genetically related the people are. Again, that is a nonexplanatory definition: according to it, whether one was a slave or not was once a highly 'heritable' trait in America: it ran in families. More generally, one acknowledges that statistical correlations do not imply anything about what causes what. But one adds the inductivist equivocation that 'they can be suggestive, though.'

Then one does the study and finds that 'happiness' is, say, 50 per cent 'heritable'. This asserts nothing about happiness itself, until the relevant explanatory theories are discovered (at some time in the future – perhaps after consciousness is understood and AIs are commonplace technology). Yet people find the result interesting, because they interpret it via everyday meanings of the words 'happiness' and 'heritable'. Under that interpretation – which the authors of the study, if they are scrupulous, will nowhere have endorsed – the result is a profound contribution to a wide class of philosophical and scientific debates about the nature of the human mind. Press reports of the discovery will reflect this. The headline will say, 'New Study Shows Happiness 50% Genetically Determined' – without quotation marks around the technical terms.

So will subsequent bad philosophy. For, suppose that someone now does dare to seek explanatory theories about the cause of human happiness. Happiness is a state of continually solving one's problems, they conjecture. Unhappiness is caused by being chronically baulked in one's attempts to do that. And solving problems itself depends on knowing how; so, external factors aside, unhappiness is caused by not knowing how. (Readers may recognize this as a special case of the principle of optimism.)

Interpreters of the study say that it has refuted that theory of happiness. *At most 50 per cent* of unhappiness can be caused by not knowing how, they say. The other 50 per cent is beyond our control: genetically determined, and hence independent of what we know or believe, pending the relevant genetic engineering. (Using the same logic on the slavery example, one could have concluded in 1860 that, say, 95 per cent of slavery is genetically determined and therefore beyond the power of political action to remedy.)

At this point – taking the step from 'heritable' to 'genetically determined' – the explanationless psychological study has transformed its correct but uninteresting result into something very exciting. For it has weighed in on a substantive philosophical issue (optimism) *and* a scientific issue about how the brain gives rise to mental states such as qualia. But it has done so without knowing anything about them.

But wait, say the interpreters. Admittedly we can't tell whether any genes *code* for happiness (or part of it). But who cares how the genes cause the effect – whether by conferring good looks or otherwise? The effect itself is real.

The effect is real, but the experiment cannot detect how much of it one can alter without genetic engineering, just by knowing how. That is because the way in which those genes affect happiness may itself depend on knowledge. For instance, a cultural change may affect what people deem to be 'good looks', and that would then change whether people tend to be made happier by virtue of having particular genes. Nothing in the study can detect whether such a change is about to happen. Similarly, it cannot detect whether a book will be written one day which will persuade some proportion of the population that all evils are due to lack of knowledge, and that knowledge is created by seeking good explanations. If some of those people consequently create more knowledge than they otherwise would have, and become happier than they otherwise would have been, then part of the 50 per cent of happiness that was 'genetically determined' in all previous studies will no longer be so.

The interpreters of the study may respond that it has proved that there can be no such book! Certainly none of them will *write* such a book, or arrive at such a thesis. And so the bad philosophy will have caused bad science, which will have stifled the growth of knowledge. Notice that this is a form of bad science that may well have conformed to all the best practices of scientific method – proper randomizing, proper controls, proper statistical analysis. All the *formal* rules of 'how to keep from fooling ourselves' may have been followed. And yet no progress could possibly be made, because *it was not being sought*: explanationless theories can do no more than entrench existing, bad explanations.

It is no accident that, in the imaginary study I have described, the outcome appeared to support a pessimistic theory. A theory that predicts how happy people will (probably) be cannot possibly take account of the effects of knowledge-creation. So, to whatever extent knowledge-creation is involved, the theory is prophecy, and will therefore be biased towards pessimism.

Behaviouristic studies of human psychology must, by their nature, lead to dehumanizing theories of the human condition. For refusing to theorize about the mind as a causative agent is the equivalent of regarding it as a non-creative automaton.

The behaviourist approach is equally futile when applied to the issue of *whether* an entity has a mind. I have already criticized it in Chapter 7, in regard to the Turing test. The same holds in regard to the controversy about animal minds – such as whether the hunting or farming of animals should be legal – which stems from philosophical disputes about whether animals experience qualia analogous to those of humans when in fear and pain, and, if so, which animals do. Now, science has little to say on this matter at present, because there is as yet no explanatory theory of qualia, and hence no way of detecting them experimentally. But this does not stop governments from trying to pass the political hot potato to the supposedly objective jurisdiction of experimental science. So, for instance, in 1997 the zoologists Patrick Bateson and Elizabeth Bradshaw were commissioned by the National Trust to determine whether stags suffer when hunted. They reported that they do, because the hunt is 'grossly stressful . . . exhausting and agonizing'. However, that *assumes* that the measurable quantities denoted there by the words 'stress' and 'agony' (such as enzyme levels in the bloodstream) signify the presence of qualia of the same names – which is precisely what the press and public assumed that the study was supposed to *discover*. The following year, the Countryside Alliance commissioned a study of the same issue, led by the veterinary physiologist Roger Harris, who concluded that the levels of those quantities are similar to those of a human who is not suffering but enjoying a sport such as football. Bateson responded – accurately – that nothing in Harris's report contradicted his own. But that is because neither study had any bearing on the issue in question.

This form of explanationless science is just bad philosophy disguised as science. Its effect is to suppress the philosophical debate about how animals should be treated, by pretending that the issue has been settled scientifically. In reality, science has, and will have, no access to this issue until explanatory knowledge about qualia has been discovered.

Another way in which explanationless science inhibits progress is that it amplifies errors. Let me give a rather whimsical example. Suppose you have been commissioned to measure the average number of people who visit the City Museum each day. It is a large building with many entrances. Admission is free, so visitors are not normally counted. You engage some assistants. They will not need any special knowledge or competence; in fact, as will become clear, the less competent they are, the better your results are going to be.

Each morning your assistants take up their stations at the doors. They mark a sheet of paper whenever someone enters through their door. After the museum closes, they count all their marks, and you add together all their counts. You do this every day for a specified period, take the average, and that is the number that you report to your client.

However, in order to claim that your count equals the number of visitors to the museum, you need some explanatory theories. For instance, you are assuming that the doors you are observing are precisely the entrances to the museum, and that they lead *only* to the museum. If one of them leads to the cafeteria or the museum shop as well, you might be making a large error if your client does not consider people who go only there to be 'visitors to the museum'. There is also the issue of museum staff – do they count as visitors? And there are visitors who leave and come back on the same day, and so on. So you need quite a sophisticated explanatory theory of what the client means by 'visitors to the museum' before you can devise a strategy for counting them.

Suppose you count the number of people coming *out* as well. If you have an explanatory theory saying that the museum is always empty at night, and that no one enters or leaves other than through the doors, and that visitors are never created, destroyed, split or merge, and so on, then one possible use for the outgoing count is to check the ingoing one: you would predict that they should be the same. Then, if they are not the same, you will have an estimate of the *accuracy* of your count. That is good science. In fact reporting your result without also making an accuracy estimate makes your report strictly meaningless. But *unless* you have an explanatory theory of the interior of the museum – which you never see – you cannot use the outgoing count, or anything else, to estimate your error.

Now, suppose you are doing your study using explanationless science instead – which really means science with unstated, uncriticized explanations, just as the Copenhagen interpretation really assumed that there was only one unobserved history connecting successive observations. Then you might analyse the results as follows. For each day, subtract the count of people entering from the count of those leaving. If the difference is not zero, then – and this is the key step in the study – call that difference the 'spontaneous-human-creation count' if it is positive, or the 'spontaneous-human-destruction count' if it is negative. If it is exactly zero, call it 'consistent with conventional physics'.

The less competent your counting and tabulating are, the more often you will find those 'inconsistencies with conventional physics'. Next, *prove* that non-zero results (the spontaneous creation or destruction of human beings) are inconsistent with conventional physics. Include this proof in your report, but also include a concession that extraterrestrial visitors would probably be able to harness physical phenomena of which we are unaware. Also, that teleportation to or from another location would be mistaken for 'destruction' (without trace) and 'creation' (out of thin air) in your experiment and that therefore this cannot be ruled out as a possible cause of the anomalies.

When headlines appear of the form 'Teleportation Possibly Observed in City Museum, Say Scientists' and 'Scientists Prove Alien

Abduction is Real,' protest mildly that you have claimed no such thing, that your results are not conclusive, merely suggestive, and that more studies are needed to determine the mechanism of this perplexing phenomenon.

You have made no false claim. Data can become 'inconsistent with conventional physics' by the mundane means of containing errors, just as genes can 'cause happiness' by countless mundane means such as affecting your appearance. The fact that your paper does not point this out does not make it false. Moreover, as I said, the crucial step consists of a definition, and definitions, provided only that they are consistent, cannot be false. You have *defined* an observation of more people entering than leaving as a 'destruction' of people. Although, in everyday language, that phrase has a connotation of people disappearing in puffs of smoke, that is not what it means in this study. For all you know, they *could* be disappearing in puffs of smoke, or in invisible spaceships: that would be consistent with your data. But your paper takes no position on that. It is entirely about the outcomes of your observations.

So you had better not name your research paper 'Errors Made When Counting People Incompetently'. Aside from being a public-relations blunder, that title might even be considered unscientific, according to explanationless science. For it would be taking a position on the 'interpretation' of the observed data, about which it provides no evidence.

In my view this is a scientific experiment in form only. The substance of scientific theories is explanation, and explanation *of errors* constitutes most of the content of the design of any non-trivial scientific experiment.

As the above example illustrates, a generic feature of experimentation is that the bigger the errors you make, either in the numbers or in your naming and interpretation of the measured quantities, the more exciting the results are, *if true*. So, without powerful techniques of error-detection and -correction – which depend on explanatory theories – this gives rise to an instability where false results drown out the true. In the 'hard sciences' – which usually do good science – false results due to all sorts of errors are nevertheless common. But they are corrected when their explanations are criticized and tested. That cannot happen in explanationless science.

Consequently, as soon as scientists allow themselves to stop demanding good explanations and consider only whether a prediction is accurate or inaccurate, they are liable to make fools of themselves. This is the means by which a succession of eminent physicists over the decades have been fooled by conjurers into believing that various conjuring tricks have been done by 'paranormal' means.

Bad philosophy cannot easily be countered by good philosophy – argument and explanation – because it holds itself immune. But it can be countered by *progress*. People want to understand the world, no matter how loudly they may deny that. And progress makes bad philosophy harder to believe. That is not a matter of refutation by logic or experience, but of explanation. If Mach were alive today I expect he would have accepted the existence of atoms once he saw them through a microscope, behaving according to atomic theory. As a matter of logic, it would still be open to him to say, 'I'm not seeing atoms, I'm only seeing a video monitor. And I'm only seeing that theory's predictions *about me*, not about atoms, come true.' But the fact that that is a general-purpose bad explanation would be borne in upon him. It would also be open to him to say, 'Very well, atoms do exist, but electrons do not.' But he might well tire of that game if a better one seems to be available – that is to say, if rapid progress is made. And then he would soon realize that it is not a game.

Bad philosophy is philosophy that denies the possibility, desirability or existence of progress. And progress is the only effective way of opposing bad philosophy. If progress cannot continue indefinitely, bad philosophy will inevitably come again into the ascendancy – for it will be true.

TERMINOLOGY

Bad philosophy   Philosophy that actively prevents the growth of knowledge.

Interpretation   The explanatory part of a scientific theory, supposedly distinct from its predictive or instrumental part.

Copenhagen interpretation   Niels Bohr's combination of instrumentalism, anthropocentrism and studied ambiguity, used to avoid understanding quantum theory as being about reality.

Positivism The bad philosophy that everything not 'derived from observation' should be eliminated from science.

Logical positivism The bad philosophy that statements not verifiable by observation are meaningless.

MEANING OF 'THE BEGINNING OF INFINITY' ENCOUNTERED IN THIS CHAPTER

– The rejection of bad philosophy.

SUMMARY

Before the Enlightenment, bad philosophy was the rule and good philosophy the rare exception. With the Enlightenment came much more good philosophy, but bad philosophy became much worse, with the descent from empiricism (merely false) to positivism, logical positivism, instrumentalism, Wittgenstein, linguistic philosophy, and the 'postmodernist' and related movements.

In science, the main impact of bad philosophy has been through the idea of separating a scientific theory into (explanationless) predictions and (arbitrary) interpretation. This has helped to legitimize dehumanizing explanations of human thought and behaviour. In quantum theory, bad philosophy manifested itself mainly as the Copenhagen interpretation and its many variants, and as the 'shut-up-and-calculate' interpretation. These appealed to doctrines such as logical positivism to justify systematic equivocation and to immunize themselves from criticism.

13

Choices

In March 1792 George Washington exercised the first presidential veto in the history of the United States of America. Unless you already know what he and Congress were quarrelling about, I doubt that you will be able to guess, yet the issue remains controversial to this day. With hindsight, one may even perceive a certain inevitability in it, for, as I shall explain, it is rooted in a far-reaching misconception about the nature of human choice, which is still prevalent.

On the face of it, the issue seems no more than a technicality: *in the US House of Representatives, how many seats should each state be allotted?* This is known as the *apportionment problem*, because the US Constitution requires seats to be 'apportioned among the several States ... according to their respective Numbers [i.e. their populations]'. So, if your state contained 1 per cent of the US population, it would be entitled to 1 per cent of the seats in the House. This was intended to implement the principle of *representative government* – that the legislature should represent the people. It was, after all, about the House of Representatives. (The US Senate, in contrast, represents the *states* of the Union, and hence each state, regardless of population, has two senators.)

At present there are 435 seats in the House of Representatives; so, if 1 per cent of the US population did live in your state, then by strict proportionality the number of representatives to which it would be entitled – known as its *quota* – would be 4.35. When the quotas are not whole numbers, which of course they hardly ever are, they have to be rounded somehow. The method of rounding is known as an *apportionment rule*. The Constitution did not specify an apportionment rule; it left such details to Congress, and that is where the centuries of controversy began.

An apportionment rule is said to 'stay within the quota' if the number of seats that it allocates to each state never differs from the state's quota by as much as a whole seat. For instance, if a state's quota is 4.35 seats, then to 'stay within the quota' a rule must assign that state either four seats or five. It may take all sorts of information into account in choosing between four and five, but if it is capable of assigning any other number it is said to 'violate quota'.

When one first hears of the apportionment problem, compromises that seem to solve it at a stroke spring easily to mind. Everyone asks, 'Why couldn't they just . . . ?' Here is what I asked: Why couldn't they just round each state's quota to the nearest whole number? Under that rule, a quota of 4.35 seats would be rounded down to four; 4.6 seats would be rounded up to five. It seemed to me that, since this sort of rounding can never add or subtract more than *half* a seat, it would keep each state within half a seat of its quota, thus 'staying within the quota' with room to spare.

I was wrong: my rule violates quota. This is easy to demonstrate by applying it to an imaginary House of Representatives with ten seats, in a nation of four states. Suppose that one of the states has just under 85 per cent of the total population, and the other three have just over 5 per cent each. The large state therefore has a quota of just under 8.5, which my rule rounds down to eight. Each of the three small states has a quota of just over half a seat, which my rule rounds up to one. But now we have allocated eleven seats, not ten. In itself that hardly matters: the nation merely has one more legislator to feed than planned. The real problem is that this apportionment is no longer representative: 85 per cent of eleven is not 8.5 but 9.35. So the large state, with only eight seats, is in fact short of its quota by well over one seat. My rule under-represents 85 per cent of the population. Because we *intended* to allocate ten seats, the exact quotas necessarily add up to ten; but the rounded ones add up to eleven. And if there are going to be eleven seats in the House, the principle of representative government – and the Constitution – requires each state to receive its fair share of those, not of the ten that we merely intended.

Again, many 'why don't they just . . . ?' ideas spring to mind. Why don't they just create three additional seats and give them to the large state, thus bringing the allocation within the quota? (Curious readers may check that no fewer than three additional seats are needed to achieve this.) Alternatively, why don't they just transfer a seat from one of the small states to the large state? Perhaps it should be from the state with the smallest population, so as to disadvantage as few people as possible. That would not only bring all the allocations within the quota, but also restore the number of seats to the originally intended ten.

Such strategies are known as *reallocation schemes*. They are indeed capable of staying within the quota. So, what is wrong with them? In the jargon of the subject, the answer is *apportionment paradoxes* – or, in ordinary language, *unfairness* and *irrationality*.

For example, the last reallocation scheme that I described is unfair by being biased against the inhabitants of the least populous state. They bear the whole cost of correcting the rounding errors. On this occasion their representation has been rounded down to zero. Yet, in the sense of minimizing the deviation from the quotas, the apportionment is almost perfectly fair: previously, 85 per cent of the population were well outside the quota, and now all are within it and 95 per cent are at the closest whole numbers to their quotas. It is true that 5 per cent now have no representatives – so they will not be able to vote in congressional elections at all – but that still leaves them within the quota, and indeed only slightly further from their exact quota than they were. (The numbers zero and one are almost equidistant from the quota of just over one half.) Nevertheless, because those 5 per cent have been completely disenfranchised, most advocates of representative government would regard this outcome as much less representative than it was before.

That must mean that the 'minimum total deviation from quota' is not the right measure of representativeness. But what is the right measure? What is the right trade-off between being slightly unfair to many people and very unfair to a few people? The Founding Fathers were aware that different conceptions of fairness, or representativeness, could conflict. For example, one of their justifications for democracy was that government was not legitimate unless everyone who was subject to the law had a representative, of equal power, among the lawmakers. This was expressed in their slogan 'No taxation without representation'. Another of their aspirations was to abolish *privilege*:

they wanted the system of government to have no built-in bias. Hence the requirement of proportional allocation. Since these two aspirations can conflict, the Constitution contains a clause that explicitly adjudicates between them: 'Each State shall have at least one Representative.' This favours the principle of representative government in the no-taxation-without-representation sense over the same principle in the abolish-privilege sense.

Another concept that frequently appeared in the Founding Fathers' arguments for representative government was 'the will of the people'. Governments are supposed to enact it. But that is a source of further inconsistencies. For in elections, only the will of *voters* counts, and not all of 'the people' are voters. At the time, voters were a fairly small minority: only free male citizens over the age of twenty-one. To address this point, the 'Numbers' referred to in the Constitution constituted the whole population of a state, including non-voters such as women, children, immigrants and slaves. In this way the Constitution attempted to treat the *population* equally by treating *voters* unequally.

So voters in states with a higher proportion of non-voters were allocated more representatives per capita. This had the perverse effect that in the states where the voters were already the most privileged *within* the state (i.e. where they were an exceptionally small minority), they now received an additional privilege relative to voters in other states: they were allocated more representation in Congress. This became a hot political issue in regard to slave-owners. Why should slave-owning states be allocated more political clout in proportion to how many slaves they had? To reduce this effect, a compromise was reached whereby a slave counted as three-fifths of a person for the purpose of apportioning seats in the House. But, even so, three-fifths of an injustice was still considered an injustice by many.* The same controversy exists today in regard to illegal immigrants, who also count as part of the population for apportionment purposes. So

---

*This rule is often misinterpreted as illustrating how slaves were regarded as less than fully human. But that has nothing to do with the issue. Black people were indeed widely regarded as being inferior to white ones, but this particular measure was designed to *reduce* the power of slave-owning states compared to what it would have been if slaves had been counted like everyone else.

states with large numbers of illegal immigrants receive extra seats in Congress, while other states correspondingly lose out.

Following the first US census, in 1790, notwithstanding the new Constitution's requirement of proportionality, seats in the House of Representatives were apportioned under a rule that violated quota. Proposed by the future president Thomas Jefferson, this rule also favoured states with higher populations, giving them more representatives per capita. So Congress voted to scrap it and substitute a rule proposed by Jefferson's arch-rival Alexander Hamilton, which is guaranteed to give a result that stays within quota as well as having no obvious bias between states.

That was the change that President Washington vetoed. The reason he gave was simply that it involved reallocation: he considered all reallocation schemes unconstitutional, because he interpreted the term 'apportioned' as meaning *divided* by a suitable numerical divisor – and then rounded, but nothing else. Inevitably, some suspected that his real reason was that he, like Jefferson, came from the most populous state, Virginia, which would have lost out under Hamilton's rule.

Ever since, Congress has continually debated and tinkered with the rules of apportionment. Jefferson's rule was eventually dropped in 1841 in favour of one proposed by Senator Daniel Webster, which does use reallocation. It also violates quota, but very rarely; and it was, like Hamilton's rule, deemed to be impartial between states.

A decade later, Webster's rule was in turn dropped in favour of Hamilton's. The latter's supporters now believed that the principle of representative government was fully implemented, and perhaps hoped that this would be the end of the apportionment problem. But they were to be disappointed. It was soon causing more controversy than ever, because Hamilton's rule, despite its impartiality and proportionality, began to make allocations that seemed outrageously perverse. For instance, it was susceptible to what came to be called the *population paradox*: a state whose population has increased since the last census can *lose* a seat to one whose population has decreased.

So, 'why didn't they just' create new seats and assign them to states that lose out under a population paradox? They did so. But unfortunately that can bring the allocation outside quota. It can also introduce another historically important apportionment paradox: the *Alabama paradox*. That happens when increasing the total number of *seats* in the House results in some state losing a seat.

And there were other paradoxes. These were not necessarily *unfair* in the sense of being biased or disproportionate. They are called 'paradoxes' because an apparently reasonable rule makes apparently unreasonable changes between one apportionment and the next. Such changes are effectively random, being due to the vagaries of rounding errors, not to any bias, and in the long run they cancel out. But impartiality *in the long run* does not achieve the intended purpose of representative government. Perfect 'fairness in the long run' could be achieved even without elections, by selecting the legislature randomly from the electorate as a whole. But, just as a coin tossed randomly one hundred times is unlikely to produce exactly fifty heads and fifty tails, so a randomly chosen legislature of 435 would in practice never be representative on any one occasion: statistically, the typical deviation from representativeness would be about eight seats. There would also be large fluctuations in how those seats were distributed among states. The apportionment paradoxes that I have described have similar effects.

The number of seats involved is usually small, but that does not make it unimportant. Politicians worry about this because votes in the House of Representatives are often very close. Bills quite often pass or fail by one vote, and political deals often depend on whether individual representatives join one faction or another. So, whenever apportionment paradoxes have caused political discord, people have tried to invent an apportionment rule that is mathematically incapable of causing that particular paradox. Particular paradoxes always make it look as though everything would be fine if only 'they' made some simple change or other. Yet the paradoxes as a whole have the infuriating property that, no matter how firmly they are kicked out of the front door, they instantly come in again at the back.

After Hamilton's rule was adopted, in 1851, Webster's still enjoyed substantial support. So Congress tried, on at least two occasions, a trick that seemed to provide a judicious compromise: adjust the number of seats in the House until the two rules agree. Surely that would please everyone! Yet the upshot was that in 1871 some states considered the result to be so unfair, and the ensuing compromise legislation was so chaotic, that it was unclear what allocation rule, if any, had been decided upon. The apportionment that was implemented – which included the last-minute creation of several additional seats for no apparent reason – satisfied neither Hamilton's rule nor Webster's. Many considered it unconstitutional.

For the next few decades after 1871, every census saw either the adoption of a new apportionment rule or a change in the number of seats, designed to compromise between different rules. In 1921 no apportionment was made at all: they kept the old one (a course of action that may well have been unconstitutional again), because Congress could not agree on a rule.

The apportionment issue has been referred several times to eminent mathematicians, including twice to the National Academy of Sciences, and on each occasion these authorities have made different recommendations. Yet none of them ever accused their predecessors of making errors *in mathematics*. This ought to have warned everyone that this problem is not really about mathematics. And on each occasion, when the experts' recommendations were implemented, paradoxes and disputes kept on happening.

In 1901 the Census Bureau published a table showing what the apportionments would be for every number of seats between 350 and 400 using Hamilton's rule. By a quirk of arithmetic of a kind that is common in apportionment, Colorado would get three seats for each of these numbers except 357, when it would get only two seats. The chairman of the House Committee on Apportionment (who was from Illinois: I do not know whether he had anything against Colorado) proposed that the number of seats be changed to 357 and that Hamilton's rule be used. This proposal was regarded with suspicion, and Congress eventually rejected it, adopting a 386-member apportionment and Webster's rule, which also gave Colorado its 'rightful' three seats. But was that apportionment really any more rightful than Hamilton's rule with 357 seats? By what criterion? Majority voting among apportionment rules?

What exactly would be wrong with working out what a large number of rival apportionment rules would do, and then allocating to each state the number of representatives that the majority of the schemes would allocate? The main thing is that that is itself an apportionment rule. Similarly, combining Hamilton's and Webster's schemes as they tried to do in 1871 just constituted adopting a third scheme. And what does such a scheme have going for it? Each of its constituent schemes was presumably designed to have some desirable properties. A combined scheme that was not designed to have those properties will not have them, except by coincidence. So it will not necessarily inherit the good features of its constituents. It will inherit some good ones and some bad ones, and have additional good and bad features of its own – but if it was not *designed* to be good, why should it be?

A devil's advocate might now ask: if majority voting among apportionment rules is such a bad idea, why is majority voting among *voters* a good idea? It would be disastrous to use it in, say, science. There are more astrologers than astronomers, and believers in 'paranormal' phenomena often point out that purported witnesses of such phenomena outnumber the witnesses of most scientific experiments by a large factor. So they demand proportionate credence. Yet science refuses to judge evidence in that way: it sticks with the criterion of good explanation. So if it would be wrong for science to adopt that 'democratic' principle, why is it right for politics? Is it just because, as Churchill put it, 'Many forms of Government have been tried and will be tried in this world of sin and woe. No one pretends that democracy is perfect or all-wise. Indeed, it has been said that democracy is the worst form of government except all those other forms that have been tried from time to time.' That would indeed be a sufficient reason. But there are cogent positive reasons as well, and they too are about explanation, as I shall explain.

Sometimes politicians have been so perplexed by the sheer perverseness of apportionment paradoxes that they have been reduced to denouncing mathematics itself. Representative Roger Q. Mills of Texas complained in 1882, 'I thought . . . that mathematics was a divine science. I thought that mathematics was the only science that spoke to inspiration and was infallible in its utterances [but] here is a new system of mathematics that demonstrates the truth to be false.' In 1901 Representative John E. Littlefield, whose own seat in Maine was under threat from the Alabama paradox, said, 'God help the State of Maine when mathematics reach for her and undertake to strike her down.'

As a matter of fact, there is no such thing as mathematical 'inspiration' (mathematical knowledge coming from an infallible source, traditionally God): as I explained in Chapter 8, our knowledge of mathematics is not infallible. But if Representative Mills meant that mathematicians are, or somehow ought to be, society's best judges of fairness, then he was simply mistaken.* The National Academy of Sciences panel that reported to Congress in 1948 included the mathematician and physicist John von Neumann. It decided that a rule invented by the statistician Joseph Adna Hill (which is the one in use today) is the most impartial between states. But the mathematicians Michel Balinski and Peyton Young have since concluded that it favours smaller states. This illustrates again that different criteria of 'impartiality' favour different apportionment rules, and which of them is the right criterion cannot be determined by mathematics. Indeed, if Representative Mills intended his complaint ironically – if he really meant that mathematics alone could not possibly be causing injustice and that mathematics alone could not cure it – then he was right.

However, there is a mathematical discovery that has changed for ever the nature of the apportionment debate: we now know that the quest for an apportionment rule that is both proportional and free from paradoxes can never succeed. Balinski and Young proved this in 1975.

Balinski and Young's Theorem
Every apportionment rule that stays within the quota
suffers from the population paradox.

This powerful 'no-go' theorem explains the long string of historical failures to solve the apportionment problem. Never mind the various other conditions that may seem essential for an apportionment to be fair: no apportionment rule can meet even the bare-bones requirements of proportionality and the avoidance of the population paradox. Balinski and Young also proved no-go theorems involving other classic paradoxes.

This work had a much broader context than the apportionment problem. During the twentieth century, and especially following the Second World War, a consensus had emerged among most major

---

*It should of course be physicists.

political movements that the future welfare of humankind would depend on an increase in society-wide (preferably worldwide) planning and decision-making. The Western consensus differed from its totalitarian counterparts in that it expected the object of the exercise to be the satisfaction of individual citizens' preferences. So Western advocates of society-wide planning were forced to address a fundamental question that totalitarians do not encounter: when society as a whole faces a choice, and citizens differ in their preferences among the options, which option is it best for society to choose? If people are unanimous, there is no problem – but no need for a planner either. If they are not, which option can be rationally defended as being 'the will of the people' – the option that society 'wants'? And that raises a second question: how should society organize its decision-making so that it does indeed choose the options that it 'wants'? These two questions had been present, at least implicitly, from the beginning of modern democracy. For instance, the US Declaration of Independence and the US Constitution both speak of the right of 'the people' to do certain things such as remove governments. Now they became the central questions of a branch of mathematical game theory known as *social-choice theory*.

Thus game theory – formerly an obscure and somewhat whimsical branch of mathematics – was suddenly thrust to the centre of human affairs, just as rocketry and nuclear physics had been. Many of the world's finest mathematical minds, including von Neumann, rose to the challenge of developing the theory to support the needs of the countless institutions of collective decision-making that were being set up. They would create new mathematical tools which, given what all the individuals in a society want or need, or prefer, would distil what that society 'wants' to do, thus implementing the aspiration of 'the will of the people'. They would also determine what systems of voting and legislating would give society what it wants.

Some interesting mathematics was discovered. But little, if any, of it ever met those aspirations. On the contrary, time and again the assumptions behind social-choice theory were proved to be incoherent or inconsistent by 'no-go' theorems like that of Balinski and Young.

Thus it turned out that the apportionment problem, which had absorbed so much legislative time, effort and passion, was the tip of an iceberg. The problem is much less parochial than it looks. For instance, rounding errors are proportionately smaller with a larger legislature. So why don't they just make the legislature very big – say, ten thousand members – so that all the rounding errors would be trivial? One reason is that such a legislature would have to organize itself internally to make any decisions. The factions within the legislature would themselves have to choose leaders, policies, strategies, and so on. Consequently, all the problems of social choice would arise within the little 'society' of a party's contingent in the legislature. So it is not really about rounding errors. Also, it is not only about people's top preferences: once we are considering the details of decision-making in large groups – how legislatures and parties and factions within parties organize themselves to contribute their wishes to 'society's wishes' – we have to take into account their second and third choices, because people still have the right to contribute to decision-making if they cannot persuade a majority to agree to their first choice. Yet electoral systems designed to take such factors into account invariably introduce more paradoxes and no-go theorems.

One of the first of the no-go theorems was proved in 1951 by the economist Kenneth Arrow, and it contributed to his winning the Nobel prize for economics in 1972. Arrow's theorem appears to deny the very existence of social choice – and to strike at the principle of representative government, and apportionment, and democracy itself, and a lot more besides.

This is what Arrow did. He first laid down five elementary axioms that any rule defining the 'will of the people' – the preferences of a group – should satisfy, and these axioms seem, at first sight, so reasonable as to be hardly worth stating. One of them is that the rule should define a group's preferences only in terms of the preferences of that group's members. Another is that the rule must not simply designate the views of one particular person to be 'the preferences of the group' regardless of what the others want. That is called the 'no-dictator' axiom. A third is that if the members of the group are unanimous about something – in the sense that they all have identical preferences about it – then the rule must deem the group to have those preferences too. Those three axioms are all expressions, in this situation, of the principle of representative government.

Arrow's fourth axiom is this. Suppose that, under a given definition of 'the preferences of the group', the rule deems the group to have a particular preference – say, for pizza over hamburger. Then it must still deem that to be the group's preference if some members who previously *disagreed* with the group (i.e. they preferred hamburger) change their minds and now prefer pizza. This constraint is similar to ruling out a population paradox. A group would be irrational if it changed its 'mind' in the opposite direction to its members.

The last axiom is that if the group has some preference, and then some members change their minds about *something else*, then the rule must continue to assign the group that original preference. For instance, if some members have changed their minds about the relative merits of strawberries and raspberries, but none of their preferences about the relative merits of pizza and hamburger have changed, then the group's preference between pizza and hamburger must not be deemed to have changed either. This constraint can again be regarded as a matter of rationality: if no members of the group change any of their opinions about a particular comparison, nor can the group.

Arrow proved that the axioms that I have just listed are, despite their reasonable appearance, logically inconsistent with each other. No way of conceiving of 'the will of the people' can satisfy all five of them. This strikes at the assumptions behind social-choice theory at an arguably even deeper level than the theorems of Balinski and Young. First, Arrow's axioms are not about the apparently parochial issue of apportionment, but about any situation in which we want to conceive of a group having preferences. Second, all five of these axioms are intuitively not just desirable to make a system fair, but essential for it to be rational. Yet they are inconsistent.

It seems to follow that a group of people jointly making decisions is necessarily irrational in one way or another. It may be a dictatorship, or under some sort of arbitrary rule; or, if it meets all three representativeness conditions, then it must sometimes change its 'mind' in a direction opposite to that in which criticism and persuasion have been effective. So it will make perverse choices, no matter how wise and benevolent the people who interpret and enforce its preferences may be – unless, possibly, one of them is a dictator (see below). So there is no such thing as 'the will of the people'. There is no way to regard 'society' as a decision-maker with self-consistent preferences. This is hardly the conclusion that social-choice theory was supposed to report back to the world.

As with the apportionment problem, there were attempts to fix the implications of Arrow's theorem with 'why don't they just . . . ?' ideas. For instance, why not take into account *how intense* people's preferences are? For, if slightly over half the electorate barely prefers X to Y, but the rest consider it a matter of life and death that Y should be done, then most intuitive conceptions of representative government would designate Y as 'the will of the people'. But intensities of preferences, and especially the differences in intensities among different people, or between the same person at different times, are notoriously difficult to define, let alone measure – like happiness. And, in any case, including such things makes no difference: there are still no-go theorems.

As with the apportionment problem, it seems that whenever one patches up a decision-making system in one way, it becomes paradoxical in another. A further serious problem that has been identified in many decision-making institutions is that they create incentives for participants to lie about their preferences. For instance, if there are two options of which you mildly prefer one, you have an incentive to register your preference as 'strong' instead. Perhaps you are prevented from doing that by a sense of civic responsibility. But a decision-making system moderated by civic responsibility has the defect that it gives disproportionate weight to the opinions of people who *lack* civic responsibility and are willing to lie. On the other hand, a society in which everyone knows everyone sufficiently well to make such lying difficult cannot have an effectively secret ballot, and the system will then give disproportionate weight to the faction most able to intimidate waverers.

One perennially controversial social-choice problem is that of devising an electoral system. Such a system is mathematically similar to an apportionment scheme, but, instead of allocating seats to states on the basis of population, it allocates them to candidates (or parties) on the basis of votes. However, it is more paradoxical than apportionment and has more serious consequences, because in the case of elections the element of *persuasion* is central to the whole exercise: an election is supposed to determine what the voters have become persuaded of.

(In contrast, apportionment is not about states trying to persuade people to migrate from other states.) Consequently an electoral system can contribute to, or can inhibit, traditions of criticism in the society concerned.

For example, an electoral system in which seats are allocated wholly or partly in proportion to the number of votes received by each party is called a 'proportional-representation' system. We know from Balinski and Young that, if an electoral system is too proportional, it will be subject to the analogue of the population paradox and other paradoxes. And indeed the political scientist Peter Kurrild-Klitgaard, in a study of the most recent eight general elections in Denmark (under its proportional-representation system), showed that every one of them manifested paradoxes. These included the 'More-Preferred-Less-Seats paradox', in which a majority of voters prefer party X to party Y, but party Y receives more seats than party X.

But that is really the least of the irrational attributes of proportional representation. A more important one – which is shared by even the mildest of proportional systems – is that they assign *dis*proportionate power in the legislature to the *third-largest party*, and often to even smaller parties. It works like this. It is rare (in any system) for a single party to receive an overall majority of votes. Hence, if votes are reflected proportionately in the legislature, no legislation can be passed unless some of the parties cooperate to pass it, and no government can be formed unless some of them form a coalition. Sometimes the two largest parties manage to do this, but the most common outcome is that the leader of the third-largest party holds the 'balance of power' and decides which of the two largest parties shall join it in government, and which shall be sidelined, and for how long. That means that it is correspondingly harder for the *electorate* to decide which party, and which policies, will be removed from power.

In Germany (formerly West Germany) between 1949 and 1998, the Free Democratic Party (FDP) was the third largest.* Though it never received more than 12.8 per cent of the vote, and usually much less, the country's proportional-representation system gave it power that was

---

*I am counting the Christian Democrat CDU and the regionally based CSU as being one party for present purposes.

insensitive to changes in the voters' opinions. On several occasions it chose which of the two largest parties would govern, twice changing sides and three times choosing to put the less popular of the two (as measured by votes) into power. The FDP's leader was usually made a cabinet minister as part of the coalition deal, with the result that for the last twenty-nine years of that period Germany had only two weeks without an FDP foreign minister. In 1998, when the FDP was pushed into fourth place by the Green Party, it was immediately ousted from government, and the Greens assumed the mantle of kingmakers. And they took charge of the Foreign Ministry as well. This disproportionate power that proportional representation gives the third-largest party is an embarrassing feature of a system whose whole *raison d'être*, and supposed moral justification, is to allocate political influence proportionately.

Arrow's theorem applies not only to collective decision-making but also to individuals, as follows. Consider a single, rational person faced with a choice between several options. If the decision requires thought, then each option must be associated with an explanation – at least a tentative one – for why it might be the best. To choose an option is to choose its explanation. So how does one decide which explanation to adopt?

Common sense says that one 'weighs' them – or weighs the evidence that their arguments present. This is an ancient metaphor. Statues of Justice have carried scales since antiquity. More recently, inductivism has cast scientific thinking in the same mould, saying that scientific theories are chosen, justified and believed – and somehow even formed in the first place – according to the 'weight of evidence' in their favour.

Consider that supposed weighing process. Each piece of evidence, including each feeling, prejudice, value, axiom, argument and so on, depending on what 'weight' it had in that person's mind, would contribute that amount to that person's 'preferences' between various explanations. Hence for the purposes of Arrow's theorem each piece of evidence can be regarded as an 'individual' participating in the decision-making process, where the person as a whole would be the 'group'.

Now, the process that adjudicates between the different explanations would have to satisfy certain constraints if it were to be rational. For instance, if, having decided that one option was the best, the person received further evidence that gave additional weight to that option, then the person's overall preference would still have to be for that option – and so on. Arrow's theorem says that those requirements are inconsistent with each other, and so seems to imply that all decision-making – all thinking – must be irrational. Unless, perhaps, one of the internal agents is a dictator, empowered to override the combined opinions of all the other agents. But this is an infinite regress: how does the 'dictator' itself choose between rival explanations about which other agents it would be best to override?

There is something very wrong with that entire conventional model of decision-making, both within single minds and for groups as assumed in social-choice theory. It conceives of decision-making as a process of selecting from existing options according to a fixed formula (such as an apportionment rule or electoral system). But in fact that is what happens only at the *end* of decision-making – the phase that does not require creative thought. In terms of Edison's metaphor, the model refers only to the perspiration phase, without realizing that decision-making is problem-solving, and that without the inspiration phase nothing is ever solved and there is nothing to choose between. At the heart of decision-making is the creation of new options and the abandonment or modification of existing ones.

To choose an option, rationally, is to choose the associated explanation. Therefore, rational decision-making consists not of weighing evidence but of explaining it, in the course of explaining the world. One judges arguments as explanations, not justifications, and one does this creatively, using conjecture, tempered by every kind of criticism. It is in the nature of good explanations – being hard to vary – that there is only one of them. Having created it, one is no longer tempted by the alternatives. They have been not outweighed, but out-argued, refuted and abandoned. During the course of a creative process, one is not struggling to distinguish between countless different explanations of nearly equal merit; typically, one is struggling to create even one good explanation, and, having succeeded, one is glad to be rid of the rest.

Another misconception to which the idea of decision-making by weighing sometimes leads is that problems can be solved by weighing – in particular, that disputes between advocates of rival explanations can be resolved by creating a weighted average of their proposals. But the fact is that a good explanation, being hard to vary at all without losing its explanatory power, is hard to mix with a rival explanation: something halfway between them is usually worse than either of them separately. Mixing two explanations to create a *better* explanation requires an additional act of creativity. That is why good explanations are discrete – separated from each other by bad explanations – and why, when choosing between explanations, we are faced with discrete options.

In complex decisions, the creative phase is often followed by a mechanical, perspiration phase in which one ties down details of the explanation that are not yet hard to vary but can be made so by non-creative means. For example, an architect whose clients ask how tall a skyscraper can be built, given certain constraints, does not just calculate that number from a formula. The decision-making process may *end* with such a calculation, but it begins creatively, with ideas for how the clients' priorities and constraints might best be met by a new design. And, before that, the clients had to decide – creatively – what those priorities and constraints should be. At the beginning of that process they would not have been aware of all the preferences that they would end up presenting to architects. Similarly, a voter may look through lists of the various parties' policies, and may even assign each issue a 'weight' to represent its importance; but one can do that only *after* one has thought about one's political philosophy, and has explained to one's own satisfaction how important that makes the various issues, what policies the various parties are likely to adopt in regard to those issues, and so on.

The type of 'decision' considered in social-choice theory is choosing from options that are known and fixed, according to preferences that are known, fixed and consistent. The quintessential example is a voter's choice, in the polling booth, not of which candidate to prefer but of which box to check. As I have explained, this is a grossly inadequate, and inaccurate, model of human decision-making. In reality, the voter is choosing between explanations, not checkboxes, and, while very few voters choose to affect the checkboxes themselves, by running for office, all rational voters create their own explanation for which checkbox they personally should choose.

So it is not true that decision-making necessarily suffers from those crude irrationalities – not because there is anything wrong with Arrow's theorem or any of the other no-go theorems, but because social-choice theory is itself based on false assumptions about what thinking and deciding consist of. It is Zeno's mistake. It is mistaking an abstract process that it has *named* decision-making for the real-life process of the same name.

Similarly, what is called a 'dictator' in Arrow's theorem is not necessarily a dictator in the ordinary sense of the word. It is simply any agent to whom the society's decision-making rules assign the sole right to make a particular decision regardless of the preferences of anyone else. Thus, every law that requires an individual's consent for something – such as the law against rape, or against involuntary surgery – establishes a 'dictatorship' in the technical sense used in Arrow's theorem. Everyone is a dictator over their own body. The law against theft establishes a dictatorship over one's own possessions. A free election is, by definition, one in which every voter is a dictator over their own ballot paper. Arrow's theorem itself assumes that all the participants are in sole control of their *contributions* to the decision-making process. More generally, the most important conditions for rational decision-making – such as freedom of thought and of speech, tolerance of dissent, and the self-determination of individuals – all require 'dictatorships' in Arrow's mathematical sense. It is understandable that he chose that term. But it has nothing to do with the kind of dictatorship that has secret police who come for you in the middle of the night if you criticize them.

Virtually all commentators have responded to these paradoxes and no-go theorems in a mistaken and rather revealing way: they *regret* them. This illustrates the confusion to which I am referring. *They wish that these theorems of pure mathematics were false.* If only mathematics permitted it, they complain, we human beings could set up a just society that makes its decisions rationally. But, faced with the impossibility of that, there is nothing left for us to do but to decide which injustices and irrationalities we like best, and to enshrine them in law. As Webster wrote, of the apportionment problem, 'That which cannot be done perfectly must be done in a manner as near perfection as can be. If exactness cannot, from the nature of things, be attained, then the nearest practicable approach to exactness ought to be made.'

But what sort of 'perfection' is a *logical contradiction*? A logical contradiction is nonsense. The truth is simpler: if your conception of justice conflicts with the demands of logic or rationality then it is unjust. If your conception of rationality conflicts with a mathematical theorem (or, in this case, with many theorems) then your conception of rationality is irrational. To stick stubbornly to logically impossible values not only guarantees failure in the narrow sense that one can never meet them, it also forces one to reject optimism ('every evil is due to lack of knowledge'), and so deprives one of the means to make progress. Wishing for something that is logically impossible is a sign that there is something better to wish for. Moreover, if my conjecture in Chapter 8 is true, an impossible wish is ultimately *uninteresting* as well.

We need something better to wish for. Something that is not incompatible with logic, reason or progress. We have already encountered it. It is the basic condition for a political system to be capable of making sustained progress: Popper's criterion that the system facilitate the removal of bad policies and bad governments without violence. That entails abandoning 'who should rule?' as a criterion for judging political systems. The entire controversy about apportionment rules and all other issues in social-choice theory has traditionally been framed by all concerned in terms of 'who should rule?': what is the right number of seats for each state, or for each political party? What does the group – presumed entitled to rule over its subgroups and individuals – 'want', and what institutions will get it what it 'wants'?

So let us reconsider collective decision-making in terms of Popper's criterion instead. Instead of wondering earnestly which of the self-evident yet mutually inconsistent criteria of fairness, representativeness and so on are the most self-evident, so that they can be entrenched, we judge such criteria, along with all other actual or proposed political institutions, according to how well they promote the removal of bad rulers and bad policies. To do this, they must embody traditions of peaceful, critical discussion – of rulers, policies and the political institutions themselves.

In this view, any interpretation of the democratic process as merely a way of consulting the people to find out who should rule or what policies to implement misses the point of what is happening. An election does not play the same role in a rational society as consulting an oracle or a priest, or obeying orders from the king, did in earlier societies. The essence of democratic decision-making is not the choice made by the system at elections, but the ideas created between elections. And elections are merely one of the many institutions whose function is to allow such ideas to be created, tested, modified and rejected. The voters are not a fount of wisdom from which the right policies can be empirically 'derived'. They are attempting, fallibly, to explain the world and thereby to improve it. They are, both individually and collectively, seeking the truth – or should be, if they are rational. And there *is* an objective truth of the matter. Problems are soluble. Society is not a zero-sum game: the civilization of the Enlightenment did not get where it is today by cleverly sharing out the wealth, votes or anything else that was in dispute when it began. It got here by creating *ex nihilo*. In particular, what voters are doing in elections is not synthesizing a decision of a superhuman being, 'Society'. They are choosing which experiments are to be attempted next, and (principally) which are to be abandoned because there is no longer a good explanation for why they are best. The politicians, and their policies, are those experiments.

When one uses no-go theorems such as Arrow's to model real decision-making, one has to assume – quite unrealistically – that none of the decision-makers in the group is able to persuade the others to modify their preferences, or to create new preferences that are easier to agree on. The realistic case is that neither the preferences nor the options need be the same at the end of a decision-making process as they were at the beginning.

Why don't they just . . . fix social-choice theory by including creative processes such as explanation and persuasion in its mathematical model of decision-making? Because it is not known how to model a creative process. Such a model would *be* a creative process: an AI.

The conditions of 'fairness' as conceived in the various social-choice problems are misconceptions analogous to empiricism: they are all about the *input* to the decision-making process – who participates, and how their opinions are integrated to form the 'preference of the group'. A rational analysis must concentrate instead on how the rules and institutions contribute to the *removal* of bad policies and rulers, and to the creation of new options.

Sometimes such an analysis does endorse one of the traditional requirements, at least in part. For instance, it is indeed important that no member of the group be privileged or deprived of representation. But this is not so that all members can contribute to the answer. It is because such discrimination entrenches in the system a preference among their potential *criticisms*. It does not make sense to *include* everyone's favoured policies, or parts of them, in the new decision; what is necessary for progress is to *exclude* ideas that fail to survive criticism, and to prevent their entrenchment, and to promote the creation of new ideas.

Proportional representation is often defended on the grounds that it leads to coalition governments and compromise policies. But compromises – amalgams of the policies of the contributors – have an undeservedly high reputation. Though they are certainly better than immediate violence, they are generally, as I have explained, bad policies. If a policy is no one's idea of what will work, then why should it work? But that is not the worst of it. The key defect of compromise policies is that when one of them is implemented and fails, no one learns anything because no one ever agreed with it. Thus compromise policies shield the underlying explanations which *do* at least seem good to some faction from being criticized and abandoned.

The system used to elect members of the legislatures of most countries in the British political tradition is that each district (or 'constituency') in the country is entitled to one seat in the legislature, and that seat goes to the candidate with the largest number of votes in that district. This is called the *plurality voting* system ('plurality' meaning 'largest number of votes') – often called the 'first-past-the-post' system, because there is no prize for any runner-up, and no second round of voting (both of which feature in other electoral systems for the sake of increasing the proportionality of the outcomes). Plurality voting typically 'over-represents' the two largest parties, compared with the proportion of votes they receive. Moreover, it is not guaranteed to avoid the population paradox, and is even capable of bringing one party to power when another has received far more votes in total.

These features are often cited as arguments against plurality voting and in favour of a more proportional system – either literal proportional representation or other schemes such as transferable-vote systems and run-off systems which have the effect of making the representation of voters in the legislature more proportional. However, under Popper's criterion, that is all insignificant in comparison with the greater effectiveness of plurality voting at removing bad governments and policies.

Let me trace the mechanism of that advantage more explicitly. Following a plurality-voting election, the usual outcome is that the party with the largest total number of votes has an overall majority in the legislature, and therefore takes sole charge. All the losing parties are removed entirely from power. This is rare under proportional representation, because some of the parties in the old coalition are usually needed in the new one. Consequently, the logic of plurality is that politicians and political parties have little chance of gaining any share in power unless they can persuade a substantial proportion of the population to vote for them. That gives all parties the incentive to find better explanations, or at least to convince more people of their existing ones, for if they fail they will be relegated to powerlessness at the next election.

In the plurality system, the winning explanations are then exposed to criticism and testing, because they can be implemented without mixing them with the most important claims of opposing agendas. Similarly, the winning *politicians* are solely responsible for the choices they make, so they have the least possible scope for making excuses later if those are deemed to have been bad choices. If, by the time of the next election, they are less convincing to the voters than they were, there is usually no scope for deals that will keep them in power regardless.

Under a proportional system, small changes in public opinion seldom count for anything, and power can easily shift in the opposite direction to public opinion. What counts most is changes in the opinion of the leader of the third-largest party. This shields not only that leader but *most* of the incumbent politicians and policies from being removed from power through voting. They are more often removed by losing support within their own party, or by shifting alliances between parties. So in that respect the system badly fails Popper's criterion. Under plurality voting, it is the other way round. The all-or-nothing nature of the constituency elections, and the consequent low representation of small parties, makes the overall outcome sensitive to small changes in opinion. When there is a small shift in opinion away from the ruling party, it is usually in real danger of losing power completely.

Under proportional representation, there are strong incentives for the system's characteristic unfairnesses to persist, or to become worse, over time. For example, if a small faction defects from a large party, it may then end up with more chance of having its policies tried out than it would if its supporters had remained within the original party. This results in a proliferation of small parties in the legislature, which in turn increases the necessity for coalitions – including coalitions with the smaller parties, which further increases their disproportionate power. In Israel, the country with the world's most proportional electoral system, this effect has been so severe that, at the time of writing, even the two largest parties combined cannot muster an overall majority. And yet, under that system – which has sacrificed all other consider-ations in favour of the supposed fairness of proportionality – even proportionality itself is not always achieved: in the election of 1992, the right-wing parties as a whole received a majority of the popular vote, but the left-wing ones had a majority of the seats. (That was because a greater proportion of the fringe parties that failed to reach the threshold for receiving even one seat were right-wing.)

In contrast, the error-correcting attributes of the plurality voting system have a tendency to avoid the paradoxes to which the system is theoretically prone, and quickly to undo them when they do happen, because all those incentives are the other way round. For instance, in the Canadian province of Manitoba in 1926, the Conservative Party received more than twice as many votes as any other party, but won *none* of the seventeen seats allocated to that province. As a result it lost power in the national Parliament despite having received the most votes nationally too. And yet, even in that rare, extreme case, the disproportionateness between the two main parties' representations in Parliament was not that great: the average Liberal voter received 1.31 times as many members of Parliament as the average Conservative one. And what happened next? In the following election the Conservative Party again had the largest number of votes nationally, but this time that gave it an overall majority in Parliament. Its vote had increased by 3 per cent of the electorate, but its representation had increased by 17 per cent of the total number of seats, bringing the parties' shares of seats back into rough proportionality and satisfying Popper's criterion with flying colours.

This is partly due to yet another beneficial feature of the plurality system, namely that elections are often very close, in terms of votes as well as in the sense that all members of the government are at serious risk of being removed. In proportional systems, elections are rarely close in either sense. What is the point of giving the party with the most votes the most seats, if the party with the third-largest number of seats can then put the second-largest party in power regardless – there to enact a compromise platform that absolutely no one voted for? The plurality voting system almost always produces situations in which a small change in the vote produces a relatively large change (in the same direction!) in who forms a government. The more proportional a system is, the less sensitive the content of the resulting government and its policies are to changes in votes.

Unfortunately there are political phenomena that can violate Popper's criterion even more strongly than bad electoral systems – for example, entrenched racial divisions, or various traditions of political violence. Hence I do not intend the above discussion of electoral systems to constitute a blanket endorsement of plurality voting as the One True System of democracy, suitable for all polities under all circumstances. Even democracy itself is unworkable under some circumstances. But in the advanced political cultures of the Enlightenment tradition the creation of knowledge can and should be paramount, and the idea that representative government depends on proportionate representation in the legislature is unequivocally a mistake.

In the United States' system of government, the Senate is required to be representative in a different sense from the House of Representatives: *states* are represented equally, in recognition of the fact that each state is a separate political entity with its own distinctive political and legal tradition. Each of them is entitled to two Senate seats, regardless of population. Because the states differ greatly in their populations (currently the most populous state, California, has nearly seventy times the population of the least populous, Wyoming), the Senate's apportionment rule creates enormous deviations from population-based proportionality – much larger than those that are so hotly disputed in regard to the House of Representatives. And yet historically, after elections, it is rare for the Senate and the House of Representatives to be controlled by different parties. This suggests that there is more going on in this vast process of apportionments and elections than merely 'representation' – the mirroring of the population by the legislature. Could it be that the problem-solving that is promoted by the plurality voting system is continually changing the *options* of the voters, and also their *preferences* among the options, through persuasion? And so opinions and preferences are, despite appearances, *converging* – not in the sense of there being less disagreement (since solutions create new problems), but in the sense of creating ever more shared knowledge.

In science, we do not consider it surprising that a community of scientists with different initial hopes and expectations, continually in dispute about their rival theories, gradually come into near-unanimous agreement over a steady stream of issues (yet still continue to disagree all the time). It is not surprising because, in their case, there are observable facts that they can use to test their theories. They converge with each other on any given issue because they are all converging on the objective truth. In politics it is customary to be cynical about that sort of convergence being possible.

But that is a pessimistic view. Throughout the West, a great deal of philosophical knowledge that is nowadays taken for granted by almost everyone – say, that slavery is an abomination, or that women should be free to go out to work, or that autopsies should be legal, or that promotion in the armed forces should not depend on skin colour – was highly controversial only a matter of decades ago, and originally the opposite positions were taken for granted. A successful truth-seeking system works its way towards broad consensus or near-unanimity – the one state of public opinion that is not subject to decision-theoretic paradoxes and where 'the will of the people' makes sense. So convergence in the broad consensus over time is made possible by the fact that all concerned are gradually eliminating errors in their positions and converging on objective truths. Facilitating that process – by meeting Popper's criterion as well as possible – is more important than which of two contending factions with near-equal support gets its way at a particular election.

In regard to the apportionment issue too, since the United States'

Constitution was instituted there have been enormous changes in the prevailing conception of what it means for a government to be 'representative'. Recognizing the right of women to vote, for instance, doubled the number of voters – and implicitly admitted that in every previous election half the population had been disenfranchised, and the other half over-represented compared with a just representation. In numerical terms, such injustices dwarf all the injustices of apportionment that have absorbed so much political energy over the centuries. But it is to the credit of the political system, and of the people of the United States and of the West in general, that, while they were fiercely debating the fairness of shifting a few percentage points' worth of representation between one state and another, they were also debating, and making, these momentous improvements. And they too became uncontroversial.

Apportionment systems, electoral systems and other institutions of human cooperation were for the most part designed, or evolved, to cope with day-to-day controversy, to cobble together ways of proceeding without violence despite intense disagreement about what would be best. And the best of them succeed as well as they do because they have, often unintentionally, implemented solutions with enormous reach. Consequently, coping with controversy in the present has become merely a means to an end. The purpose of deferring to the majority in democratic systems should be to approach unanimity in the future, by giving all concerned the incentive to abandon bad ideas and to conjecture better ones. Creatively *changing the options* is what allows people in real life to cooperate in ways that no-go theorems seem to say are impossible; and it is what allows individual minds to choose at all.

The growth of the body of knowledge about which there is unanimous agreement does not entail a dying-down of controversy: on the contrary, human beings will never disagree any less than they do now, and that is a very good thing. If those institutions do, as they seem to, fulfil the hope that it is possible for changes to be for the better, on balance, then human life can improve without limit as we advance from misconception to ever better misconception.

TERMINOLOGY

Representative government   A system of government in which the composition or opinions of the legislature reflect those of the people.

Social-choice theory   The study of how the 'will of society' can be defined in terms of the wishes of its members, and of what social institutions can cause society to enact its will, thus defined.

Popper's criterion   Good political institutions are those that make it as easy as possible to detect whether a ruler or policy is a mistake, and to remove rulers or policies without violence when they are.

MEANINGS OF 'THE BEGINNING OF INFINITY' ENCOUNTERED IN THIS CHAPTER

- Choice that involves creating new options rather than weighing existing ones.
- Political institutions that meet Popper's criterion.

SUMMARY

It is a mistake to conceive of choice and decision-making as a process of selecting from existing options according to a fixed formula. That omits the most important element of decision-making, namely the creation of new options. Good policies are hard to vary, and therefore conflicting policies are discrete and cannot be arbitrarily mixed. Just as rational thinking does not consist of weighing the justifications of rival theories, but of using conjecture and criticism to seek the best explanation, so coalition governments are not a desirable objective of electoral systems. They should be judged by Popper's criterion of how easy they make it to remove bad rulers and bad policies. That designates the plurality voting system as best in the case of advanced political cultures.

14

Why are Flowers Beautiful?

My daughter Juliet, then aged six . . . pointed out some flowers by the wayside. I asked her what she thought wildflowers were for. She gave a rather thoughtful answer. 'Two things,' she said. 'To make the world pretty, and to help the bees make honey for us.' I was touched by this and sorry I had to tell her that it wasn't true.

Richard Dawkins, *Climbing Mount Improbable* (1996)

'Displace one note and there would be diminishment. Displace one phrase and the structure would fall.' That is how Mozart's music is described in Peter Shaffer's 1979 play *Amadeus*. This is reminiscent of the remark by John Archibald Wheeler with which this book begins, speaking of a hoped-for unified theory of fundamental physics: 'an idea so simple, so beautiful, that when we grasp it . . . how could it have been otherwise?'

Shaffer and Wheeler were describing the same attribute: being hard to vary while still doing the job. In the first case it is an attribute of aesthetically good music, and in the second of good scientific explanations. And Wheeler speaks of the scientific theory as being *beautiful* in the same breath as describing it as hard to vary.

Scientific theories are hard to vary because they correspond closely with an objective truth, which is independent of our culture, our personal preferences and our biological make-up. But what made Peter Shaffer think that Mozart's music is hard to vary? The prevailing view among both artists and non-artists is, I think, that there is nothing objective about artistic standards. Beauty, says the adage, is in the eye of the beholder. The very phrase 'It's a matter of taste' is used interchangeably with 'There is no objective truth of the matter.' Artistic standards are, in this view, nothing more than artefacts of fashion and other cultural accidents, or of individual whim, or of biological predisposition. Many are willing to concede that in science and mathematics one idea can be objectively truer than another (though, as we have seen, some deny even that), but most insist that there is no such thing as one object being objectively more beautiful than another. Mathematics has its proofs (so the argument goes), and science has its experimental tests; but if you choose to believe that Mozart was an inept and cacophonous composer then neither logic nor experiment nor anything else objective will ever contradict you.

However, it would be a mistake to dismiss the possibility of objective beauty for that sort of reason, for it is none other than the relic of empiricism that I discussed in Chapter 9 – the assertion that philosophical knowledge in general cannot exist. It is true that, just as one cannot *deduce* moral maxims from scientific theories, likewise nor can one deduce aesthetic values. But that would not prevent aesthetic truths from being linked to physical facts through explanations, as moral ones are. Wheeler was very nearly asserting such a link in that quotation.

Facts can be used to criticize aesthetic theories, as they can moral theories. For instance, there is the criticism that, since most arts depend on parochial properties of human senses (such as which range of colours and sounds they can detect), they cannot be attaining anything objective. Extraterrestrial people whose senses detected radio waves but not light or sound would have art that was inaccessible to us, and vice versa. And the reply to that criticism might be, first, that perhaps our arts are merely scratching the surface of what is possible: they are indeed parochial, but they are a first approximation to something universal. Or, second, that deaf composers on Earth have composed, and appreciated, great music; why could deaf extraterrestrials (or humans who were born deaf) not learn to do the same – if by no other means than by downloading a set of deaf-composer aesthetics into their brains? Or, third, what is the difference between using radio telescopes to understand the physics of quasars and using prosthetic senses (wired into the brain to create new qualia) to appreciate extra-terrestrial art?

Experience may also provide artistic *problems*. Our ancestors had eyes and paint, which may have led them to wonder how paint could be used in a way that would look more beautiful.

Just as Bronowski pointed out that scientific discovery depends on a commitment to certain moral values, might it not also entail the appreciation of certain forms of beauty? It is a fact – often mentioned but seldom explained – that deep truth is often beautiful. Mathematicians and theoretical scientists call this form of beauty 'elegance'. Elegance is the beauty in explanations. It is by no means synonymous with how good, or how true, an explanation is. The poet John Keats' assertion (which I think was ironic) that 'Beauty is truth, truth beauty' is refuted by what the evolutionist Thomas Huxley called 'the great tragedy of Science – the slaying of a beautiful hypothesis by an ugly fact – which is so constantly being enacted under the eyes of philosophers'. (By 'philosophers' he meant 'scientists'.) I think Huxley, too was being ironic in calling this process a great tragedy, especially since he was referring to the refutation of spontaneous-generation theories. But it is true that some important mathematical proofs, and some scientific theories, are far from elegant. Yet the truth so often *is* elegant that elegance is, at least, a useful heuristic when searching for fundamental truths. And when a 'beautiful hypothesis' is slain, it is more often than not replaced, as the spontaneous-generation theory was, by a more beautiful one. Surely this is not coincidence: it is a regularity in nature. So it must have an explanation.

The processes of science and art can look rather different: a new artistic creation rarely proves an old one wrong; artists rarely look at a scene through microscopes, or understand a sculpture through equations. Yet scientific and artistic creation do sometimes look remarkably alike. Richard Feynman once remarked that the only equipment a theoretical physicist needs is a stack of paper, a pencil and a waste-paper basket, and some artists, when they are at work, closely resemble that picture. Before the invention of the typewriter, novelists used exactly the same equipment.

Composers like Ludwig van Beethoven agonized through change after change, apparently seeking something that they knew was there to be created, apparently meeting a standard that could be met only after much creative effort and much failure. Scientists often do the same. In both science and art there are the exceptional creators like Mozart or the mathematician Srinivasa Ramanujan, who reputedly made brilliant contributions without any such effort. But from what we know of knowledge-creation we have to conclude that in such cases the effort, and the mistakes, did happen, invisibly, inside their brains.

Are these resemblances only superficial? Was Beethoven fooling himself when he thought that the sheets in his waste-paper basket contained *mistakes*: that they were *worse* than the sheets he would eventually publish? Was he merely meeting the arbitrary standards of his culture, like the twentieth-century women who carefully adjusted their hemlines each year to conform to the latest fashions? Or is there a real meaning to saying that the music of Beethoven and Mozart was as far above that of their Stone Age ancestors banging mammoth bones together as Ramanujan's mathematics was above tally marks?

Is it an illusion that the *criteria* that Beethoven and Mozart were trying to meet were better too? Or is there no such thing as better? Is there only 'I know what I like,' or what tradition or authority designates as good? Or what our genes predispose us to like? The psychologist Shigeru Watanabe has found that sparrows prefer harmonious to discordant music. Is that all that human artistic appreciation is?

All these theories assume – with little or no argument – that for each logically possible aesthetic standard there could exist, say, a culture in which people would enjoy and be deeply moved by art that met that standard. Or that a genetic predisposition could exist with the same properties. But is it not much more plausible that only very exceptional aesthetic standards could possibly end up as the norm of any culture, or be the objective towards which some great artist, creating a new artistic style, spent a lifetime working? Quite generally, cultural relativism (about art or morality) has a very hard time explaining what people are doing when they think they are improving a tradition.

Then there is the equivalent of instrumentalism: is art no more than a means to non-artistic ends? For instance, artistic creations can deliver information – a painting can depict something, and a piece of music can represent an emotion. But their beauty is not primarily in that content. It is in the form. For instance, here is a boring picture:

and here is another picture with much the same content:

yet with greater aesthetic value. One can see that someone thought about the second picture. In its composition, framing, cropping, lighting, focus – it has the *appearance of design* by the photographer. But design for what? Unlike Paley's watch, it does not seem to have a function – it only seems to be more beautiful than the first picture. But what does that mean?

One possible instrumental purpose of beauty is *attraction*. A beautiful object can be attractive to people who appreciate the beauty. Attractiveness (to a given audience) can be functional, and is a down-to-earth, scientifically measurable quantity. Art can be literally attractive in the sense of causing people to move towards it. Visitors to an art gallery can see a painting and be reluctant to leave, and then later be caused, by the painting, to return to it. People may travel great distances to hear a musical performance – and so on. If you see a work of art that

357

you appreciate, that means that you want to dwell on it, to give it your attention, in order to appreciate more in it. If you are an artist, and halfway through creating a work of art you see something in it that you want to bring out, then you are being attracted by a beauty that you have not yet experienced. You are being attracted by the *idea* of a piece of art before you have created it.

Not all attractiveness has anything to do with aesthetics. You lose your balance and fall off a log because we are all attracted to the planet Earth. That may seem merely a play on the word 'attraction': our attraction to the Earth is due not to aesthetic appreciation but to a law of physics, which affects artists no more than it does aardvarks. A red traffic light may induce us to stop and stare at it so long as it remains red. But that is not artistic appreciation either, even though it is attraction. It is mechanical.

But, when analysed in sufficient detail, *everything* is mechanical. The laws of physics are sovereign. So, can one draw the conclusion that beauty cannot have an objective meaning other than 'that which we are attracted to by processes in our brains and hence by the laws of physics'? One cannot, because by that argument the physical world would not exist objectively either, since the laws of physics also determine what a scientist or mathematician wants to call true. Yet one cannot *explain* what a mathematician does – or what Hofstadter's dominoes do – without referring to the objective truths of mathematics.

New art is unpredictable, like new scientific discoveries. Is that the unpredictability of randomness, or the deeper unknowability of knowledge-creation? In other words, is art truly creative, like science and mathematics? That question is usually asked the other way round, because the idea of creativity is still rather confused by various misconceptions. Empiricism miscasts science as an automatic, non-creative process. And art, though acknowledged as 'creative', has often been seen as the antithesis of science, and hence irrational, random, inexplicable – and hence unjudgeable, and non-objective. But if beauty *is* objective, then a new work of art, like a newly discovered law of nature or mathematical theorem, adds something irreducibly new to the world.

We stare at the red traffic light because doing so will allow us to continue our journey with the least possible delay. An animal can be attracted towards another animal in order to mate with it, or to eat it. Once the predator has taken a bite, it is attracted to take another – unless the bite tastes bad, in which case it will be repelled. So there we have a literal matter of taste. And that matter of taste is indeed caused by the laws of physics in the form of the laws of chemistry and biochemistry. We can guess that there is no higher-level explanation of the resulting behaviour than the zoological level, because the behaviour is predictable. It is repetitive, and where it is not repetitive it is random.

Art does not consist of repetition. But in human tastes there can be genuine novelty. Because we are universal explainers, we are not simply obeying our genes. For instance, humans often act in ways that are contrary to any preferences that might plausibly have been built into our genes. People fast – sometimes for aesthetic reasons. Some abstain from sex. People act in very diverse ways for religious reasons or for any number of other reasons, philosophical or scientific, practical or whimsical. We have an inborn aversion to heights and to falling, yet people go skydiving – not in spite of this feeling, but because of it. It is that very feeling of inborn aversion that humans can reinterpret into a larger picture which to them is attractive – they want more of it; they want to appreciate it more deeply. To a skydiver, the vista from which we were born to recoil is beautiful. The whole activity of skydiving is beautiful, and part of that beauty is in the very sensations that evolved to deter us from trying it. The conclusion is inescapable: that attraction is not inborn, just as the contents of a newly discovered law of physics or mathematical theorem are not inborn.

Could it be purely cultural? We pursue beauty as well as truth, and in both cases we can be fooled. Perhaps we see a face as beautiful because it really is, or perhaps it is only because of a combination of our genes and our culture. A beetle is attracted to another beetle that you and I may see as hideous. But not if you are an entomologist. People can *learn* to see many things as beautiful or ugly. But, there again, people can also learn to see false scientific theories as true, and true ones as false, yet there is such a thing as objective scientific truth. So that still does not tell us whether there is such a thing as objective beauty.

Now, why is a flower the shape that it is? Because the relevant genes evolved to make it attractive to insects. Why would they do this? Because when insects visit a flower they are dusted with pollen, which they then deposit in other flowers of the same species, and so the genes in the DNA in that pollen are spread far and wide. This is the reproductive mechanism that flowering plants evolved and which most still use today: before there were insects, there were no flowers on Earth. But the mechanism could work only because insects, at the same time, evolved genes that attracted them to flowers. Why did they? Because flowers provide nectar, which is food. Just as there is co-evolution between the genes to coordinate mating behaviours in males and females of the same species, so genes for making flowers and giving them their shapes and colours co-evolved with genes in insects for recognizing flowers with the best nectar.

During that biological co-evolution, just as in the history of art, *criteria* evolved, and *means of meeting those criteria* co-evolved with them. That is what gave flowers the knowledge of how to attract insects, and insects the knowledge of how to recognize those flowers and the propensity to fly towards them. But what is surprising is that these same flowers *also attract humans*.

This is so familiar a fact that it is hard to see how amazing it is. But think of all the countless hideous animals in nature, and think also that all of them who find their mates by sight have evolved to find that appearance attractive. And therefore it is not surprising that we do not. With predators and prey there is a similar co-evolution, but in a competitive sense rather than a cooperative one: each has genes that evolved to enable it to recognize the other and to make it run towards or away from it respectively, while other genes evolve to make their organism hard to recognize against the relevant background. That is why tigers are striped.

Occasionally it happens by chance that the parochial criteria of attractiveness that evolved within a species produce something that looks beautiful to us: the peacock's tail is an example. But that is a rare anomaly: in the overwhelming majority of species, we do not share any of their criteria for finding something attractive. Yet with flowers – most flowers – we do. Sometimes a leaf can be beautiful; even a puddle of water can. But, again, only by rare chance. With flowers it is reliable.

It is another regularity in nature. What is the explanation? Why are flowers beautiful?

Given the prevailing assumptions in the scientific community – which are still rather empiricist and reductionist – it may seem plausible that flowers are not objectively beautiful, and that their attractiveness is merely a cultural phenomenon. But I think that that fails closer inspection. We find flowers beautiful that we have never seen before, and which have not been known to our culture before – and quite reliably, for most humans in most cultures. The same is not true of the *roots* of plants, or the leaves. Why only the flowers?

One unusual aspect of the flower–insect co-evolution is that it involved the creation of a complex code, or language, for signalling information between *species*. It had to be complex because the genes were facing a difficult communication problem. The code had to be, on the one hand, easily recognizable by the right insects, and, on the other, difficult to forge by other species of flower – for if other species could cause their pollen to be spread by the same insects without having to manufacture nectar for them, which requires energy, they would have a selective advantage. So the criterion that was evolving in the insects had to be discriminating enough to pick the right flowers and not crude imitations; and the flowers' design had to be such that no design that other flower species could easily evolve could be mistaken for it. Thus both the criterion and the means of meeting it had to be hard to vary.

When genes are facing a similar problem *within* a species, notably in the co-evolution of criteria and characteristics for choosing mates, they already have a large amount of shared genetic knowledge to draw on. For instance, even before any such co-evolution begins, the genome may already contain adaptations for recognizing fellow members of the species, and for detecting various attributes of them. Moreover, the attributes that a mate is searching for may initially be objectively useful ones – such as neck length in a giraffe. One theory of the evolution of giraffe necks is that it began as an adaptation for feeding, but then continued through sexual selection. However, there is no such shared knowledge to build on across the gap between distant species. They are starting from scratch.

And therefore my guess is that the easiest way to signal across such a gap with hard-to-forge patterns designed to be recognized by hard-to-emulate pattern-matching algorithms is to use *objective* standards of beauty. So flowers have to create objective beauty, and insects have to recognize objective beauty. Consequently the only species that are attracted by flowers are the insect species that co-evolved to do so – and humans.

If true, this means that Dawkins' daughter was partly right about the flowers after all. They *are* there to make the world pretty; or, at least, prettiness is no accidental side effect but is what they specifically evolved to have. Not because anything intended the world to be pretty, but because the best-replicating genes depend on embodying *objective prettiness* to get themselves replicated. The case of honey, for instance, is very different. The reason that honey – which is sugar water – is easy for flowers and bees to make, and why its taste is attractive to humans and insects alike, is that we *do* all have a shared genetic heritage going back to our common ancestor and before, which includes biochemical knowledge about many uses of sugar, and the means to recognize it.

Could it be that what humans find attractive in flowers – or in art – is indeed objective, but it is not objective *beauty*? Perhaps it is something more mundane – something like a liking for bright colours, strong contrasts, symmetrical shapes. Humans seem to have an inborn liking for symmetry. It is thought to be a factor in sexual attractiveness, and it may also be useful in helping us to classify things and to organize our environment physically and conceptually. So a side effect of these inborn preferences might be a liking for flowers, which happen to be colourful and symmetrical. However, some flowers are white (at least to us – they may have colours that we cannot see and insects can), but we still find their shapes beautiful. All flowers do contrast with their background in some sense – that is a precondition for being used for signalling – but a spider in the bath contrasts with its background even more, and there is no widespread consensus that such a sight is beautiful. As for symmetry: again, spiders are quite symmetrical, while some flowers, such as orchids, are very unsymmetrical, yet we do not find them any less attractive for that. So I do not think that symmetry, colour and contrast are all that we are seeing in flowers when we imagine that we are seeing beauty.

A sort of mirror image of that objection is that there are other things in nature that we also find beautiful – things that are not results of either human creativity or co-evolution across a gap: the night sky; waterfalls; sunsets. So why not flowers too? But the cases are not alike. Those things may be attractive to look at, but they have no appearance of design. They are analogous not to Paley's watch, but to the sun as a timekeeper. One cannot explain why the watch is as it is without referring to timekeeping, because it would be useless for timekeeping if it had been made slightly differently. But, as I mentioned, the sun would still be useful for keeping time even if the solar system were altered. Similarly, Paley might have found a stone that looked attractive. He might well have taken it home to use as an ornamental paperweight. But he would not have sat down to write a monograph about how changing any detail of the stone would have made it incapable of serving that function, because that would not have been so. The same is true of the night sky, waterfalls and almost all other natural phenomena. But flowers do have the appearance of design for beauty: if they looked like leaves, or roots, they would lose their universal appeal. Displace even one petal, and there would be diminishment.

We know what the watch was designed for, but we do not know what beauty is. We are in a similar position to an archaeologist who finds inscriptions in an unknown language in an ancient tomb: they look like writing and not just meaningless marks on the walls. Conceivably this is mistaken, but they look as though they were inscribed there for a purpose. Flowers are like that: they have the appearance of having been evolved for a purpose which we call 'beauty', which we can (imperfectly) recognize, but whose nature is poorly understood.

In the light of these arguments I can see only one explanation for the phenomenon of flowers being attractive to humans, and for the various other fragments of evidence I have mentioned. It is that the attribute we call beauty is of two kinds. One is a parochial kind of attractiveness, local to a species, to a culture or to an individual. The other is unrelated to any of those: it is universal, and as objective as the laws of physics. Creating either kind of beauty requires knowledge; but the second kind requires knowledge with universal reach. It reaches all the way from the flower genome, with its problem of competitive pollination, to human minds which appreciate the resulting flowers as art. Not great art – human artists are far better, as is to be expected. But with the hard-to-fake appearance of design for beauty.

Now, why do *humans* appreciate objective beauty, if there has been no equivalent of that co-evolution in our past? At one level the answer is simply that we are universal explainers and can create knowledge about anything. But still, why did we want to create aesthetic knowledge in particular? It is because we *did* face the same problem as the flowers and the insects. Signalling across the gap between two humans is analogous to signalling across the gap between two entire species. A human being, in terms of knowledge content and creative individuality, is like a species. All the individuals of any other species have virtually the same programming in their genes and use virtually the same criteria for acting and being attracted. Humans are quite unlike that: the amount of information in a human mind is more than that in the genome of any species, and overwhelmingly more than the genetic information unique to one person. So human artists are trying to signal across the same scale of gap between humans as the flowers and insects are between species. They can use some species-specific criteria; but they can also reach towards objective beauty. Exactly the same is true of all our other knowledge: we can communicate with other people by sending predetermined messages determined by our genes or culture, or we can invent something new. But in the latter case, to have any chance of communicating, we had better strive to rise above parochialism and seek universal truths. This may be the proximate reason that humans ever began to do so.

One amusing corollary of this theory is, I think, that it is quite possible that human *appearance*, as influenced by human sexual selection, satisfies standards of objective beauty as well as species-specific ones. We may not be very far along that path yet, because we diverged from apes only a few hundred thousand years ago, so our appearance isn't yet all that different from that of apes. But I guess that when beauty is better understood it will turn out that most of the differences have been in the direction of making humans objectively more beautiful than apes.

The two types of beauty are usually created to solve two types of problem – which could be called pure and applied. The applied kind is that of signalling information, and is usually solved by creating the parochial type of beauty. Humans have problems of that type too: the beauty of, say, the graphical user interface of a computer is created primarily to promote comfort and efficiency in the machine's use. Sometimes a poem or song may be written for a similar practical purpose: to give more cohesiveness to a culture, or to advance a political agenda, or even to advertise beverages. Again, sometimes these purposes can also be met by creating *objective* beauty, but usually the parochial kind is used because it is easier to create.

The other kind of problem, the pure kind, which has no analogue in biology, is that of creating beauty for its own sake – which includes creating improved criteria for beauty: new artistic standards or styles. This is the analogue of pure scientific research. The states of mind involved in that sort of science and that sort of art are fundamentally the same. Both are seeking universal, objective truth.

And both, I believe, are seeking it through good explanations. This is most straightforwardly so in the case of art forms that involve stories – fiction. There, as I mentioned in Chapter 11, a good story has a good explanation of the fictional events that it portrays. But the same is true in all art forms. In some, it is especially hard to express in words the explanation of the beauty of a particular work of art, even if one knows it, because the relevant knowledge is itself not expressed in words – it is *inexplicit*. No one yet knows how to translate musical explanations into natural language. Yet when a piece of music has the attribute 'displace one note and there would be diminishment' there is an explanation: it was known to the composer, and it is known to the listeners who appreciate it. One day it will be expressible in words.

This, too, is not as different from science and mathematics as it looks: poetry and mathematics or physics share the property that they develop a language different from ordinary language in order to state things efficiently that it would be very inefficient to state in ordinary language. And both do this by constructing variants of ordinary language: one has to understand the latter first in order to understand explanations of, and in, the former.

Applied art and pure art 'feel' the same. And, just as we need sophisticated knowledge to tell the difference between the motion of a bird across the sky, which is happening objectively, and the motion of the sun across the sky, which is just a subjective illusion caused by our own motion, and the motion of the moon, which is a bit of each, so pure and applied art, universal and parochial beauty, are mixed together in our subjective appreciation of things. It will be important to discover which is which. For it is only in the objective direction that we can expect to make unlimited progress. The other directions are inherently finite. They are circumscribed by the finite knowledge inherent in our genes and our existing traditions.

That has a bearing on various existing theories of what art is. Ancient fine art, for instance in Greece, was initially concerned with the skill of reproducing the shapes of human bodies and other objects. That is not the same as the pursuit of objective beauty, because, among other things, it is perfectible (in the bad sense that it can reach a state that cannot be much improved on). But it is a skill that can allow artists to pursue pure art as well, and they did so in the ancient world, and then again during the revival of that tradition in the Renaissance.

There are utilitarian theories of the purpose of art. These theories deprecate pure art, just as pure science and mathematics are deprecated by the same arguments. But one has no choice about what constitutes an artistic improvement any more than one has a choice as to what is true and false in mathematics. And if one tries to tune one's scientific theories or philosophical positions to meet a political agenda, or a personal preference, then one is at cross purposes. Art can be *used* for many purposes. But artistic values are not subordinate to, or derived from, anything else.

The same critique applies to the theory that art is self-expression.

Expression is conveying something that is already there, while objective progress in art is about creating something new. Also, *self*-expression is about expressing something subjective, while pure art is objective. For the same reason, any kind of art that consists solely of spontaneous or mechanical acts, such as throwing paint on to canvas, or of pickling sheep, lacks the means of making artistic progress, because real progress is difficult and involves many errors for every success.

If I am right, then the future of art is as mind-boggling as the future of every other kind of knowledge: art of the future can create unlimited increases in beauty. I can only speculate, but we can presumably expect new kinds of unification too. When we understand better what elegance really is, perhaps we shall find new and better ways to seek truth using elegance or beauty. I guess that we shall also be able to design new senses, and design new qualia, that can encompass beauty of new kinds literally inconceivable to us now. 'What is it like to be a bat?' is a famous question asked by the philosopher Thomas Nagel. (More precisely, what would it be like for a person to have the echo-location senses of a bat?) Perhaps the full answer is that in future it will be not so much the task of philosophy to discover what that is like, but the task of technological art to give us the experience itself.

TERMINOLOGY

Aesthetics   The philosophy of beauty.
Elegance   The beauty in explanations, mathematical formulae and so on.
Explicit   Expressed in words or symbols.
Inexplicit   Not explicit.
Implicit   Implied or otherwise contained in other information.

MEANINGS OF 'THE BEGINNING OF INFINITY' ENCOUNTERED IN THIS CHAPTER

– The fact that elegance is a heuristic guide to truth.
– The need to create objective knowledge in order to allow different people to communicate.

SUMMARY

There are objective truths in aesthetics. The standard argument that there cannot be is a relic of empiricism. Aesthetic truths are linked to factual ones by explanations, and also because artistic problems can emerge from physical facts and situations. The fact that flowers reliably seem beautiful to humans when their designs evolved for an apparently unrelated purpose is evidence that beauty is objective. Those convergent criteria of beauty solve the problem of creating hard-to-forge signals where prior shared knowledge is insufficient to provide them.

15

The Evolution of Culture

Ideas that survive

A *culture* is a set of ideas that cause their holders to behave alike in some ways. By 'ideas' I mean any information that can be stored in people's brains and can affect their behaviour. Thus the shared values of a nation, the ability to communicate in a particular language, the shared knowledge of an academic discipline and the appreciation of a given musical style are all, in this sense, 'sets of ideas' that define cultures. Many of them are inexplicit; in fact all ideas have some inexplicit component, since even our knowledge of the meanings of words is held largely inexplicitly in our minds. Physical skills, such as the ability to ride a bicycle, have an especially high inexplicit content, as do philosophical concepts such as freedom and knowledge. The distinction between explicit and inexplicit is not always sharp. For instance, a poem or a satire may be explicitly about one subject, while the audience in a particular culture will reliably, and without being told, interpret it as being about a different one.

The world's major cultures – including nations, languages, philosophical and artistic movements, social traditions and religions – have been created incrementally over hundreds or even thousands of years. Most of the ideas that define them, including the inexplicit ones, have a long history of being passed from one person to another. That makes these ideas *memes* – ideas that are replicators.

Nevertheless, cultures change. People modify cultural ideas in their minds, and sometimes they pass on the modified versions. Inevitably, there are unintentional modifications as well, partly because of straightforward error, and partly because inexplicit ideas are hard to convey accurately: there is no way to download them directly from one brain

369

to another like computer programs. Even native speakers of a language will not give identical definitions of every word. So it can be only rarely, if ever, that two people hold precisely the same cultural idea in their minds. That is why, when the founder of a political or philosophical movement or a religion dies, or even before, schisms typically happen. The movement's most devoted followers are often shocked to discover that they disagree about what its doctrines 'really' are. It is not much different when the religion has a holy book in which the doctrines are stated explicitly: then there are disputes about the meanings of the words and the interpretation of the sentences.

Thus a culture is in practice defined not by a set of strictly identical memes, but by a set of variants that cause slightly different characteristic behaviours. Some variants tend to have the effect that their holders are eager to enact or talk about them, others less so. Some are easier than others for potential recipients to replicate in their own minds. These factors and others affect how likely each variant of a meme is to be passed on faithfully. A few exceptional variants, once they appear in one mind, tend to spread throughout the culture with very little change in meaning (as expressed in the behaviours that they cause). Such memes are familiar to us because long-lived cultures are composed of them; but, nevertheless, in another sense they are a very unusual type of idea, for most ideas are short-lived. A human mind considers many ideas for every one that it ever acts upon, and only a small proportion of those cause behaviour that anyone else notices – and, of those, only a small proportion are ever replicated by anyone else. So the over-whelming majority of ideas disappear within a lifetime or less. The behaviour of people in a long-lived culture is therefore determined partly by recent ideas that will soon become extinct, and partly by *long-lived memes*: exceptional ideas that have been accurately replicated many times in succession.

A fundamental question in the study of cultures is: what is it about a long-lived meme that gives it this exceptional ability to resist change throughout many replications? Another – central to the theme of this book – is: when such memes do change, what are the conditions under which they can change for the better?

The idea that cultures evolve is at least as old as that of evolution in biology. But most attempts to understand how they evolve have been based on misunderstandings of evolution. For example, the communist thinker Karl Marx believed that his theory of history was evolutionary because it spoke of a progression through historical stages determined by economic 'laws of motion'. But the real theory of evolution has nothing to do with predicting the attributes of organisms from those of their ancestors. Marx also thought that Darwin's theory of evolution 'provides a basis in natural science for the historical class struggle'. He was comparing his idea of inherent conflict between socio-economic classes with the supposed competition between biological species. Fascist ideologies such as Nazism likewise used garbled or inaccurate evolutionary ideas, such as 'the survival of the fittest', to justify violence. But in fact the competition in biological evolution is not between different species, but between *variants of genes within a species* – which does not resemble the supposed 'class struggle' at all. It *can* give rise to violence or other competition between species, but it can also produce cooperation (such as the symbiosis between flowers and insects) and all sorts of intricate combinations of the two.

Although Marx and the fascists assumed false theories of biological evolution, it is no accident that analogies between society and the biosphere are often associated with grim visions of society: the biosphere is a grim place. It is rife with plunder, deceit, conquest, enslavement, starvation and extermination. Hence those who think that cultural evolution is like that end up either opposing it (advocating a static society) or condoning that kind of immoral behaviour as necessary or inevitable.

Arguments by analogy are fallacies. Almost any analogy between any two things contains some grain of truth, but one cannot tell what that is until one has an independent explanation for what is analogous to what, and why. The main danger in the biosphere–culture analogy is that it encourages one to conceive of the human condition in a reductionist way that obliterates the high-level distinctions that are essential for understanding it – such as those between mindless and creative, determinism and choice, right and wrong. Such distinctions are meaningless at the level of biology. Indeed, the analogy is often drawn for the very purpose of debunking the common-sense idea of human beings as causal agents with the ability to make moral choices and to create new knowledge for themselves.

As I shall explain, although biological and cultural evolution are described by the same underlying theory, the mechanisms of transmission, variation and selection are all very different. That makes the resulting 'natural histories' different too. There is no close cultural analogue of a species, or of an organism, or a cell, or of sexual or asexual reproduction. Genes and memes are about as different as can be at the level of mechanisms, and of outcomes; they are similar only at the lowest level of explanation, where they are both *replicators* that embody *knowledge* and are therefore conditioned by the same fundamental principles that determine the conditions under which knowledge can or cannot be preserved, can or cannot improve.

Meme evolution

In the classic 1956 science-fiction story 'Jokester', by Isaac Asimov, the main character is a scientist studying jokes. He finds that, although most people do sometimes make witty remarks that are original, they never invent what he considers to be a fully fledged joke: a story with a plot and a punchline that causes listeners to laugh. Whenever they tell such a joke, they are merely repeating one that they have heard from someone else. So, where do jokes come from originally? Who creates them? The fictional answer given in 'Jokester' is far-fetched and need not concern us here. But the premise of the story is not so absurd: it really is plausible that some jokes were not created by anyone – that they evolved.

People tell each other amusing stories – some fictional, some factual. They are not jokes, but some become memes: they are interesting enough for the listeners to retell them to other people, and some of those people retell them in turn. But they rarely recite them word for word; nor do they preserve every detail of the content. Hence an often-retold story will come to exist in different versions. Some of those versions will be retold more often than others – in some cases because people find them amusing. When that is the main reason for retelling them, successive versions that remain in circulation will tend to be ever more amusing. So the conditions are there for evolution: repeated cycles of imperfect copying of information, alternating with selection. Eventually the story becomes amusing enough to make people laugh, and a fully fledged joke has evolved.

It is conceivable that a joke could evolve through variations that were not intended to improve upon the funniness. For example, people who hear a story can mishear or misunderstand aspects of it, or change it for pragmatic reasons, and in a small proportion of cases, by sheer luck, that will produce a funnier version of the story, which will then propagate better. If a joke has evolved in that way from a non-joke, it truly has no author. Another possibility is that most of the people who altered the amusing story on its way to becoming a joke *designed* their contributions, using creativity to make it funnier intentionally. In such cases, although the joke was indeed created by variation and selection, its funniness was the result of human creativity. In that case it would be misleading to say that 'no one created it.' It had many co-authors, each of whom contributed creative thought to the outcome. But it may still be that literally no one understands why the joke is as funny as it is, and hence that no one could create another joke of similar quality at will.

Although we do not know exactly how creativity works, we do know that it is itself an evolutionary process within individual brains. For it depends on conjecture (which is variation) and criticism (for the purpose of selecting ideas). So, somewhere inside brains, blind variations and selections are adding up to creative thought at a higher level of emergence.

The idea of memes has come in for a great deal of radical, and in my view mistaken, criticism to the effect that it is vague and pointless, or else tendentious. For example, when the ancient Greek religion was suppressed, but the stories of its gods continued to be told, though now only as fiction, were those stories still the same memes despite now causing new behaviours? When Newton's laws were translated into English from the original Latin, they caused different words to be spoken and written. Were they the same memes? But in fact such questions cast no doubt on the existence of memes, nor on the usefulness of the concept. It is like the controversy about which objects in the solar system should be called 'planets'. Is Pluto a 'real' planet even though it is smaller than some of the moons in our solar system? Is Jupiter really not a planet but an un-ignited star? It is not important. What is important is what is really there. And memes are really there, regardless of what we call them or how we classify them. Just as the basic theory of genes was developed long before the discovery of DNA, so today, without knowing *how* ideas are stored in brains, we do know that some ideas can be passed from one person to another and affect people's behaviour. Memes are those ideas.

Another line of criticism is that memes, unlike genes, are not stored in identical physical forms in every holder. But, as I shall explain, that does not necessarily make it impossible for memes to be transmitted 'faithfully' in the sense that matters for evolution. It is indeed meaningful to think of memes as retaining their identity as they pass from one holder to the next.

Just as genes often work together in groups to achieve what we might think of as a single adaptation, so there are memeplexes consisting of several ideas which can, alternatively, be thought of as a single more complex idea, such as quantum theory or neo-Darwinism. So it does not matter if we refer to a memeplex as a meme, just as it does not matter if we refer to quantum theory as a single theory or a group of theories. However, ideas, including memes, cannot be indefinitely analysed into sub-memes, because there comes a point where replacing a meme by part of itself would result in its not being copied. So, for instance, '2 + 3 = 5' is not a meme, because it does not have what it takes to cause itself reliably to be copied, except under circumstances which would also copy some theory of arithmetic with universal reach, which itself could not be transmitted without also transmitting the knowledge that $2 + 3 = 5$.

Laughing at a joke and retelling it are both behaviours caused by the joke, but we often do not know why we are enacting them. That reason is objectively there in the meme, but we do not know it. We may try to guess, but our guess will not necessarily be true. For instance, we may guess that the humour in a particular joke lay in the unexpectedness of its punchline. But further experience with the same joke may reveal that it remains funny when we hear it again. In such a case, we are in the counter-intuitive (but common) position of having been *mistaken about the reason for our own behaviour*.

The same sort of thing happens with rules of grammar. We say, 'I am learning to play *the* piano' (in British English), but never 'I am learning to play *the* baseball.' We know how to form such sentences correctly, but, until we think about it, very few of us know that the inexplicit grammatical rule we are following even exists, let alone what it is. In American English the rule is slightly different, so the phrase 'learning to play piano' is acceptable. We may wonder why, and guess that the British are more fond of the definite article. But, again, that is not the explanation: in British English a patient is 'in hospital', and in American English 'in *the* hospital'.

The same is true of memes in general: they implicitly contain information that is not known to the holders, but which nevertheless causes the holders to behave alike. Hence, just as native English speakers may be mistaken about why they have said 'the' in a given sentence, people enacting all sorts of other memes often give false explanations, even to themselves, of why they are behaving in that way.

Like genes, all memes contain knowledge (often inexplicit) of how to cause their own replication. This knowledge is encoded in strands of DNA or remembered by brains respectively. In both cases, the knowledge is *adapted* to causing itself to be replicated: it causes that more reliably than nearly all its variants do. In both cases, this adaptation is the outcome of alternating rounds of variation and selection.

However, the logic of the copying mechanism is very different for genes and memes. In organisms that reproduce by dividing, either all the genes are copied into the next generation or (if the individual fails to reproduce) none are. In sexual reproduction, a full complement of genes randomly chosen from both parents is copied, or none are. In all cases, the DNA duplication process is automatic: genes are copied indiscriminately. One consequence is that some genes can be replicated for many generations without ever being 'expressed' (causing any behaviour) at all. Whether your parents ever broke a bone or not, genes for repairing broken bones will (barring unlikely mutations) be passed on to you and your descendants.

The situation faced by memes is utterly different. Each meme has to be expressed as behaviour every time it is replicated. For it is that behaviour, and only that behaviour (given the environment created by all the other memes), that effects the replication. That is because a recipient cannot see the representation of the meme in the holder's mind. A meme cannot be downloaded like a computer program. If it is not enacted, it will not be copied.

The upshot of this is that memes necessarily become embodied in two different physical forms alternately: as memories in a brain, and as behaviour:

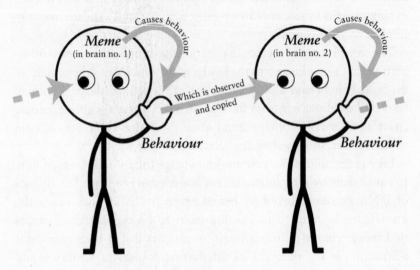

A meme exists in a brain form and a behaviour form, and each is copied to the other.

Each of the two forms has to be copied (specifically, translated into the other form) in each meme generation. (Meme 'generations' are simply successive instances of copying to another individual.) Technology can add further stages to a meme's life cycle. For instance, the behaviour may be to write something down – thus embodying the meme in a third physical form, which may later cause a person who reads it to enact other behaviour, which then causes the meme to appear in someone's brain. But all memes must have at least two physical forms.

In contrast, for genes the replicator exists in only one physical form – the DNA strand (of a germ cell). Even though it may be copied to other locations in the organism, translated into RNA, and expressed as behaviour, none of those forms is a replicator. The idea that the behaviour might be a replicator is a form of Lamarckism, since it implies that behaviours that had been modified by circumstances would be inherited.

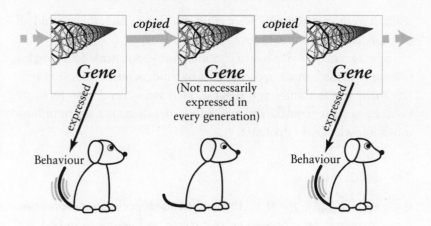

A gene exists in only one physical form, which is copied.

Because of the alternating physical forms of a meme, it has to survive two different, and potentially unrelated, mechanisms of selection in every generation. The brain-memory form has to cause the holder to enact the behaviour; and the behaviour form has to cause the new recipient to remember it – and to enact it.

So, for example, although religions prescribe behaviours such as educating one's children to adopt the religion, the mere intention to transmit a meme to one's children or anyone else is quite insufficient to make that happen. That is why the overwhelming majority of attempts to start a new religion fail, even if the founder members try hard to propagate it. In such cases, what has happened is that an idea that people have adopted has succeeded in causing them to enact various behaviours including ones intended to cause their children and others to do the same – but the behaviour has failed to cause the same idea to be stored in the minds of those recipients.

The existence of long-lived religions is sometimes explained from the premise that 'children are gullible', or that they are 'easily frightened' by tales of the supernatural. But that is not the explanation. The overwhelming majority of ideas simply do not have what it takes to persuade (or frighten or cajole or otherwise cause) children or anyone else into doing the same to other people. If establishing a faithfully replicating meme were that easy, the whole adult population in our society would be proficient at algebra, thanks to the efforts made to teach it to them when they were children. To be exact, they would all be proficient algebra *teachers*.

To be a meme, an idea has to contain quite sophisticated knowledge of how to cause humans to do at least two independent things: assimilate the meme faithfully, and enact it. That some memes can replicate themselves with great fidelity for many generations is a token of how much knowledge they contain.

The selfish meme

If a gene is in a genome at all, then, when suitable circumstances arise, it will definitely be expressed as an enzyme, as I described in Chapter 6, and will then cause its characteristic effects. Nor can it be left behind if the rest of its genome is successfully replicated. But merely being present in a mind does not automatically get a meme expressed as behaviour: the meme has to compete for that privilege with other ideas – memes and non-memes, about all sorts of subjects – in the same mind. And merely being expressed as behaviour does not automatically get the meme copied into a recipient along with other memes: it has to compete for the recipients' attention and acceptance with all sorts of behaviours by other people, and with the recipient's own ideas. All that is in addition to the analogue of the type of selection that genes face, each meme competing with rival versions of itself across the population, perhaps by containing the knowledge for some useful function.

Memes are subject to all sorts of random and intentional variation in addition to all that selection, and so they evolve. So to this extent the same logic holds as for genes: memes are 'selfish'. They do not necessarily evolve to benefit their holders, or their society – or, again, even themselves, except in the sense of replicating better than other memes. (Though now *most* other memes are their rivals, not just variants of themselves.) The successful meme variant is the one that changes the behaviour of its holders in such a way as to make itself best at displacing other memes from the population. This variant may well benefit its holders, or their culture, or the species as a whole. But if it harms them, or destroys them, it will spread anyway. Memes that harm society are a familiar phenomenon. You need only consider the harm done by adherents of political views, or religions, that you especially abhor. Societies have been destroyed because some of the memes that were best at spreading through the population were bad for a society. I shall discuss one example in Chapter 17. And countless individuals have been harmed or killed by adopting memes that were bad for them – such as irrational political ideologies or dangerous fads. Fortunately, in the case of memes, that is not the whole story. To understand the rest of the story, we have to consider the basic strategies by which memes cause themselves to be faithfully replicated.

Static societies

As I have explained, a human brain – quite unlike a genome – is itself an arena of intense variation, selection and competition. Most ideas within a brain are created by it for the very purpose of trying them out in imagination, criticizing them, and varying them until they meet the person's preferences. In other words, meme replication itself involves evolution, within individual brains. In some cases there can be thousands of cycles of variation and selection before any of the variants is ever enacted. Then, even after a meme has been copied into a new holder, it has not yet completed its life cycle. It still has to survive a further selection process, namely the holder's choice of whether to enact it or not.

Some of the criteria that a mind uses to make such choices are themselves memes. Some are ideas that it has created for itself (by altering memes, or otherwise), and which will never exist in any other mind. Such ideas are potentially highly variable between different people, yet they can decisively affect whether any given meme does or does not survive via a given person.

Since a person can enact and transmit a meme soon after receiving it, a meme generation can be much shorter than a human generation. And many cycles of variation and selection can take place inside the minds concerned even during one meme generation. Also, memes can be passed to people other than the holders' biological descendants. Those factors make meme evolution enormously faster than gene evolution, which partly explains how memes can contain so much knowledge. Hence the frequently cited metaphor of the history of life on Earth, in which human civilization occupies only the final 'second' of the 'day' during which life has so far existed, is misleading. In reality, a substantial proportion of all evolution on our planet to date has occurred in human brains. And it has barely begun. The whole of biological evolution was but a preface to the main story of evolution, the evolution of memes.

But, for the same reason, on the face of it meme replication is inherently less reliable than gene replication. Since the inexplicit content of memes cannot be literally copied but has to be guessed from the holders' behaviour, and since a meme can be subjected to large intentional variations inside every holder, it could be considered something of a miracle that any meme manages to be transmitted faithfully even once. And indeed the survival strategies of all long-lived memes are dominated by this problem.

Another way of stating the problem is that people think and try to improve upon their ideas – which entails changing them. A long-lived meme is an idea that runs that gauntlet again and again, and survives. How is that possible?

The post-Enlightenment West is the only society in history that for more than a couple of lifetimes has ever undergone change rapid enough for people to notice. Short-lived rapid changes have always happened: famines, plagues and wars have begun and ended; maverick kings have attempted radical change. Occasionally empires were rapidly created or whole civilizations were rapidly destroyed. But, while a society lasted, all important areas of life seemed changeless to the participants: they could expect to die under much the same moral values, personal lifestyles, conceptual framework, technology and pattern of economic production as they were born under. And, of the changes that did occur, few were for the better. I shall call such societies 'static societies': societies changing on a timescale unnoticed by the inhabitants. Before we can understand our unusual, dynamic sort of society, we must understand the usual, static sort.

For a society to be static, all its memes must be unchanging or changing too slowly to be noticed. From the perspective of our rapidly changing society, such a state of affairs is hard even to imagine. For instance, consider an isolated, primitive society that has, for whatever reason, remained almost unchanged for many generations. Why? Quite possibly no one in the society even wants it to change, because they can conceive of no other way of life. Nevertheless, its members are not immune from pain, hunger, grief, fear or other forms of physical and mental suffering. They try to think of ideas to alleviate some of that suffering. Some of those ideas are original, and occasionally one of them would actually help. It need be only a small, tentative improvement: a way of hunting or growing food with slightly less effort, or of making slightly better tools; a better way of recording debts or laws; a subtle change in the relationship between husband and wife, or between parent and child; a slightly different attitude towards the society's rulers or gods. What will happen next?

The person with that idea may well want to tell other people. Those who believe the idea will see that it could make life a little less nasty, brutish and short. They will tell their families and friends, and they theirs. This idea will be competing in people's minds with other ideas about how to make life better, most of them presumably false. But suppose, for the sake of argument, that this particular true idea happens to be believed, and spreads through the society.

Then the society will have been changed. It may not have changed very much, but this was merely the change caused by a single person, thinking of a single idea. So multiply all that by the number of thinking minds in the society, and by a lifetime's worth of thought in each of them, and let this continue for only a few generations, and the result is an exponentially increasing, revolutionary force transforming every aspect of the society.

But in a static society that beginning of infinity never happens. Despite the fact that I have assumed nothing other than that people try to improve their lives, and that they cannot transmit their ideas perfectly, and that information subject to variation and selection evolves, I have entirely failed to imagine a static society in this story.

For a society to be static, something else must be happening as well. One thing my story did not take into account is that static societies have customs and laws – taboos – that prevent their memes from changing. They enforce the enactment of the existing memes, forbid the enactment of variants, and suppress criticism of the status quo. However, that alone could not suppress change. First, *no* enactment of a meme is completely identical to that of the previous generation. It is infeasible to specify every aspect of acceptable behaviour with perfect precision. Second, it is impossible to tell in advance which small deviations from traditional behaviour would initiate further changes. Third, once a variant idea has begun to spread to even one more person – which means that people are preferring it – preventing it from being transmitted further is extremely difficult. Therefore no society could remain static solely by suppressing new ideas once they have been created.

That is why the enforcement of the status quo is only ever a secondary method of preventing change – a mopping-up operation. The primary method is always – and can only be – to disable the source of new ideas, namely human creativity. So static societies always have traditions of bringing up children in ways that disable their creativity and critical faculties. That ensures that most of the new ideas that would have been capable of changing the society are never thought of in the first place.

How is this done? The details are variable and not relevant here, but the sort of thing that happens is that people growing up in such a society acquire a set of values for judging themselves and everyone else which amounts to ridding themselves of distinctive attributes and seeking only conformity with the society's constitutive memes. They not only enact those memes: they see themselves as existing only in order to enact them. So, not only do such societies enforce qualities such as obedience, piety and devotion to duty, their members' sense of their own selves is invested in the same standards. People know no others. So they feel pride and shame, and form all their aspirations and opinions, by the criterion of how thoroughly they subordinate them-selves to the society's memes.

How do memes 'know' how to achieve all such complex, reproducible effects on the ideas and behaviour of human beings? They do not, of course, *know*: they are not sentient beings. They merely contain that knowledge implicitly. How did they come by that knowledge? It evolved. The memes exist, at any instant, in many variant forms, and those are subject to selection in favour of *faithful replication*. For every long-lived meme of a static society, millions of variants of it will have fallen by the wayside because they lacked that tiny extra piece of information, that extra degree of ruthless efficiency in preventing rivals from being thought of or acted upon, that slight advantage in psychological leverage, or whatever it took to make it spread through the population better than its rivals and, once it was prevalent, to get it copied and enacted with just that extra degree of fidelity. If ever a variant happened to be a little better at inducing behaviour with those self-replicating properties, it soon became prevalent. As soon as it did, there were again many variants of that variant, which were again subject to the same evolutionary pressure. Thus, successive versions of the meme accumulated knowledge that enabled them ever more reliably to inflict their characteristic style of damage on their human victims. Like genes, they may also confer benefits, though, even then, they are unlikely to do so optimally. Just as genes for the eye implicitly 'know' the laws of optics, so the long-lived memes of a static society implicitly possess knowledge of the human condition, and use it mercilessly to evade the defences and exploit the weaknesses of the human minds that they enslave.

A remark about timescales: Static societies, by this definition, are not perfectly unchanging. They are static on the timescale that humans can notice; but memes cannot prevent changes that are slower than that. So meme evolution still occurs in static societies, but too slowly for most members of the society to notice, most of the time. For instance, palaeontologists examining tools from the Old Stone Age cannot date them, by their shapes, to an accuracy better than many thousands of years, because tools at that time simply did not improve any faster than that. (Note that this is still much faster than biological evolution.) Examining a tool from the static society of ancient Rome or Egypt, one may be able to date it by its technology alone to the nearest century, say. But historians in the future examining cars and other technological artefacts of today will easily be able to date them to the nearest decade – and in the case of computer technology to the nearest year or less.

Meme evolution tends towards making *memes* static, but not necessarily whole societies. Like genes, memes do not evolve to benefit the group. Nevertheless, just as gene evolution can create long-lasting organisms and confer some benefits on them, so it is not surprising that meme evolution can sometimes create static societies, cooperate to keep them static, and help them to function by embodying truths. It is also not surprising that memes are often useful (though seldom optimally) to their holders. Just as organisms are the tools of genes, so individuals are used by memes to achieve their 'purpose' of spreading themselves through the population. And, to do this, memes sometimes confer benefits. One difference from the biological case, however, is that, while organisms are *nothing but* the slaves of all their genes, memes only ever control part of a person's thinking, even in the most slavishly static of societies. That is why some people use the metaphor of memes as *viruses* – which control part of the functionality of cells to propagate themselves. Some viruses do just install themselves into the host's DNA and do little else except participate in being copied from then on – but that is unlike memes, which *must* cause their distinctive behaviours and use knowledge to cause their own copying. Other viruses destroy their host cell – just as some memes destroy their holders: when someone commits suicide in a newsworthy way, there is often a spate of 'copycat suicides'.

The overarching selection pressure on memes is towards being faithfully replicated. But, within that, there is also pressure to do as little damage to the holder's mind as possible, because that mind is what the human uses to be long-lived enough to be able to enact the meme's behaviours as much as possible. This pushes memes in the direction of causing a *finely tuned* compulsion in the holder's mind: ideally, this would be just the inability to refrain from enacting that particular meme (or memeplex). Thus, for example, long-lived religions typically cause fear of specific supernatural entities, but they do not cause general fearfulness or gullibility, because that would both harm the holders in general and make them more susceptible to rival memes. So the evolutionary pressure is for the psychological damage to be confined to a relatively narrow area of the recipients' thinking, but to be deeply entrenched, so that the recipients find themselves facing a large emotional cost if they subsequently consider deviating from the meme's prescribed behaviours.

A static society forms when there is no escape from this effect: all significant behaviour, all relationships between people, and all thoughts are subordinated to causing faithful replication of the memes. In all areas *controlled* by the memes, no critical faculties are exercised. No innovation is tolerated, and almost none is attempted. This destruction of human minds makes static societies almost unimaginable from our perspective. Countless human beings, hoping throughout lifetimes, and for generations, for their suffering to be relieved, not only fail to make progress in realizing any such hope: they largely fail even to try to make any, or even to think about trying. If they do see an opportunity, they reject it. The spirit of creativity with which we are all born is systematically extinguished in them before it can ever create anything new.

A static society involves – in a sense *consists* of – a relentless struggle to prevent knowledge from growing. But there is more to it than that. For there is no reason to expect that a rapidly spreading idea, if one did happen to arise in a static society, would be true or useful. That is another aspect missing from my story of the static society above. I *assumed* that the change would be for the better. It might not have been, especially as the lack of critical sophistication in a static society would leave people vulnerable to false and harmful ideas from which their taboos did not protect them. For instance, when the Black Death plague destabilized the static societies of Europe in the fourteenth century, the new ideas for plague-prevention that spread best were extremely bad ones. Many people decided that this was the end of the world, and that therefore attempting any further earthly improvements was pointless. Many went out to kill Jews or 'witches'. Many crowded together in churches and monasteries to pray (thus unwittingly facilitating the spread of the disease, which was carried by fleas). A cult called the Flagellants arose, whose members devoted their lives to flogging them- selves, and to preaching all the above measures, in order to prove to God that his children were sorry. All these ideas were functionally harmful as well as factually false, and were eventually suppressed by the authorities in their drive to return to stasis.

Thus, ironically, there is much truth in the typical static-society fear that any change is much more likely to do harm than good. A static society is indeed in constant danger of being harmed or destroyed by a newly arising dysfunctional meme. However, in the aftermath of the Black Death a few true and functional ideas did also spread, and may well have contributed to ending that particular static society in an unusually good way (with the Renaissance).

Static societies survive by effectively eliminating the type of evolution that is unique to memes, namely creative variation intended to meet the holders' individual preferences. In the absence of that, meme evolution resembles gene evolution more closely, and some of the grim conclusions of the naive analogies between them apply after all. Static societies do tend to settle issues by violence, and they do tend to sacrifice the welfare of individuals for the 'good' of (that is to say, for the prevention of changes in) society. I mentioned that people who rely on such analogies end up either advocating a static society or condoning violence and oppression. We now see that those two responses are essentially the same: oppression is what it takes to keep a society static; oppression of a given kind will not last long unless the society is static.

Since the sustained, exponential growth of knowledge has unmistakable effects, we can deduce without historical research that every society on Earth before the current Western civilization has either been static or has been destroyed within a few generations. The golden ages of Athens and Florence are examples of the latter, but there may have been many others. This directly contradicts the widely held belief that individuals in primitive societies were happy in a way that has not been possible since – that they were unconstrained by social convention and other imperatives of civilization, and hence were able to achieve self-expression and fulfilment of their needs and desires. But primitive societies (including tribes of hunter-gatherers) must all have been static societies, because if ever one ceased to be static it would soon cease to be primitive, or else destroy itself by losing its distinctive knowledge. In the latter case, the growth of knowledge would still be inhibited by the raw violence which would immediately replace the static society's institutions. For once violence is mediating changes, they will typically not be for the better. Since static societies cannot exist without effectively extinguishing the growth of knowledge, they cannot allow their members much opportunity to pursue happiness. (Ironically, creating knowledge is itself a natural human need and desire, and static societies, however primitive, 'unnaturally' suppress it.) From the point of view of every individual in such a society, its creativity-suppressing mechanisms are catastrophically harmful. Every static society must leave its members chronically baulked in their attempts to achieve anything positive for themselves as people, or indeed anything at all, other than their meme-mandated behaviours. It can perpetuate itself only by suppressing its members' self-expression and breaking their spirits, and its memes are exquisitely adapted to doing this.

Dynamic societies

But our society (the West) is not a static society. It is the only known instance of a long-lived dynamic (rapidly changing) society. It is unique in history for its ability to mediate long-term, rapid, peaceful change and improvement, including improvements in the broad consensus about values and aims, as I described in Chapter 13. This has been made possible by the emergence of a radically different class of memes which, though still 'selfish', are not necessarily harmful to individuals.

To explain the nature of these new memes, let me pose the question: what sort of meme can cause itself to be replicated for long periods *in a rapidly changing environment*? In such an environment, people are continually being faced with unpredictable problems and opportunities. Hence their needs and wishes are changing unpredictably too. How can a meme remain unchanged under such a regime? The memes of a static society remain unchanged by effectively eliminating all the individuals' choices: people choose neither which ideas to acquire nor which to enact. Those memes also combine to make the society static, so that people's circumstances vary as little as possible. But once the stasis has broken down, and people are choosing, they will choose, in part, according to their individual circumstances and ideas, in which case memes will face selection criteria that vary unpredictably from recipient to recipient as well as over time.

To be transferred to a single person, a meme need seem useful only to that person. To be transferred to a group of similar people under unchanging circumstances, it need be only a parochial truth. But what sort of idea is best suited to getting itself adopted many times in succession by many people who have diverse, unpredictable objectives? A *true* idea is a good candidate. But not just any truth will do. It must seem useful to *all* those people, for it is they who will be choosing whether to enact it or not. 'Useful' in this context does not necessarily mean functionally useful: it refers to any property that can make people want to adopt an idea and enact it, such as being interesting, funny, elegant, easily remembered, morally right and so on. And the best way to *seem* useful to diverse people under diverse, unpredictable circumstances is to *be* useful. Such an idea is, or embodies, a truth in the broadest sense: factually true if it is an assertion of fact, beautiful if it is an artistic value or behaviour, objectively right if it is a moral value, funny if it is a joke, and so on.

The ideas with the best chance of surviving through many generations of change are truths with reach – deep truths. People are fallible; they often have preferences for false, shallow, useless or morally wrong ideas. But *which* false ideas they prefer differs from one person to another, and changes with time. Under changed circumstances, a specious falsehood or parochial truth can survive only by luck. But a true, deep idea has an objective reason to be considered useful by people with diverse purposes over long periods. For instance, Newton's laws are useful for building better cathedrals, but also for building better bridges and designing better artillery. Because of this reach, they get themselves remembered and enacted by all sorts of people, many of them vehemently opposed to each other's objectives, over many generations. This is the kind of idea that has a chance of becoming a long-lived meme in a rapidly changing society.

In fact such memes are not merely *capable* of surviving under rapidly changing criteria of criticism, they positively rely on such criticism for their faithful replication. Unprotected by any enforcement of the status quo or suppression of people's critical faculties, they are criticized, *but so are their rivals*, and the rivals fare worse, and are not enacted. In the absence of such criticism, true ideas no longer have that advantage and can deteriorate or be superseded.

Rational and anti-rational memes

Thus, memes of this new kind, which are created by rational and critical thought, subsequently also depend on such thought to get themselves replicated faithfully. So I shall call them *rational memes*. Memes of the older, static-society kind, which survive by disabling their holders' critical faculties, I shall call *anti-rational memes*. Rational and anti-rational memes have sharply differing properties, originating in their fundamentally different replication strategies. They are about as different from each other as they both are from genes.

If a certain type of hobgoblin has the property that, if children fear it, they will grow up to make *their* children fear it, then the behaviour of telling stories about that type of hobgoblin is a meme. Suppose it is a rational meme. Then criticism, over generations, will cast doubt on the story's truth. Since in reality there are no hobgoblins, the meme might evolve away to extinction. Note that it does not 'care' if it goes extinct. Memes do what they have to do: they have no intentions, even about themselves. But there are also other paths that it might evolve down. It might become overtly fictional. Because rational memes must be seen as beneficial by the holders, those that evoke unpleasant emotions are at a disadvantage, so it may also evolve away from evoking terror and towards, for instance, being pleasantly thrilling – or else (if it settled on a genuine danger) exploring practicalities for the present and optimism for the future.

Now suppose it is an anti-rational meme. Evoking unpleasant emotions will then be useful in doing the harm that it needs to do – namely disabling the listener's ability to be rid of the hobgoblin and entrenching the compulsion to think and therefore speak of it. The more accurately the hobgoblin's attributes exploit genuine, widespread vulnerabilities of the human mind, the more faithfully the anti-rational meme will propagate. If the meme is to survive for many generations, it is essential that its implicit knowledge of these vulnerabilities be true and deep. But its overt content – the idea of the hobgoblin's existence – need contain no truth. On the contrary, the non-existence of the hobgoblin helps to make the meme a better replicator, because the story is then unconstrained by the mundane attributes of any genuine menace, which are always finite and to some degree combatable. And that will be all the more so if the story can also manage to undermine the principle of optimism. Thus, just as rational memes evolve towards deep truths, anti-rational memes evolve away from them.

As usual, mixing the above two replication strategies does no good. If a meme contains true and beneficial knowledge for the recipient, but disables the recipient's critical faculties in regard to itself, then the recipient will be less able to correct errors in that knowledge, and so will reduce the faithfulness of transmission. And if a meme relies on the recipients' belief that it is beneficial, but it is not in fact beneficial, then that increases the chance that the recipient will reject it or refuse to enact it.

Similarly, a rational meme's natural home is a dynamic society – more or less any dynamic society – because there the tradition of criticism (optimistically directed at problem-solving) will suppress variants of the meme with even slightly less truth. Moreover, the rapid progress will subject these variants to continually varying criteria of criticism, which again only deeply true memes have a chance of surviving. An anti-rational meme's natural home is a static society – not any static society, but preferably the one in which it evolved – for all the converse reasons. And therefore each type of meme, when present in a society that is broadly of the *opposite* kind, is less able to cause itself to be replicated.

The Enlightenment

Our society in the West became dynamic not through the sudden failure of a static society, but through generations of static-society-type evolution. Where and when the transition began is not very well defined, but I suspect that it began with the philosophy of Galileo and perhaps became irreversible with the discoveries of Newton. In meme terms, Newton's laws replicated themselves as rational memes, and their fidelity was very high – because they were so useful for so many purposes. This success made it increasingly difficult to ignore the philosophical implications of the fact that nature had been understood in unprecedented depth, and of the methods of science and reason by which this had been achieved.

In any case, following Newton, there was no way of missing the fact that rapid progress was under way. (Some philosophers, notably Jean-Jacques Rousseau, did try – but only by arguing that reason was harmful, civilization bad and primitive life happy.) There was such an avalanche of further improvements – scientific, philosophical and political – that the possibility of resuming stasis was swept away. Western society would become the beginning of infinity or be destroyed. Nations beyond the West today are also changing rapidly, sometimes through the exigencies of warfare with their neighbours, but more often and even more powerfully by the peaceful transmission of Western memes. Their cultures, too, cannot become static again. They must either become 'Western' in their mode of operation or lose all their knowledge and thus cease to exist – a dilemma which is becoming increasingly significant in world politics.

Even in the West, the Enlightenment today is nowhere near complete. It is relatively advanced in a few, vital areas: the physical sciences and Western political and economic institutions are prime examples. In those areas ideas are now fairly open to criticism and experimentation, and to choice and change. But in many other areas memes are still replicated in the old manner, by means that suppress the recipients' critical faculties and ignore their preferences. When girls strive to be ladylike and to meet culturally defined standards of shape and appearance, and when boys do their utmost to look strong and not to cry when distressed, they are struggling to replicate ancient 'gender-stereotyping' memes that are still part of our culture – despite the fact that explicitly endorsing them has become a stigmatized behaviour. Those memes have the effect of preventing vast ranges of ideas about what sort of life one should lead from ever crossing the holders' minds. If their thoughts ever wander in the forbidden directions, they feel uneasiness and embarrassment, and the same sort of fear and loss of centredness as religious people have felt since time immemorial at the thought of betraying their gods. And their world views and critical faculties are left disabled in precisely such a way that they will in due course draw the next generation into the same pattern of thought and behaviour.

That anti-rational memes are still, today, a substantial part of our culture, and of the mind of every individual, is a difficult fact for us to accept. Ironically, it is harder for us than it would have been for the profoundly closed-minded people of earlier societies. They would not have been troubled by the proposition that most of their lives were spent enacting elaborate rituals rather than making their own choices and pursuing their own goals. On the contrary, the degree to which a person's life was controlled by duty, obedience to authority, piety, faith and so on was the very measure by which people judged themselves and others. Children who asked why they were required to enact onerous behaviours that did not seem functional would be told 'Because I say so', and in due course they would give their children the same reply to the same question, never realizing that they were giving the full explanation. (This is a curious type of meme whose explicit content is true though its holders do not believe it.) But today, with our eagerness for change and our unprecedented openness to new ideas and to self-criticism, it conflicts with most people's self-image that we are still, to a significant degree, the slaves of anti-rational memes. Most of us would admit to having a hang-up or two, but in the main we consider our behaviour to be determined by our own decisions, and our decisions by our reasoned assessment of the arguments and evidence about what is in our rational self-interest. This rational self-image is itself a recent development of our society, many of whose memes explicitly promote, and implicitly give effect to, values such as reason, freedom of thought, and the inherent value of individual human beings. We naturally try to explain ourselves in terms of meeting those values.

Obviously there is truth in this; but it is not the whole story. One need look no further than our clothing styles, and the way we decorate our homes, to find evidence. Consider how you would be judged by other people if you went shopping in pyjamas, or painted your home with blue and brown stripes. That gives a hint of the narrowness of the conventions that govern even these objectively trivial and inconsequential choices about style, and the steepness of the social costs of violating them. Is the same thing true of the more momentous patterns in our lives, such as careers, relationships, education, morality, political outlook and national identity? Consider what we should *expect* to happen when a static society is gradually switching from anti-rational to rational memes.

Such a transition is necessarily gradual, because keeping a dynamic society stable requires a great deal of knowledge. Creating that knowledge, starting with only the means available in a static society – namely small amounts of creativity and knowledge, many misconceptions, the blind evolution of memes, and trial and error – must necessarily take time.

Moreover, the society has to continue to function throughout. But the coexistence of rational and anti-rational memes makes this transition unstable. Memes of each type cause behaviours that impede the faithful replication of the other: to replicate faithfully, anti-rational memes need people to avoid thinking critically about their choices, while rational memes need people to think as critically as possible. That means that no memes in our society replicate as reliably as the most successful memes of either a very static society or an (as yet hypothetical) fully dynamic society. This causes a number of phenomena that are peculiar to our transitional era.

One of them is that some anti-rational memes evolve against the grain, towards rationality. An example is the transition from an autocratic monarchy to a 'constitutional monarchy', which has played a positive role in some democratic systems. Given the instability that I have described, it is not surprising that such transitions often fail.

Another is the formation within the dynamic society of anti-rational subcultures. Recall that anti-rational memes suppress criticism selectively and cause only finely tuned damage. This makes it possible for the members of an anti-rational subculture to function normally in other respects. So such subcultures can survive for a long time, until they are destabilized by the haphazard effects of reach from other fields. For example, racism and other forms of bigotry exist nowadays almost entirely in subcultures that suppress criticism. Bigotry exists not because it benefits the bigots, but despite the harm they do to themselves by using fixed, non-functional criteria to determine their choices in life.

Present-day methods of education still have a lot in common with their static-society predecessors. Despite modern talk of encouraging critical thinking, it remains the case that teaching by rote and inculcating standard patterns of behaviour through psychological pressure are integral parts of education, even though they are now wholly or partly renounced in explicit theory. Moreover, in regard to academic knowledge, it is still taken for granted, in practice, that the main purpose of education is to transmit a standard curriculum faithfully. One consequence is that people are acquiring scientific knowledge in an anaemic and instrumental way. Without a critical, discriminating approach to what they are learning, most of them are not effectively replicating the memes of science and reason into their minds. And so we live in a society in which people can spend their days conscientiously using laser technology to count cells in blood samples, and their evenings sitting cross-legged and chanting to draw supernatural energy out of the Earth.

Living with memes

Existing accounts of memes have neglected the all-important distinction between the rational and anti-rational modes of replication. Consequently they end up missing most of what is happening, and why. Moreover, since the most obvious examples of memes are long-lived anti-rational memes and short-lived arbitrary fads, the tenor of such accounts is usually anti-meme, even when these accounts formally accept that the best and most valuable knowledge also consists of memes.

For example, the psychologist Susan Blackmore, in her book *The Meme Machine*, attempts to provide a fundamental explanation of the human condition in terms of meme evolution. Now, memes are indeed integral to the explanation for the existence of our species – though, as I shall explain in the next chapter, I believe that the specific mechanism she proposes would not have been possible. But, crucially, Blackmore downplays the element of creativity both in the replication of memes and in their origin. This leads her, for example, to doubt that technological progress is best explained as being due to individuals as the conventional narrative would have it. She regards it instead as meme evolution. She cites the historian George Basalla, whose book *The Evolution of Technology* denies 'the myth of the heroic inventor'.

But that distinction between 'evolution' and 'heroic inventors' as being the agents of discovery makes sense only in a static society. There, most change is indeed brought about in the way that I guessed jokes might evolve, with no great creativity being exercised by any individual participant. But in a dynamic society, scientific and technological innovations are generally made creatively. That is to say, they emerge from individual minds as novel ideas, having acquired significant adaptations inside those minds. Of course, in both cases, ideas are built from previous ideas by a process of variation and selection, which constitutes evolution. But when evolution takes place largely within an individual mind, it is not meme evolution. It is creativity by a heroic inventor.

Worse, in regard to progress, Blackmore denies that there has been 'progress towards anything in particular' – that is to say, no progress towards anything objectively better. She recognizes only increasing complexity. Why? Because *biological* evolution does not have a 'better' or 'worse'. This despite her own warning that memes and genes evolve differently. Again, her claim is largely true of static societies, but not of ours.

How *should* we understand the existence of the distinctively human emergent phenomena such as creativity and choice, in the light of the fact that part of our behaviour is caused by autonomous entities whose content we do not know? And, worse, given that we are liable to be systematically misled by those entities about the reasons for our own thoughts, opinions and behaviour?

The basic answer is that it should not come as a surprise that we can be badly mistaken in any of our ideas, even about ourselves, and even when we feel strongly that we are right. So we should respond no differently, in principle, from how we respond to the possibility of being in error for any other reason. We are fallible, but through conjecture, criticism and seeking good explanations we may correct some of our errors. Memes hide, but, just as with the optical blind spot, there is nothing to prevent our using a combination of explanation and observation to detect a meme and discover its implicit content indirectly.

For example, whenever we find ourselves enacting a complex or narrowly defined behaviour that has been accurately repeated from one holder to the next, we should be suspicious. If we find that enacting this behaviour thwarts our efforts to attain our personal objectives, or is faithfully continued when the ostensible justifications for it disappear, we should become more suspicious. If we then find ourselves explaining our own behaviour with bad explanations, we should become still more suspicious. Of course, at any given point we may fail either to notice these things or to discover the true explanation of them. But failure need not be permanent in a world in which all evils are due to lack of knowledge. We failed at first to notice the non-existence of a force of gravity. Now we understand it. Locating hang-ups is, in the last analysis, easier.

Another thing that should make us suspicious is the presence of the *conditions* for anti-rational meme evolution, such as deference to authority, static subcultures and so on. Anything that says 'Because I say so' or 'It never did me any harm,' anything that says 'Let us suppress criticism of our idea because it is true,' suggests static-society thinking. We should examine and criticize laws, customs and other institutions with an eye to whether they set up conditions for anti-rational memes to evolve. Avoiding such conditions is the essence of Popper's criterion.

The Enlightenment is the moment at which explanatory knowledge is beginning to assume its soon-to-be-normal role as the most important determinant of physical events. At least it could be: we had better remember that what we are attempting – the sustained creation of knowledge – has never worked before. Indeed, everything that we shall ever try to achieve from now on will never have worked before. We have, so far, been transformed from the victims (and enforcers) of an eternal status quo into the mainly passive recipients of the benefits of relatively rapid innovation in a bumpy transition period. We now have to accept, and rejoice in bringing about, our next transformation: to active agents of progress in the emerging rational society – and universe.

TERMINOLOGY

Culture A set of shared ideas that cause their holders to behave alike in some ways.

Rational meme An idea that relies on the recipients' critical faculties to cause itself to be replicated.

Anti-rational meme An idea that relies on disabling the recipients' critical faculties to cause itself to be replicated.

Static culture/society One whose changes happen on a timescale longer than its members can notice. Such cultures are dominated by anti-rational memes.

Dynamic culture/society One that is dominated by rational memes.

MEANINGS OF 'THE BEGINNING OF INFINITY' ENCOUNTERED IN THIS CHAPTER

– Biological evolution was merely a finite preface to the main story of evolution, the unbounded evolution of memes.
– So was the evolution of anti-rational memes in static societies.

SUMMARY

Cultures consist of memes, and they evolve. In many ways memes are analogous to genes, but there are also profound differences in the way they evolve. The most important differences are that each meme has to include its own replication mechanism, and that a meme exists alternately in two different physical forms: a mental representation and a behaviour. Hence also a meme, unlike a gene, is separately selected, at each replication, for its ability to cause behaviour and for the ability of that behaviour to cause new recipients to adopt the meme. The holders of memes typically do not know why they are enacting them: we enact the rules of grammar, for instance, much more accurately than we are able to state them. There are only two basic strategies of meme replication: to help prospective holders or to disable the holders' critical faculties. The two types of meme – rational memes and anti-rational memes – inhibit each other's replication and the ability of the culture as a whole to propagate itself. Western civilization is in an unstable transitional period between stable, static societies consisting of anti-rational memes and a stable dynamic society consisting of rational memes. Contrary to conventional wisdom, primitive societies are unimaginably unpleasant to live in. Either they are static, and survive only by extinguishing their members' creativity and breaking their spirits, or they quickly lose their knowledge and disintegrate, and violence takes over. Existing accounts of memes fail to recognize the significance of the rational/anti-rational distinction and hence tend to be implicitly anti-meme. This is tantamount to mistaking Western civilization for a static society, and its citizens for the crushed, pessimistic victims of memes that the members of static societies are.

16

The Evolution of Creativity

What use was creativity?

Of all the countless biological adaptations that have evolved on our planet, creativity is the only one that can produce scientific or mathematical knowledge, art or philosophy. Through the resulting technology and institutions, it has had spectacular physical effects – most noticeably near human habitations, but also further afield: a substantial portion of the Earth's land area is now used for human purposes. Human choice – itself a product of creativity – determines which other species to exclude and which to tolerate or cultivate, which rivers to divert, which hills to level, and which wildernesses to preserve. In the night sky, a bright, fast-moving spot may well be a space station carrying humans higher and faster than any biological adaptation can carry anything. Or it may be a satellite through which humans communicate across distances that biological communication has never spanned, using phenomena such as radio waves and nuclear reactions, which biology has never harnessed. The unique effects of creativity dominate our experience of the world.

Nowadays that includes the experience of rapid innovation. By the time you read these words, the computer on which I am writing them will be obsolete: there will be functionally better computers that will require less human effort to build. Other books will have been written, and innovative buildings and other artefacts will be constructed, some of which will be quickly superseded while others will stand for longer than the pyramids have so far. Surprising scientific discoveries will be made, some of which will change the standard textbooks for ever. All these consequences of creativity make for an ever-changing way of life, which is possible only in a long-lived dynamic society – itself a phenomenon that nothing other than creative thought could possibly bring about.

However, as I pointed out in the previous chapter and Chapter 1, it was only recently in the history of our species that creativity has had any of those effects. In prehistoric times it would not have been obvious to a casual observer (say, an explorer from an extraterrestrial civilization) that humans were capable of creative thought at all. It would have seemed that we were doing no more than endlessly repeating the lifestyle to which we were genetically adapted, just like all the other billions of species in the biosphere. Clearly, we were tool-users – but so were many other species. We were communicating using symbolic language – but, again, that was not unusual: even bees do that. We were domesticating other species – but so do ants. Closer observation would have revealed that human languages and the knowledge for human tool use were being transmitted through memes and not genes. That made us fairly unusual, but still not obviously creative: several other species have memes. But what they do not have is the means of improving them other than through random trial and error. Nor are they capable of sustained improvement over many generations. Today, the creativity that humans use to improve ideas is what pre-eminently sets us apart from other species. Yet for most of the time that humans have existed it was not noticeably in use.

Creativity would have been even less noticeable in the predecessor of our species. Yet it must already have been evolving in that species, or ours would never have been the result. In fact the advantage conferred by successive mutations that gave our predecessors' brains slightly more creativity (or, more precisely, more of the ability that *we now think of as creativity*) must have been quite large, for by all accounts modern humans evolved from ape-like ancestors very rapidly by gene-evolution standards. Our ancestors must have been continually out-breeding their cousins who had slightly less ability to create new knowledge. Why? What were they using this knowledge for?

If we did not know better, the natural answer would be that they were using it as we do today, for innovation and for understanding the world, in order to improve their lives. For instance, individuals who could improve stone tools would have ended up with better tools, and hence with better food and more surviving offspring. They would also have been able to make better weapons, thus denying the holders of rival genes access to food and mates – and so on. Yet if that had happened, the palaeontological record would show those improvements happening on a timescale of generations. But it does not.

Moreover, during the period when creativity was evolving, the ability to replicate memes was evolving too. It is believed that some members of the species *Homo erectus* living 500,000 years ago knew how to make camp fires. That knowledge was in their memes, not in their genes. And, once creativity and meme transmission are both present, they greatly enhance each other's evolutionary value, for then anyone who improves something also has the means to bequeath the innovation to all future generations, thus multiplying the benefit to the relevant genes. And memes can be improved much faster by creativity than by random trial and error. Since there is no upper limit to the value of ideas, the conditions would have been there for a runaway co-evolution between the two adaptations: creativity and the ability to use memes.

Yet, again, there is something wrong with that scenario. The two adaptations presumably did co-evolve, but the driving force behind that evolution cannot have been that people were improving on ideas and passing the improvements on to their children, because, again, if they had been, they would have been making cumulative improvements on a timescale of generations. Before the beginning of agriculture, about 12,000 years ago, many thousands of years passed between noticeable changes. It is as though each small genetic improvement in creativity produced just one noticeable innovation and then nothing more – rather like today's experiments in 'artificial evolution'. But how can that be? Unlike present-day artificial-evolution and AI research, our ancestors were evolving *real* creativity, which is the capacity to create an endless stream of innovations.

Their ability to innovate was increasing rapidly, but they were barely innovating. This is a puzzle not because it is odd behaviour, but because, if innovation was that rare, how could there have been a differential effect on the reproduction of individuals with more or less ability to innovate? That there were thousands of years between noticeable changes presumably means that in most generations even the most creative individuals in the population would not have been making any innovations. Hence their greater ability to innovate would have caused no selection pressure in their favour. Why did tiny improvements in that ability keep spreading rapidly through the population? Our ancestors must have been using their creativity – and using it to its limits, and frequently – for *something*. But evidently not for innovation. What else could it have been used for?

One theory is that it did not evolve to provide any functional advantage, but merely through sexual selection: people used it to create displays to attract mates – colourful clothing, decorations, story-telling, wit and the like. A preference to mate with the individuals with the most creative displays co-evolved with the creativity to meet that preference in an evolutionary spiral – so the theory goes – just like peahens' preferences and peacocks' tails.

But creativity is an unlikely target for sexual selection. It is a sophisticated adaptation which, to this day, we are unable to reproduce artificially. So it is presumably much harder to evolve than attributes like coloration or the size and shape of body parts – some of which, it is thought, did indeed evolve by sexual selection in humans and many other animals. Creativity, as far as we know, evolved only once. Moreover, its most visible effects are cumulative: it would be hard to detect small differences in the creativity of potential mates on any one occasion, especially if that creativity was not being used for practical purposes. (Consider how hard it would be, today, to detect tiny *genetic* differences in people's artistic abilities by means of an art competition. In practice, any such differences would be swamped by other factors.) So why did we not evolve multi-coloured hair or fingernails instead of the capacity to create new knowledge, or any one of countless other attributes that would have been far easier to evolve, and far easier to assess reliably?

A more plausible variant of the sexual-selection theory is that people chose mates according to social status, rather than favouring creativity directly. Perhaps the most creative individuals were able to gain status more effectively through intrigue or other social manipulation. This could have given them an evolutionary advantage without producing any progress of which we would see evidence. However, all such theories still face the problem of explaining why, if creativity was being used intensively for any purpose, it was not *also* used for functional purposes. Why would a chief who had gained power through creative intrigue not be thinking about better spears for hunting? Why wouldn't a subordinate who invented such a thing have been favoured? Similarly, wouldn't potential mates who were impressed by artistic displays *also* have been impressed by practical innovations? In any case, some practical innovations would themselves have helped the discoverers to produce better displays. And innovations sometimes have reach: a new skill of making a string of decorative beads in one generation might become the skill of making a slingshot in the next. So why were practical innovations originally so rare?

From the discussion in the previous chapter, one might guess that it was because the tribes or families in which people were living were static societies, in which any noticeable innovation would reduce one's status and hence presumably one's eligibility to mate. So how does one gain status, specifically by exercising more creativity than anyone else, without becoming noticeable as a taboo-violator?

I think there is only one way: it is to enact that society's memes more faithfully than the norm. To display exceptional conformity and obedience. To refrain exceptionally well from innovation. A static society has no choice but to reward that sort of conspicuousness. So – can enhanced creativity help one to be *less* innovative than other people? That turns out to be a pivotal question, to which I shall return below. But first I must address a second puzzle.

How do you replicate a meaning?

Meme replication is often characterized (for example by Blackmore) as *imitation*. But that cannot be so. A meme is an idea, and we cannot observe ideas inside other people's brains. Nor do we have the hardware to download them from one brain to another like computer programs, nor to replicate them like DNA molecules. So we cannot literally copy or imitate memes. The only access we have to their content is through their holders' behaviour (including their speech, and consequences of their behaviour such as their writings).

Meme replication always follows this pattern: one observes the holders' behaviour, directly or indirectly. Then, later – sometimes immediately, sometimes after years of such observation – memes from the holders' brains are present in one's own brain. How do they get there? It looks a bit like induction, does it not? But induction is impossible.

The process often seems to involve imitating the holders. For instance, we learn words by imitating their sounds; we learn how to wave by being waved to and imitating what we see. Thus, outwardly, and even to our own introspection, we appear to be copying aspects of what other people do, and remembering what they say and write. This common-sense misconception is even corroborated by the fact that our species' closest living relatives, the great apes, also have a (much more limited, but nevertheless striking) ability to imitate. But, as I shall explain, the truth is that imitating people's actions and remembering their utterances could not possibly be the basis of human meme replication. In reality these play only a small – and for the most part inessential – role.

Meme acquisition comes so naturally to us that it is hard to see what a miraculous process it is, or what is really happening. It is especially hard to see where the *knowledge* is coming from. There is a great deal of knowledge in even the simplest of human memes. When we learn to wave, we learn not only the gesture but also which aspects of the situation made it appropriate to wave, and how, and to whom. We are not told most of this, yet we learn it anyway. Similarly, when we learn a word, we also learn its meaning, including highly inexplicit subtleties. How do we acquire that knowledge?

Not by imitating the holders. Popper used to begin his lecture course on the philosophy of science by asking the students simply to 'observe'. Then he would wait in silence for one of them to ask *what* they were supposed to observe. This was his way of demonstrating one of many flaws in the empiricism that is still part of common sense today. So he would explain to them that scientific observation is impossible without pre-existing knowledge about what to look at, what to look for, how to look, and how to interpret what one sees. And he would explain that, therefore, theory has to come first. It has to be conjectured, not derived.

Popper could have made the same point by asking his audience to *imitate*, rather than merely to observe. The logic would have been the same: under what explanatory theory should they 'imitate'? *Whom* should they imitate? Popper? In that case, should they walk to the podium, push him out of the way, and stand where he had been standing? If not, should they at least turn to face the rear of the room, to imitate where he was facing? Should they imitate his heavy Austrian accent, or should they speak in their normal voices, because he was speaking in his normal voice? Or should they do nothing special at the time, but merely include such demonstrations in their lectures when they themselves became professors of philosophy? There are infinitely many possible interpretations of 'imitate Popper', each defining a different behaviour for the imitator. Many of those ways would look very different from each other. Each way corresponds to a different theory of what ideas, in Popper's mind, were causing the observed behaviour.

So there is no such thing as 'just imitating the behaviour' – still less, therefore, can one discover those *ideas* by imitating it. One needs to know the ideas *before* one can imitate the behaviour. So imitating behaviour cannot be how we acquire memes.

The hypothetical genes that caused meme replication by imitation would also have to specify *whom* to imitate. Blackmore, for instance, suggests that the criterion may be 'imitate the best imitators'. But this is impossible for the same reason. One can only judge how well someone is imitating if one already knows, or has guessed, *what* (which aspect of behaviour, and whose) they are imitating, and which of the circumstances they are taking into account and how.

The same holds if the behaviour consists of *stating* the memes. As Popper remarked, 'It is impossible to speak in such a way that you cannot be misunderstood.' One can only state the explicit content, which is insufficient to define the meaning of a meme or anything else. Even the most explicit of memes – such as laws – have inexplicit content without which they cannot be enacted. For example, many laws refer to what is 'reasonable'. But no one can define that attribute accurately enough for, say, a person from a different culture to be able to apply the definition in judging a criminal case. Hence we certainly do not learn what 'reasonable' means by hearing its meaning *stated*. But we do learn it, and the versions of it that are learned by people in the same culture are sufficiently close for laws based on it to be practicable.

In any case, as I remarked in the previous chapter, we do not explicitly know the rules by which we behave. We know the rules, meanings and patterns of speech of our native language largely inexplicitly, yet we pass its rules on with remarkable fidelity to the next generation – including the ability to apply them in situations the new holder has never experienced, and including patterns of speech that people explicitly try to prevent the next generation from replicating.

The real situation is that people need inexplicit knowledge to understand laws and other explicit statements, not vice versa. Philosophers and psychologists work hard to discover, and to make explicit, the assumptions that our culture tacitly makes about social institutions, human nature, right and wrong, time and space, intention, causality, freedom, necessity and so on. But we do not acquire those assumptions by reading the results of such research: it is entirely the other way round.

If behaviour is impossible to imitate without prior knowledge of the theory causing the behaviour, how it is that apes, famously, can ape? They have memes: they can learn a new way of opening a nut by watching another ape that already knows that way. How is it that apes are not confused by the infinite ambiguity of what it means to imitate? Even parrots, famously, parrot: they can commit to memory dozens of sounds that they have heard, and repeat them later. How do they cope with the ambiguity of which sounds to imitate, and when to repeat them?

They cope with it by knowing the relevant inexplicit theories in advance. Or, rather, their genes know them. Evolution has built into the genes of parrots an implicit definition of what 'imitating' means: to them, it means recording sequences of sounds that meet some inborn criterion, and later replaying them under conditions that meet some other inborn criterion. An interesting fact follows, about parrot physiology: the parrot's brain must also contain a translation system that analyses incoming nerve signals from the ears and generates outgoing ones that will cause the parrot's vocal cords to play the same sounds. That translation requires some quite sophisticated computation, which is encoded in genes, not memes. It is thought to be achieved in part by a system based on 'mirror neurons'. These are neurons that fire when an animal performs a given action, and also when the animal perceives the same action being performed by another. These neurons have been identified experimentally in animals that have the capacity to imitate. Scientists who believe that human meme replication is a sophisticated form of imitation tend to believe that mirror neurons are a key to understanding all sorts of functions of the human mind. Unfortunately, that cannot possibly be so.

It is not known *why* parroting evolved. It is a fairly common adaptation in birds, and may play more than one role. But, whatever the reason, the important thing for present purposes is that parrots never have a choice about which sounds to imitate, or about what constitutes imitating them. A ringing doorbell and a barking dog may happen to provide conditions that meet the inborn criterion that initiates parroting behaviour, and, when they do, the parrot will always mimic exactly the same aspects of them: their sounds. So, it resolves the infinite ambiguity by making no choices. It does not occur to it to ignore the dog under those conditions, or to imitate the wagging of its tail, because it is incapable of conceiving of any other criterion for imitation than the one built into its mirror-neuron system. It is devoid of creativity and *relies* on its lack of creativity to replicate the sounds faithfully. This is reminiscent of humans in static societies – except for a crucial difference which I shall explain below.

Now, imagine that a parrot had been present at Popper's lectures, and learned to parrot some of Popper's favourite sentences. It would, in a sense, have 'imitated' some of Popper's ideas: in principle, an interested student could later learn the ideas by listening to the parrot. But the parrot would merely be transmitting those memes from one place to another – which is no more than the air in the lecture theatre does. The parrot could not be said to have acquired the memes, because it would be reproducing only one of the countless behaviours that they could produce. The parrot's subsequent behaviour as a result of having learned the sounds by heart – such as its responses to questions – would not resemble Popper's. The sound of the meme would be there, but its meaning would not. And it is the meaning – the knowledge – that is the replicator.

The parrot is oblivious to the human meanings of the sounds that it parrots. Had those lectures been not about philosophy but about recipes for fried parrot, it would have been just as eager to quote from them to anyone who would listen. But it is not *oblivious* to the content of the sound – it is not like a mechanical recorder. Quite the contrary: parrots neither record sounds indiscriminately nor replay them randomly. Their inborn criteria do implicitly attribute meaning to sounds that they hear; it is just that the meaning is always drawn from the same, narrow set of possibilities: if the evolutionary function of parroting is, for instance, to create identifying calls, then every sound it hears is either a potential identifying call or not.

Apes are capable of recognizing a much larger set of possible meanings. Some of them are so complex that aping has often been misinterpreted as evidence of human-like understanding. For example, when an ape learns a new method of cracking nuts by hitting them with rocks, it does not then play the movements back blindly in a fixed sequence like a parrot does. The movements required to crack the nut are never the same twice: the ape has to *aim* the rock at the nut; it may have to *chase* the nut and fetch it back if it rolls away; it has to keep hitting it until it cracks, rather than a fixed number of times; and so on. During some parts of the procedure the ape's two hands must cooperate, each performing a different sub-task. Before it can even begin, it must be able to recognize a nut as being suitable for the procedure; it must look for a rock and, again, recognize a suitable one.

Such activities may seem to depend on explanation – on understanding how and why each action within the complex behaviour has to fit in with the other actions in order to achieve the overall purpose. But recent discoveries have revealed how apes are able to imitate such behaviours without ever creating any explanatory knowledge. In a remarkable series of observational and theoretical studies, the evolutionary psychologist and animal-behaviour researcher Richard Byrne has shown how they achieve this by a process that he calls *behaviour parsing* (which is analogous to the grammatical analysis or 'parsing' of human speech or computer programs).

Humans and computers separate continuous streams of sounds or characters into individual elements such as words, and then interpret those elements as being connected by the logic of a larger sentence or program. Similarly, in behaviour parsing (which evolved millions of years before human language parsing), an ape parses a continuous stream of behaviour that it witnesses into individual elements, each of which it already knows – genetically – how to imitate. The individual elements can be inborn behaviours, such as biting; or behaviours learned by trial and error, such as grasping a nettle without being stung; or previously learned memes. As for connecting these elements together in the right way without knowing why, it turns out that, in every known case of complex behaviours in non-humans, the necessary information can be obtained merely by watching the behaviour many times and looking out for simple statistical patterns – such as which right-hand behaviour often goes with which left-hand behaviour, and which elements are often omitted. It is a very inefficient method, requiring a lot of watching of behaviours that a human could mimic almost immediately by understanding their purpose. Also, it allows only a few fixed options for connecting the behaviours together, so only relatively simple memes can be replicated. Apes can copy certain individual actions instantly – the ones of which they have pre-existing knowledge through their mirror-neuron system – but it takes them years to learn a repertoire of memes that involve combinations of actions. Yet those memes – trivially simple tricks by human standards – are enormously valuable: using them, apes have privileged access to sources of food that are closed to all other animals; and meme evolution gives them the ability to switch to other sources far faster than gene evolution would allow.

So, an ape knows (inexplicitly) that another ape is 'picking up a rock', and not performing any of the countless other possible interpretations of the same actions, such as 'picking up an object in a given relative position', because picking up a rock is in its inborn repertoire of copiable behaviours while the other possibilities are not. Indeed, it may well be that apes *cannot* imitate the behaviour of 'picking up an object in a given relative position'. Note, in this connection, that apes are unable to imitate sounds. They cannot even parrot sounds (repeat them blindly), despite having a complex inborn repertoire of calls that they can make, recognize and act upon in genetically predetermined ways. Their behaviour-parsing system simply did not evolve a predetermined translation mechanism from hearing sounds to uttering them, so they cannot ape them. Consequently there are no customized sounds in any of the apes' memetically controlled behaviours.

Thus, in the crucial respect that is relevant to meme replication, aping has the same logic as parroting: like the parrot, the ape avoids the infinite ambiguity of copying by already knowing, inexplicitly, the meaning of every action that it is capable of copying. And it is only capable of associating one meaning with each action that it can copy – one definition of how to perform the 'same' action under various circumstances. That is how ape memes can be replicated without the impossible step of literally copying knowledge from another ape. The recipient of the meme instantly recognizes the meaning of each element of the behaviour; and it relates the elements by statistical analysis, not by discovering how they support each other's functioning.

Human beings acquiring human memes are doing something profoundly different. When an audience is watching a lecture, or a child is learning language, their problem is almost the opposite of that of parroting or aping: the meaning of the behaviour that they are observing is precisely what they are striving to discover and do not know in advance. The actions themselves, and even the logic of how they are connected, are largely secondary and are often entirely forgotten afterwards. For example, as adults we remember few of the actual sentences from which we learned to speak. If a parrot had copied snatches of Popper's voice at a lecture, it would certainly have copied them with his Austrian accent: parrots are incapable of copying an utterance without its accent. But a human student might well be unable to copy it *with* the accent. In fact a student might well acquire a complex meme at a lecture without being able to repeat a single sentence spoken by the lecturer, even immediately afterwards. In such a case the student has replicated the meaning – which is the whole content – of the meme without imitating any actions at all. As I said, imitation is not at the heart of human meme replication.

Suppose that the lecturer had repeatedly returned to a certain key idea, and had expressed it with different words and gestures each time. The parrot's (or ape's) job would be that much harder than imitating only the first instance; the student's much easier, because to a human observer each different way of putting the idea would convey additional knowledge. Alternatively, suppose that the lecturer had consistently misspoken in a way that altered the meaning, and had then made one correction at the end. The parrot would copy the wrong version. The student would not. Even if the lecturer never corrected the error at all, a human listener might still have a good chance of understanding the idea that was in the lecturer's mind – and, again, without imitating any behaviour. If someone else reported the lecture but in a way that contained severe misconceptions, a human listener might *still* be able to detect what the lecturer meant, by explaining the reporter's misconceptions as well as the lecturer's intention – just as a conjuring expert might be able to detect what really happened during a trick given only a false account from the audience of what they saw.

Rather than imitating behaviour, a human being tries to explain it – to understand the ideas that caused it – which is a special case of the general human objective of explaining the world. When we succeed in explaining someone's behaviour, and we approve of the underlying intention, we may subsequently behave 'like' that person in the relevant sense. But if we disapprove, we might behave unlike that person. Since creating explanations is second nature (or, rather, first nature) to us, we can easily misconstrue the process of acquiring a meme as 'imitating what we see'. Using our explanations, we 'see' right through the behaviour to the meaning. Parrots copy distinctive sounds; apes copy purposeful movements of a certain limited class. But humans do not especially copy any behaviour. They use conjecture, criticism and experiment to create good explanations of the meaning of things – other people's behaviour, their own, and that of the world in general. That is what creativity does. And if we end up behaving like other people, it is because we have rediscovered the same idea.

That is why the audience at a lecture, when striving to assimilate the lecturer's memes, are not tempted to face the rear wall of the lecture room, or to imitate the lecturer in any one of infinitely many other ways. They reject such interpretations of what is worth copying about the lecturer not because they are genetically incapable of conceiving of them, as other animals are, but because they are bad explanations of what the lecturer is doing, and bad ideas by the audience's own values.

Both puzzles have the same solution

In this chapter I have presented two puzzles. The first is why human creativity was evolutionarily advantageous at a time when there was almost no innovation. The second is how human memes can possibly be replicated, given that they have content that the recipient never observes.

I think that both those puzzles have the same solution: what replicates human memes is creativity; and creativity was used, while it was evolving, *to replicate memes*. In other words, it was used to acquire existing knowledge, not to create new knowledge. But *the mechanism to do both things is identical*, and so in acquiring the ability to do the former, we automatically became able to do the latter. It was a momentous example of reach, which made possible everything that is uniquely human.

A person acquiring a meme faces the same logical challenge as a scientist. Both must discover a hidden explanation. For the former, it is an idea in the minds of other people; for the latter, a regularity or law of nature. Neither person has direct access to this explanation. But both have access to evidence with which explanations can be tested: the observed behaviour of people who hold the meme, and physical phenomena conforming to the law.

The puzzle of how one can possibly translate behaviour back into a theory that contains its meaning is therefore the same puzzle as where scientific knowledge comes from. And the idea that memes are copied by imitating their holders' behaviour is the same mistake as empiricism or inductivism or Lamarckism. They all depend on there being a way of automatically translating *problems* (like the problem of planetary motions, or of how to reach leaves on tall trees or to be invisible to one's prey) into their solutions. In other words, they assume that the environment (in the form of an observed phenomenon, or a tall tree, say) can 'instruct' minds or genomes in how to meet its challenges. Popper wrote:

> The inductivist or Lamarckian approach operates with the idea of instruction from without, or from the environment. But the critical or Darwinian approach only allows instruction from within – from within the structure itself . . .
>
> I contend that there is no such thing as instruction from without the structure. We do not discover new facts or new effects by copying them, or by inferring them inductively from observation, or by any other method of instruction by the environment. We use, rather, the method

of trial and the elimination of error. As Ernst Gombrich says, 'making comes before matching': the active production of a new trial structure comes before its exposure to eliminating tests.

The Myth of the Framework

Popper could just as well have written, 'We do not *acquire new memes* by copying them, or by inferring them inductively from observation, or by any other method of imitation of, or instruction by, the environment.' The transmission of human-type memes – memes whose meaning is not mostly predefined within the receiver – cannot be other than a creative activity on the part of the receiver.

Memes, like scientific theories, are not derived from anything. They are created afresh by the recipient. They are conjectural explanations, which are then subjected to criticism and testing before being tentatively adopted.

This same pattern of creative conjecture, criticism and testing generates inexplicit as well as explicit ideas. In fact all creativity does, for no idea can be represented entirely explicitly. When we make an explicit conjecture, it has an inexplicit component whether we are aware of it or not. And so does all criticism.

Thus, as has so often happened in the history of universality, the human capacity for universal explanation did not evolve to have a universal function. It evolved simply to increase the volume of memetic information that our ancestors could acquire, and the speed and accuracy with which they could acquire it. But since the easiest way for evolution to do that was to give us a universal ability to explain, through creativity, that is what it did. This epistemological fact provides not only the solution of the two puzzles I mentioned, but also the reason for the evolution of human creativity – and therefore the human species – in the first place.

It must have happened something like this. In early pre-human societies, there were only very simple memes – the kind that apes now have, though perhaps with a wider repertoire of copiable elementary behaviours. Those memes were about practical things like how to get food that was otherwise inaccessible. The value of such knowledge must have been high, so this created a ready-made niche for any adaptation that would reduce the effort required to replicate memes. Creativity was the ultimate adaptation to fill that niche. As it increased, further adaptations co-evolved, such as an increase in memory capacity (to store more memes), finer motor control, and specialized brain structures for dealing with language. As a result, the meme bandwidth (the amount of memetic information that could be passed from each generation to the next) increased too. Memes also became more complex and sophisticated.

This is why and how our species evolved, and why it evolved rapidly – at first. Memes gradually came to dominate our ancestors' behaviour. Meme evolution took place, and, like all evolution, this was always in the direction of greater faithfulness. This meant becoming ever more anti-rational. At some point, meme evolution achieved static societies – presumably they were tribes. Consequently, all those increases in creativity never produced streams of innovations. Innovation remained imperceptibly slow, even as the capacity for it was increasing rapidly.

Even in a static society, memes still evolve, due to imperceptible errors of replication. They just evolve more slowly than anyone can notice: imperceptible errors cannot be suppressed. They would generally evolve towards greater fidelity of replication, as usual with evolution, and hence to greater staticity of the society.

Status in such a society is reduced by transgressing people's expectations of proper behaviour, and is improved by meeting them. There would have been the expectations of parents, priests, chiefs and potential mates (or whoever controlled mating in that society) – who were themselves conforming to the wishes and expectations of the society at large. Those people's opinions would determine one's ability to eat, thrive and reproduce, and hence the fate of one's genes.

But how does one discover the wishes and expectations of other people? They might issue commands, but they could never specify every detail of what they expected, let alone every detail of how to achieve it. When one is commanded to do something (or expected to, as a condition for being considered worthy of food or mating, for instance), one might remember seeing an already-respected person doing the same thing, and one might try to emulate that person. To do that effectively, one would have to understand what the point of doing it was, and to try to achieve *that* as best one could. One would impress one's chief, priest, parent or potential mate by replicating, and following, their standards of what one should strive for. One would impress the tribe as a whole by replicating their idea (or the ideas of the most influential among them) of what was worthy, and acting accordingly.

Hence, paradoxically, it requires creativity to thrive in a static society – creativity that enables one to be *less* innovative than other people. And that is how primitive, static societies, which contained pitifully little knowledge and existed only by suppressing innovation, constituted environments that strongly favoured the evolution of an ever-greater ability to innovate.

From the perspective of those hypothetical extraterrestrials observing our ancestors, a community of advanced apes with memes before the evolution of creativity began would have looked superficially similar to their descendants after the jump to universality. The latter would merely have had many more memes. But the mechanism keeping those memes replicating faithfully would have changed profoundly. The animals of the earlier community would have been relying on their lack of creativity to replicate their memes; the people, despite living in a static society, would be relying entirely on their creativity.

As with all jumps to universality, the way in which the jump emerged out of gradual changes is interesting to think about. Creativity is a property of *software*. As I said, we could be running AI programs on our laptop computers today if we knew how to write (or evolve) such programs. Like all software, it would require the computer to have certain hardware specifications in order to be able to process the required amount of data in the required time. It so happened that the hardware specifications that would make creativity practicable were included in those that were being heavily favoured for pre-creative meme replication. The principal one would have been memory capacity: the more one could remember, the more memes one could enact, and the more accurately one could enact them. But there may also have been hardware abilities such as mirror neurons for imitating a wider range of elementary actions than apes could ape – for instance, the elementary sounds of a language. It would have been natural for such hardware assistance for language abilities to be evolving at the same time as the increased meme bandwidth. So, by the time creativity was evolving, there would already have been significant co-evolution between genes and memes: genes evolving hardware to support more and better memes, and memes evolving to take over ever more of what had previously been genetic functions such as choice of mate, and methods of eating, fighting and so on. Therefore, my speculation is that the creativity program is not entirely inborn. It is a combination of genes and memes. The hardware of the human brain would have been *capable* of being creative (and sentient, conscious and all those other things) long before any creative program existed. Considering a sequence of brains during this period, the earliest ones *capable* of supporting creativity would have required very ingenious programming to fit the capacity into the barely suitable hardware. As the hardware improved, creativity could have been programmed more easily, until the moment when it became easy enough actually to be done by evolution. We do not know *what* was being gradually increased in that approach to a universal explainer. If we did, we could program one tomorrow.

The future of creativity

Before Blackmore and others realized the significance of memes in human evolution, all sorts of root causes had been suggested for what propelled a normal-looking lineage of apes into rapidly becoming a species that can explain and control the universe. Some proposed that it was the adaptation of walking upright, which freed the front limbs, with their opposable thumbs, to specialize in manipulation. Some proposed that climate change favoured adaptations that would make our ancestors more able to exploit diverse habitats. And, as I have mentioned, sexual selection is always a candidate for explaining rapid evolution. Then there is the 'Machiavellian hypothesis' that human intelligence evolved in order to predict the behaviour of others, and to fool them. There is also the hypothesis that human intelligence is an enhanced version of the apes' aping adaptation – which, as I have argued, could not be true. Nevertheless, Blackmore's 'meme machine' idea, that human brains evolved in order to replicate memes, must be true. The reason it must be true is that, *whatever* had set off the evolution of any of those attributes, creativity would have had to evolve as well. For no human-level mental achievements would be possible without human-type (explanatory) memes, and the laws of epistemology dictate that no such memes are possible without creativity.

Not only is creativity necessary for human meme replication, it is also sufficient. Deaf people and blind people and paralysed people are still able to acquire and create human ideas to a more or less full extent. Hence, neither upright walking nor fine motor control nor the ability to parse sounds into words nor any of those other adaptations, though they might have played a role historically in creating the conditions for human evolution, were functionally necessary to allow humans to become creative. Nor, therefore, are they philosophically significant in understanding what humans are today, namely *people*: creative, universal explainers.

It was specifically creativity that made the difference between ape memes – expensive in terms of the time and effort required to replicate them, and inherently limited in the knowledge that they were capable of expressing – and human memes, which are efficiently transmitted and universal in their expressive power. The beginning of creativity was, in that sense, the beginning of infinity. We have no way of telling, at present, how likely it was for creativity to begin to evolve in apes. But, once it began to, there would automatically have been evolutionary pressure for it to continue, and for other meme-facilitating adaptations to follow in its wake. This increase must have continued through all the static societies of prehistory.

The horror of static societies, which I described in the previous chapter, can now be seen as a hideous practical joke that the universe played on the human species. Our creativity, which evolved in order to increase the amount of knowledge that we could use, and which would immediately have been capable of producing an endless stream of useful innovations as well, was from the outset prevented from doing so by the very knowledge – the memes – that that creativity preserved. The strivings of individuals to better themselves were, from the outset, perverted by a superhumanly evil mechanism that turned their efforts to exactly the opposite end: to thwart all attempts at improvement; to keep sentient beings locked in a crude, suffering state for eternity. Only the Enlightenment, hundreds of thousands of years later, and after who knows how many false starts, may at last have made it practical to escape from that eternity into infinity.

TERMINOLOGY

Imitation   Copying behaviour. This is different from human meme replication, which copies the knowledge that is causing the behaviour.

MEANINGS OF 'THE BEGINNING OF INFINITY' ENCOUNTERED IN THIS CHAPTER

– The evolution of creativity.
– The reassignment of creativity from its original function of preserving memes faithfully, to the function of creating new knowledge.

SUMMARY

On the face of it, creativity cannot have been useful during the evolution of humans, because knowledge was growing much too slowly for the more creative individuals to have had any selective advantage. This is a puzzle. A second puzzle is: how can complex memes even exist, given that brains have no mechanism to download them from other brains? Complex memes do not mandate specific bodily actions, but *rules*. We can see the actions, but not the rules, so how do we replicate them? We replicate them by creativity. That solves both problems, for replicating memes unchanged is the function for which creativity evolved. And that is why our species exists.

17

Unsustainable

Easter Island in the South Pacific is famous mainly – let's face it, *only* – for the large stone statues that were built there many centuries ago by the islanders. The purpose of the statues is unknown, but is thought to be connected with an ancestor-worshipping religion. The first settlers may have arrived on the island as early as the fifth century CE. They developed a complex Stone Age civilization, which suddenly collapsed over a millennium later. By some accounts there was starvation, war and perhaps cannibalism. The population fell to a small fraction of what it had been, and their culture was lost.

The prevailing theory is that the Easter Islanders brought disaster upon themselves, in part by chopping down the forest which had originally covered most of the island. They eliminated the most useful species of tree altogether. This is not a wise thing to do if you rely on timber for shelter, or if fish form a large part of your diet and your boats and nets are made of wood. And there were knock-on effects such as soil erosion, precipitating the destruction of the environment on which the islanders had depended.

Some archaeologists dispute this theory. For example, Terry Hunt has concluded that the islanders arrived only in the thirteenth century, and that their civilization continued to function throughout the deforestation (which he attributes to rats, not tree-felling) until it was destroyed by epidemics, caused by contact with Europeans. However, I do not want to discuss whether the prevailing theory is accurate, but only to use it as an example of a common fallacy – an argument by analogy about issues far less parochial.

Easter Island is 2,000 kilometres from the nearest habitation, namely Pitcairn Island (where the *Bounty*'s crew took refuge after their famous mutiny). Both islands are far from anywhere, even by today's standards. Nevertheless, in 1972 Jacob Bronowski made his way to Easter Island to film part of his magnificent television series *The Ascent of Man*. He and his film crew travelled by ship all the way from California, a round trip of some 14,000 kilometres. He was in poor health, and the crew had literally to carry him to the location for filming. But he persevered because those distinctive statues were the perfect setting for him to deliver the central message of his series – which is also a theme of this book – that our civilization is unique in history for its capacity to make progress. He wanted to celebrate its values and achievements, and to attribute the latter to the former, and to contrast our civilization with the alternative as epitomized by ancient Easter Island.

The Ascent of Man had been commissioned by the naturalist David Attenborough, then controller of the British television channel BBC2. A quarter of a century later Attenborough – who had by then become the doyen of natural-history film-making – led another film crew to Easter Island, to film another television series, *The State of the Planet*. He too chose those grim-faced statues as a backdrop, for his closing scene. Alas, his message was almost exactly the opposite of Bronowski's.

The philosophical difference between these two great broadcasters – so alike in their infectious sense of wonder, their clarity of exposition, and their humanity – was immediately evident in their different attitudes towards those statues. Attenborough called them 'astonishing stone sculptures . . . vivid evidence of the technological and artistic skills of the people who once lived here'. Now, I wonder whether Attenborough was really all that impressed by the islanders' skills, which had been exceeded millennia earlier in other Stone Age societies. I expect he was being polite, for it is de rigueur in our culture to heap praise upon any achievement of a primitive society. But Bronowski refused to conform to that convention. He remarked, 'People often ask about Easter Island, How did men come here? They came here by accident: that is not in question. The question is, Why could they not get off?' And why, he might have added, did others not follow to trade with them (there was a great deal of trade among Polynesians other than Easter Islanders), or to rob them, or to learn from them? Because they did not know how.

As for the statues being 'vivid evidence of . . . artistic skills', Bronowski was having none of that either. To him they were vivid evidence of failure, not success:

> The critical question about these statues is, Why were they all made *alike*? You see them sitting there, like Diogenes in their barrels, looking at the sky with empty eye-sockets, and watching the sun and the stars go overhead without ever trying to understand them. When the Dutch discovered this island on Easter Sunday in 1722, they said that it had the makings of an earthly paradise. But it did not. An earthly paradise is not made by this empty repetition . . . These frozen faces, these frozen frames in a film that is running down, mark a civilization which failed to take the first step on the ascent of rational knowledge.
>
> *The Ascent of Man* (1973)

The statues were all made alike because Easter Island was a static society. It never took that first step in the ascent of man – the beginning of infinity.

Of the hundreds of statues on the island, built over the course of several centuries, fewer than half are at their intended destinations. The rest, including the largest, are in various stages of completion, with as many as 10 per cent already in transit on specially built roads. Again there are conflicting explanations, but, according to the prevailing theory, it is because there was a large increase in the rate of statue-building just before it stopped for ever. In other words, as disaster loomed, the islanders diverted ever more effort not into addressing the problem – for they did not know how to do that – but into making ever more and bigger (but very rarely better) monuments to their ancestors. And what were those roads made of? Trees.

When Bronowski made his documentary, there were as yet no detailed theories of how the Easter Island civilization fell. But, unlike Attenborough, he was not interested in that, because his whole purpose in going to Easter Island was to point out the profound *difference* between our civilization and civilizations like the one that built those statues. *We are not like them* was his message. We have taken the step that they did not. Attenborough's argument rests on the opposite claim: *we are like them* and are following headlong in their footsteps. And so he drew an extended analogy between the Easter Island civilization and ours, feature for feature, and danger for danger:

> A warning of what the future could hold can be seen on one of the remotest places on Earth ... When the first Polynesian settlers landed here they found a miniature world that had ample resources to sustain them. They lived well ...
>
> *The State of the Planet* (BBC TV, 2000)

A miniature world: there, in three words, is Attenborough's reason for travelling all the way to Easter Island and telling its story. He believed that it holds a warning for the world because Easter Island was itself a miniature world – a Spaceship Earth – that went wrong. It had 'ample resources' to sustain its population, just as the Earth has seemingly ample resources to sustain us. (Imagine how amazed Malthus would have been had he known that the Earth's resources would still be called 'ample' by pessimists in the year 2000.) Its inhabitants 'lived well', just as we do. And yet they were doomed, just as we are doomed unless we change our ways. If we do not, here is 'what the future could hold':

> The old culture that had sustained them was abandoned and the statues toppled. What had been a rich, fertile world in miniature had become a barren desert.

Again, Attenborough puts in a good word for the old culture: it 'sustained' the islanders (just as the ample resources did, until the islanders failed to use them *sustainably*). He uses the toppling of the statues to symbolize the fall of that culture, as if to warn of future disaster for ours, and he reiterates his world-in-miniature analogy between the society and technology of ancient Easter Island and that of our whole planet today.

Thus Attenborough's Easter Island is a variant of Spaceship Earth: humans are sustained *jointly* by the 'rich, fertile' biosphere *and* the cultural knowledge of a static society. In this context, 'sustain' is an interestingly ambiguous word. It can mean providing someone with what they need. But it can also mean preventing things from changing – which can be almost the opposite meaning, for the suppression of change is seldom what human beings need.

The knowledge that currently sustains human life in Oxfordshire does so only in the first sense: it does not make us enact the same, traditional way of life in every generation. In fact it prevents us from doing so. For comparison: if your way of life merely makes you build a new, giant statue, you can continue to live afterwards exactly as you did before. That is *sustainable*. But if your way of life leads you to invent a more efficient method of farming, and to cure a disease that has been killing many children, that is *unsustainable*. The population grows because children who would have died survive; meanwhile, fewer of them are needed to work in the fields. And so there is no way to continue as before. You have to live the solution, and to set about solving the new problems that this creates. It is because of this unsustainability that the island of Britain, with a far less hospitable climate than the subtropical Easter Island, now hosts a civilization with at least three times the population density that Easter Island had at its zenith, and at an enormously higher standard of living. Appropriately enough, this civilization has knowledge of how to live well without the forests that once covered much of Britain.

The Easter Islanders' culture sustained them in both senses. This is the hallmark of a functioning static society. It provided them with a way of life; but it also inhibited change: it sustained their determination to enact and re-enact the same behaviours for generations. It sustained the values that placed forests – literally – beneath statues. And it sustained the shapes of those statues, and the pointless project of building ever more of them.

Moreover, the portion of the culture that sustained them in the sense of providing for their needs was not especially impressive. Other Stone Age societies have managed to take fish from the sea and sow crops without wasting their efforts in endless monument-building. And, if the prevailing theory is true, the Easter Islanders started to starve *before* the fall of their civilization. In other words, even after it had stopped providing for them, it retained its fatal proficiency at sustaining a fixed pattern of behaviour. And so it remained effective at preventing them from addressing the problem by the only means that could possibly have been effective:

creative thought and innovation. Attenborough regards the culture as having been very valuable and its fall as a tragedy. Bronowski's view was closer to mine, which is that since the culture never improved, its *survival* for many centuries was a tragedy, like that of all static societies.

Attenborough is not alone in drawing frightening lessons from the history of Easter Island. It has become a widely adduced version of the Spaceship Earth metaphor. But what exactly is the analogy behind the lesson? The idea that civilization depends on good *forest* management has little reach. But the broader interpretation, that survival depends on good *resource* management, has almost no content: *any* physical object can be deemed a 'resource'. And, since problems are soluble, all disasters are caused by 'poor resource management'. The ancient Roman ruler Julius Caesar was stabbed to death, so one could summarize his mistake as 'imprudent iron management, resulting in an excessive build-up of iron in his body'. It is true that if he had succeeded in keeping iron away from his body he would not have died in the (exact) way he did, yet, as an explanation of how and why he died, that ludicrously misses the point. The interesting question is not what he was stabbed with, but how it came about that other politicians plotted to remove him violently from office and that they succeeded. A Popperian analysis would focus on the fact that Caesar had taken vigorous steps to ensure that he could not be removed *without* violence. And then on the fact that his removal did not rectify, but actually entrenched, this progress-suppressing innovation. To understand such events and their wider significance, one has to understand the politics of the situation, the psychology, the philosophy, sometimes the theology. Not the cutlery. The Easter Islanders may or may not have suffered a forest-management fiasco. But, if they did, the explanation would not be about why they made mistakes – problems are inevitable – but why they failed to correct them.

I have argued that the laws of nature cannot possibly impose any bound on progress: by the argument of Chapters 1 and 3, denying this is tantamount to invoking the supernatural. In other words, progress is *sustainable*, indefinitely. But only by people who engage in a particular kind of thinking and behaviour – the problem-solving and problem-creating kind characteristic of the Enlightenment. And that requires the optimism of a dynamic society.

One of the consequences of optimism is that one expects to learn from failure – one's own and others'. But the idea that our civilization has something to learn from the Easter Islanders' alleged forestry failure is not derived from any structural resemblance between our situation and theirs. For they failed to make progress in practically every area. No one expects the Easter Islanders' failures in, say, medicine to explain our difficulties in curing cancer, or their failure to understand the night sky to explain why a quantum theory of gravity is elusive to us. The Easter Islanders' errors, both methodological and substantive, were simply too elementary to be relevant to us, and their imprudent forestry, if that is really what destroyed their civilization, would merely be typical of their lack of problem-solving ability across the board. We should do much better to study their many small successes than their entirely commonplace failures. If we could discover their rules of thumb (such as 'stone mulching' to help grow crops on poor soil), we might find valuable fragments of historical and ethnological knowledge, or perhaps even something of practical use. But one cannot draw general conclusions from rules of thumb. It would be astonishing if the details of a primitive, static society's collapse had any relevance to hidden dangers that may be facing our open, dynamic and scientific society, let alone what we should do about them.

The knowledge that would have saved the Easter Islanders' civilization has already been in our possession for centuries. A sextant would have allowed them to explore their ocean and bring back the seeds of new forests and of new ideas. Greater wealth, and a written culture, would have enabled them to recover after a devastating plague. But, most of all, they would have been better at solving problems of all kinds if they had known some of our ideas about how to do that, such as the rudiments of a scientific outlook. Such knowledge would not have guaranteed their welfare, any more than it guarantees ours. Nevertheless, the fact that their civilization failed for lack of what ours discovered long ago cannot be an ominous 'warning of what the future could hold' for us.

This knowledge-based approach to explaining human events follows from the general arguments of this book. We know that achieving arbitrary physical transformations that are not forbidden by the laws of physics (such as replanting a forest) can only be a matter of knowing how. We know that finding out how is a matter of seeking good explanations. We also know that whether a particular attempt to make progress will succeed or not is profoundly unpredictable. It can be understood in retrospect, but not in terms of factors that could have been known in advance. Thus we now understand why alchemists never succeeded at transmutation: because they would have had to understand some nuclear physics first. But this could not have been known at the time. And the progress that they did make – which led to the science of chemistry – depended strongly on how individual alchemists *thought*, and only peripherally on factors like which chemicals could be found nearby. The conditions for a beginning of infinity exist in almost every human habitation on Earth.

In his book *Guns, Germs and Steel*, the biogeographer Jared Diamond takes the opposite view. He proposes what he calls an 'ultimate explanation' of why human history was so different on different continents. In particular, he seeks to explain why it was Europeans who sailed out to conquer the Americas, Australasia and Africa and not vice versa. In Diamond's view, the psychology and philosophy and politics of historical events are no more than ephemeral ripples on the great river of history. Its course is set by factors independent of human ideas and decisions. Specifically, he says, the continents on our planet had different natural resources – different geographies, plants, animals and micro-organisms – and, details aside, that is what explains the broad sweep of history, including which human ideas were created and what decisions were made, politics, philosophy, cutlery and all.

For example, part of his explanation of why the Americas never developed a technological civilization before the advent of Europeans is that there were no animals there suitable for domestication as beasts of burden.

Llamas are native to South America, and have been used as beasts of burden since prehistoric times, so Diamond points out that they are not native to the continent as a whole, but only to the Andes mountains. Why did no technological civilization arise in the Andes mountains? Why did the Incan Empire not have an Enlightenment? Diamond's position is that other biogeographical factors were unfavourable.

The communist thinker Friedrich Engels proposed the same ultimate

explanation of history, and made the same proviso about llamas, in 1884:

> The Eastern Hemisphere ... possessed nearly all the animals adaptable to domestication ... The Western Hemisphere, America, had no mammals that could be domesticated except the llama, which, moreover, was only found in one part of South America ... Owing to these differences in natural conditions, the population of each hemisphere now goes on its own way ...
>
> *The Origin of the Family, Private Property and the State*
> (Friedrich Engels, based on notes by Karl Marx)

But why did llamas *continue* to be 'only found in one part of South America', if they could have been useful elsewhere? Engels did not address that issue. But Diamond realized that it 'cries out for explanation'. Because, unless the reason that llamas were not exported was itself biogeographical, Diamond's 'ultimate explanation' is false. So he proposed a biogeographical reason: he pointed out that a hot, lowland region, unsuitable for llamas, separates the Andes from the highlands of Central America where llamas would have been useful in agriculture.

But, again, *why* must such a region have been a barrier to the spread of domesticated llamas? Traders travelled between South and Central America for centuries, perhaps overland and certainly by sea. Where there are long-range traders, it is not necessary for an idea to be useful in an unbroken line of places for it to be able to spread. As I remarked in Chapter 11, knowledge has the unique ability to take aim at a distant target and utterly transform it while having scarcely any effect on the space between. So, what would it have taken for some of those traders to take some llamas north for sale? Only the idea: the leap of imagination to guess that if something is useful here, it might be useful there too. And the boldness to take the speculative and physical risk. Polynesian traders did exactly that. They ranged further, across a more formidable natural barrier, carrying goods including livestock. Why did none of the South American traders ever think of selling llamas to the Central Americans? We may never know – but why should it have had anything to do with geography? They may simply have been too set in their ways. Perhaps innovative uses for animals were taboo. Perhaps such a trade was attempted, but failed every time because of sheer bad luck. But, whatever the reason was, it cannot have been that the hot region constituted a physical barrier, for it did not.

Those are the parochial considerations. The bigger picture is that the spread of llamas *can only* have been prevented by people's ideas and outlook. Had the Andeans had a Polynesian outlook instead, llamas might have spread all over the Americas. Had the ancient Polynesians not had that outlook, they might never have settled Polynesia in the first place, and biogeographical explanations would now be referring to the great ocean barrier as the 'ultimate explanation' for that. If the Polynesians had been even better at long-range trading, they might have managed to transport horses from Asia to their islands and thence to South America – a feat perhaps no more impressive than Hannibal's transporting elephants across the Alps. If the ancient Greek enlightenment had continued, Athenians might have been the first to settle the Pacific islands and *they* would now be the 'Polynesians'. Or, if the early Andeans had worked out how to breed giant war llamas and had ridden out to explore and conquer before anyone else had even thought of domesticating the horse, South American biogeographers might now be explaining that their ancestors colonized the world because no other continent had llamas.

Moreover, the Americas had not always lacked large quadrupeds. When the first humans arrived there, many species of 'mega-fauna' were common, including wild horses, mammoths, mastodons and other members of the elephant family. According to some theories, the humans hunted them to extinction. What would have happened if one of those hunters had had a different idea: to ride the beast before killing it? Generations later, the knock-on effects of that bold conjecture might have been tribes of warriors on horses and mammoths pouring back through Alaska and re-conquering the Old World. Their descendants would now be attributing this to the geographical distribution of mega-fauna. But the real cause would have been that one idea in the mind of that one hunter.

In early prehistory, populations were tiny, knowledge was parochial, and history-making ideas were millennia apart. In those days, a meme spread only when one person observed another enacting it nearby, and (because of the staticity of cultures) rarely even then. So at that time human behaviour resembled that of other animals, and much of what happened was indeed explained by biogeography. But developments such as abstract language, explanation, wealth above the level of subsist-ence, and long-range trade all had the potential to erode parochialism and hence to give causal power to ideas. By the time history began to be recorded, it had long since become the history of ideas far more than anything else – though unfortunately the ideas were still mainly of the self-disabling, anti-rational variety. As for subsequent history, it would take considerable dedication to insist that biogeographical explanations account for the broad sweep of events. Why, for instance, did the societies in North America and Western Europe, rather than Asia and Eastern Europe, win the Cold War? Analysing climate, minerals, flora, fauna and diseases can teach us nothing about that. The explanation is that the Soviet system lost because its ideology wasn't true, and all the biogeography in the world cannot explain what was false about it.

Coincidentally, one of the things that was most false about the Soviet ideology was the very idea that there is an ultimate explanation of history in mechanical, non-human terms, as proposed by Marx, Engels and Diamond. Quite generally, mechanical reinterpretations of human affairs not only lack explanatory power, they are morally wrong as well, for in effect they deny the humanity of the participants, casting them and their ideas merely as side effects of the landscape. Diamond says that his main reason for writing *Guns, Germs and Steel* was that, unless people are convinced that the relative success of Europeans was caused by biogeography, they will for ever be tempted by *racist* explanations. Well, not readers of this book, I trust! Presumably

Diamond can look at ancient Athens, the Renaissance, the Enlighten-ment – all of them the quintessence of causation through the power of abstract ideas – and see no way of attributing those events to ideas and to people; he just takes it for granted that the only alternative to one reductionist, dehumanizing reinterpretation of events is another.

In reality, the difference between Sparta and Athens, or between Savonarola and Lorenzo de' Medici, had nothing to do with their genes; nor did the difference between the Easter Islanders and the imperial British. They were all people – universal explainers and constructors. But their *ideas* were different. Nor did landscape cause the Enlightenment. It would be much truer to say that the landscape we live in is the *product* of ideas. The primeval landscape, though packed with evidence and therefore opportunity, contained not a single idea. It is knowledge alone that converts landscapes into resources, and humans alone who are the authors of explanatory knowledge and hence of the uniquely human behaviour called 'history'.

Physical resources such as plants, animals and minerals afford opportunities, which may inspire new ideas, but they can neither create ideas nor cause people to have particular ideas. They also cause problems, but they do not prevent people from finding ways to solve those problems. Some overwhelming natural event like a volcanic eruption might have wiped out an ancient civilization regardless of what the victims were thinking, but that sort of thing is exceptional. Usually, if there are human beings left alive to think, there are ways of thinking that can improve their situation, and then improve it further. Unfortunately, as I have explained, there are also ways of thinking that can prevent all improvement. Thus, since the beginning of civilization and before, both the principal opportunities for progress and the principal obstacles to progress have consisted of ideas alone. These are the determinants of the broad sweep of history. The primeval distribution of horses or llamas or flint or uranium can affect only the details, and then only *after* some human being has had an idea for how to use those things. The effects of ideas and decisions almost entirely determine which biogeographical factors have a bearing on the next chapter of human history, and what that effect will be. Marx, Engels and Diamond have it the wrong way round.

A thousand years is a long time for a static society to survive. We think of the great centralized empires of antiquity which lasted even longer; but that is a selection effect: we have no record of most static societies, and they must have been much shorter-lived. A natural guess is that most were destroyed by the first challenge that would have required the creation of a significantly new pattern of behaviour. The isolated location of Easter Island, and the relatively hospitable nature of its environment, might have given its static society a longer lifespan than it would have had if it had been exposed to more tests by nature and by other societies. But even those factors are still largely human, not biogeographical: if the islanders had known how to make long-range ocean voyages, the island would not have been 'isolated' in the relevant sense. Likewise, how 'hospitable' Easter Island is depends on what the inhabitants know. If its settlers had known as little about survival techniques as I do, then they would not have survived their first week on the island. And, on the other hand, today thousands of people live on Easter Island without starving and without a forest – though now they are planting one because they want to and know how.

The Easter Island civilization collapsed because no human situation is free of new problems, and static societies are inherently unstable in the face of new problems. Civilizations rose and collapsed on other South Pacific islands too – including Pitcairn Island. That was part of the broad sweep of history in the region. And, in the big picture, the cause was that they all had problems that they failed to solve. The Easter Islanders failed to navigate their way off the island, just as the Romans failed to solve the problem of how to change governments peacefully. If there was a forestry disaster on Easter Island, that was not what brought its inhabitants down: it was that they were chronically unable to solve the problem that this raised. If that problem had not dispatched their civilization, some other problem eventually would have. Sustaining their civilization in its static, statue-obsessed state was never an option. The only options were whether it would collapse suddenly and painfully, destroying most of what little knowledge they had, or change slowly and for the better. Perhaps they would have chosen the latter if only they had known how.

We do not know what horrors the Easter Island civilization per-petrated in the course of preventing progress. But apparently its fall did not improve anything. Indeed, the fall of tyranny is never enough.

The sustained creation of knowledge depends also on the presence of certain kinds of idea, particularly optimism, and an associated tradition of criticism. There would have to be social and political institutions that incorporated and protected such traditions: a society in which some degree of dissent and deviation from the norm was tolerated, and whose educational practices did not entirely extinguish creativity. None of that is trivially achieved. Western civilization is the current consequence of achieving it – which is why, as I said, it already has what it takes to avoid an Easter Island disaster. If it really is facing a crisis, it must be some other crisis. If it ever collapses, it will be in some other way and if it needs to be saved, it will have to be by its own, unique methods.

In 1971, while I was still at school, I attended a lecture for high-school students entitled 'Population, Resources, Environment'. It was given by the population scientist Paul Ehrlich. I do not remember what I was expecting – I don't think I had ever heard of 'the environment' before – but nothing had prepared me for such a bravura display of raw pessimism. Ehrlich starkly described to his young audience the living hell we would be inheriting. Half a dozen varieties of resource-management catastrophe were just around the corner, and it was already too late to avoid some of them. People would be starving to death by the billion in ten years, twenty at best. Raw materials were running out: the Vietnam War, then in progress, was a last-ditch struggle for the region's tin, rubber and petroleum. (Notice how his biogeographical explanation blithely shrugged off the political dis-agreements that were in fact causing the conflict.) The troubles of the day in American inner cities, rising crime, mental illness – all were part of the same great catastrophe. All were linked by Ehrlich to over-population, pollution and the reckless overuse of finite resources: we had created too many power stations and factories, and mines, and intensive farms – too much economic growth, far more than the planet could sustain. And, worst of all, too many people – the ultimate source of all the other ills. In this respect, Ehrlich was following in the footsteps of Malthus, making the same error: setting *predictions* of one process against *prophecies* of another. Thus he calculated that, if the United States was to sustain even its 1971 standard of living, it would have to reduce its population by three-quarters, to 50 million – which was of course impossible in the time available. The planet as a whole was overpopulated by a factor of seven, he said. Even Australia was nearing its maximum sustainable population. And so on.

We had little basis for doubting what the professor was telling us about the field he was studying. Yet for some reason our conversation afterwards was not that of a group of students who had just had their futures stolen. I do not know about the others, but I can remember when I stopped worrying. At the end of the lecture a girl asked Ehrlich a question. I have forgotten the details, but it had the form 'What if we solve [one of the problems that Ehrlich had described] within the next few years? Wouldn't that affect your conclusion?' Ehrlich's reply was brisk. How could we possibly solve it? (She did not know.) And, even if we did, how could that do more than briefly delay the catastrophe? And what would we do *then*?

What a relief! Once I realized that Ehrlich's prophesies amounted to saying, 'If we stop solving problems, we are doomed,' I no longer found them shocking, for how could it be otherwise? Quite possibly that girl went on to solve the very problem she asked about, *and* the one after it. At any rate, someone must have, because the catastrophe scheduled for 1991 has still not materialized. Nor have any of the others that Ehrlich foretold.

Ehrlich thought that he was investigating a planet's physical resources and predicting their rate of decline. In fact he was prophesying the content of future knowledge. And, by envisaging a future in which only the best knowledge of 1971 was deployed, he was implicitly assuming that only a small and rapidly dwindling set of problems would ever be solved again. Furthermore, by casting problems in terms of 'resource depletion', and ignoring the human level of explanation, he missed all the important determinants of what he was trying to predict, namely: did the relevant people and institutions have what it takes to solve problems? And, more broadly, what *does* it take to solve problems?

A few years later, a graduate student in the then new subject of environmental science explained to me that *colour television* was a sign of the imminent collapse of our 'consumer society'. Why? Because, first of all, he said, it served no useful purpose. All the useful functions of television could be performed just as well in monochrome. Adding colour, at several times the cost, was merely 'conspicuous consumption'. That term had been coined by the economist Thorstein Veblen in 1902, a couple of decades before even monochrome television was invented; it meant wanting new possessions in order to show off to the neighbours. That we had now reached the physical limit of conspicuous consumption could be proved, said my colleague, by analysing the resource constraints scientifically. The cathode-ray tubes in colour televisions depended on the element *europium* to make the red phosphors on the screen. Europium is one of the rarest elements on Earth. The planet's total known reserves were only enough to build a few hundred million more colour televisions. After that, it would be back to monochrome. But worse – think what this would mean. From then on there would be two kinds of people: those with colour televisions and those without. And the same would be true of everything else that was being consumed. It would be a world with permanent class distinction, in which the elites would hoard the last of the resources and live lives of gaudy display, while, to sustain that illusory state through its final years, everyone else would be labouring on in drab resentment. And so it went on, nightmare built upon nightmare.

I asked him how he knew that no new source of europium would be discovered. He asked how I knew that it would. And, even if it were, what would we do *then*? I asked how he knew that colour cathode-ray tubes could not be built without europium. He assured me that they could not: it was a miracle that there existed even one element with the necessary properties. After all, why should nature supply elements with properties to suit our convenience?

I had to concede the point. There aren't that many elements, and each of them has only a few energy levels that could be used to emit light. No doubt they had all been assessed by physicists. If the bottom line was that there was no alternative to europium for making colour televisions, then there was no alternative.

Yet something deeply puzzled me about that 'miracle' of the red phosphor. If nature provides only one pair of suitable energy levels, why does it provide *even* one? I had not yet heard of the fine-tuning problem (it was new at the time), but this was puzzling for a similar reason. Transmitting accurate images in real time is a natural thing for people to want to do, like travelling fast. It would not have been puzzling if the laws of physics forbade it, just as they do forbid faster-than-light travel. For them to allow it but only if one knew how would be normal too. But for them *only just* to allow it would be a fine-tuning coincidence. Why would the laws of physics draw the line so close to a point that happened to have significance for human technology? It would be as if the centre of the Earth had turned out to be within a few kilometres of the centre of the universe. It seemed to violate the Principle of Mediocrity.

What made this even more puzzling was that, as with the real fine-tuning problem, my colleague was claiming that there were *many* such coincidences. His whole point was that the colour-television problem was just one representative instance of a phenomenon that was happening simultaneously in many areas of technology: the ultimate limits were being reached. Just as we were using up the last stocks of the rarest of rare-earth elements for the frivolous purpose of watching soap operas in colour, so everything that looked like progress was actually just an insane rush to exploit the last resources left on our planet. The 1970s were, he believed, a unique and terrible moment in history.

He was right in one respect: no alternative red phosphor has been discovered to this day. Yet, as I write this chapter, I see before me a superbly coloured computer display that contains not one atom of europium. Its pixels are liquid crystals consisting entirely of common elements, and it does not require a cathode-ray tube. Nor would it matter if it did, for by now enough europium has been mined to supply every human being on earth with a dozen europium-type screens, and the known reserves of the element comprise several times that amount.

Even while my pessimistic colleague was dismissing colour television technology as useless and doomed, optimistic people were discovering new ways of achieving it, and new uses for it – uses that he thought he had ruled out by considering for five minutes how well colour televisions could do the existing job of monochrome ones. But what stands out, for me, is not the failed prophecy and its underlying fallacy, nor relief that the nightmare never happened. It is the contrast between two different conceptions of what *people* are. In the pessimistic conception, they are wasters: they take precious resources and madly convert them into useless coloured pictures. This is *true* of static societies: those statues really were what my colleague thought colour televisions are – which is why comparing our society with the 'old culture' of Easter Island is exactly wrong. In the optimistic conception – the one that was unforeseeably vindicated by events – people are problem-solvers: creators of the unsustainable solution and hence also of the next problem. In the pessimistic conception, that distinctive ability of people is a disease for which sustainability is the cure. In the optimistic one, sustainability is the disease and people are the cure.

Since then, whole new industries have come into existence to harness great waves of innovation, and in many of those – from medical imaging to video games to desktop publishing to nature documentaries like Attenborough's – colour television proved to be very useful after all. And, far from there being a permanent class distinction between monochrome- and colour-television users, the monochrome technology is now practically extinct, as are cathode-ray televisions. Colour displays are now so cheap that they are being given away free with magazines as advertising gimmicks. And all those technologies, far from being divisive, are inherently egalitarian, sweeping away many formerly entrenched barriers to people's access to information, opinion, art and education.

Optimistic opponents of Malthusian arguments are often – rightly – keen to stress that all evils are due to lack of knowledge, and that problems are soluble. Prophecies of disaster such as the ones I have described do illustrate the fact that the prophetic mode of thinking, no matter how plausible it seems prospectively, is fallacious and inherently biased. However, to expect that problems will *always* be solved in time to avert disasters would be the same fallacy. And, indeed, the deeper and more dangerous mistake made by Malthusians is that they claim to have a way of *averting* resource-allocation disasters (namely, sustainability). Thus they also deny that other great truth that I suggested we engrave in stone: *problems are inevitable.*

A solution may be problem-free for a period, and in a parochial application, but there is no way of identifying in advance which problems will have such a solution. Hence there is no way, short of stasis, to avoid unforeseen problems arising from new solutions. But stasis is itself unsustainable, as witness every static society in history.

Malthus could not have known that the obscure element uranium, which had just been discovered, would eventually become relevant to the survival of civilization, just as my colleague could not have known that, within his lifetime, colour televisions would be saving lives every day.

So there is no resource-management strategy that can prevent disasters, just as there is no political system that provides only good leaders and good policies, nor a scientific method that provides only true theories. But there are ideas that reliably *cause* disasters, and one of them is, notoriously, the idea that the future can be scientifically planned. The only rational policy, in all three cases, is to judge institutions, plans and ways of life according to how good they are at correcting mistakes: removing bad policies and leaders, superseding bad explanations, and recovering from disasters.

For example, one of the triumphs of twentieth-century progress was the discovery of antibiotics, which ended many of the plagues and endemic illnesses that had caused suffering and death since time immemorial. However, it has been pointed out almost from the outset by critics of 'so-called progress' that this triumph may only be temporary, because of the evolution of antibiotic-resistant pathogens. This is often held up as an indictment of – to give it its broad context – Enlightenment hubris. We need lose only one battle in this war of science against bacteria and their weapon, evolution (so the argument goes), to be doomed, because our other 'so-called progress' – such as cheap world-wide air travel, global trade, enormous cities – makes us more vulnerable than ever before to a global pandemic that could exceed the Black Death in destructiveness and even cause our extinction.

But *all* triumphs are temporary. So to use this fact to reinterpret progress as 'so-called progress' is bad philosophy. The fact that reliance on specific antibiotics is unsustainable is only an indictment from the point of view of someone who expects a sustainable lifestyle. But in reality there is no such thing. Only progress is sustainable.

The prophetic approach can see only what one might do to *postpone* disaster, namely improve sustainability: drastically reduce and disperse the population, make travel difficult, suppress contact between different geographical areas. A society which did this would not be able to afford the kind of scientific research that would lead to new antibiotics. Its members would hope that their lifestyle would protect them instead. But note that this lifestyle did not, when it was tried, prevent the Black Death. Nor would it cure cancer.

Prevention and delaying tactics are useful, but they can be no more than a minor part of a viable strategy for the future. Problems are inevitable, and sooner or later survival will depend on being able to cope when prevention and delaying tactics have failed. Obviously we need to work towards cures. But we can do that only for diseases that we already know about. So we need the capacity to deal with unforeseen, unforeseeable failures. For this we need a large and vibrant research community, interested in explanation and problem-solving. We need the wealth to fund it, and the technological capacity to implement what it discovers.

This is also true of the problem of climate change, about which there is currently great controversy. We face the prospect that carbon-dioxide emissions from technology will cause an increase in the average temperature of the atmosphere, with harmful effects such as droughts, sea-level rises, disruption to agriculture, and the extinctions of some species. These are forecast to outweigh the beneficial effects, such as an increase in crop yields, a general boost to plant life, and a reduction in the number of people dying of hypothermia in winter. Trillions of dollars, and a great deal of legislation and institutional change, intended to reduce those emissions, currently hang on the outcomes of simulations of the planet's climate by the most powerful supercomputers, and on projections by economists about what those computations imply about the economy in the next century. In the light of the above discussion, we should notice several things about the controversy and about the underlying problem.

First, we have been lucky so far. Regardless of how accurate the prevailing climate models are, it is uncontroversial from the laws of physics, without any need for supercomputers or sophisticated modelling, that such emissions *must*, *eventually*, increase the temperature, which must, eventually, be harmful. Consider, therefore: what if the relevant parameters had been just slightly different and the moment of disaster had been in, say, 1902 – Veblen's time – when carbon-dioxide emissions were already orders of magnitude above their pre-Enlightenment values. Then the disaster would have happened before anyone could have predicted it or known what was happening. Sea levels would have risen, agriculture would have been disrupted, millions would have begun to die, with worse to come. And the great issue of the day would have been not how to prevent it but what could be done about it.

They had no supercomputers then. Because of Babbage's failures and the scientific community's misjudgements – and, perhaps most importantly, their lack of wealth – they lacked the vital technology of automated computing altogether. Mechanical calculators and roomfuls of clerks would have been insufficient. But, much worse: they had almost no atmospheric physicists. In fact the total number of physicists of all kinds was a small fraction of the number who today work on climate change alone. From society's point of view, physicists were a luxury in 1902, like colour televisions were in the 1970s. Yet, to recover from the disaster, society would have needed more scientific knowledge, and better technology, and more of it – that is to say, more wealth. For instance, in 1900, building a sea wall to protect the coast of a low-lying island would have required resources so enormous that the only islands that could have afforded it would have been those with either large concentrations of cheap labour or exceptional wealth, as in the Netherlands, much of whose population already lived below sea level thanks to the technology of dyke-building.

This is a challenge that is highly susceptible to automation. But people were in no position to address it in that way. All relevant machines were underpowered, unreliable, expensive, and impossible to produce in large numbers. An enormous effort to construct a Panama canal had just failed with the loss of thousands of lives and vast amounts of money, due to inadequate technology and scientific knowledge. And, to compound those problems, the world as a whole had very little wealth by today's standards. Today, a coastal defence project would be well within the capabilities of almost any coastal nation – and would add decades to the time available to find other solutions to rising sea levels.

If none are found, what would we do *then*? That is a question of a wholly different kind, which brings me to my second observation on the climate-change controversy. It is that, while the supercomputer simulations make (conditional) *predictions*, the economic forecasts make almost pure *prophecies*. For we can expect the future of human responses to climate to depend heavily on how successful people are at creating new knowledge to address the problems that arise. So comparing predictions with prophecies is going to lead to that same old mistake.

Again, suppose that disaster had already been under way in 1902. Consider what it would have taken for scientists to forecast, say, carbon-dioxide emissions for the twentieth century. On the (shaky) assumption that energy use would continue to increase by roughly the same exponential factor as before, they could have estimated the resulting increase in emissions. But that estimate would not have included the effects of nuclear power. It could not have, because radioactivity itself had only just been discovered, and would not be harnessed for power until the middle of the century. But suppose that somehow they had been able to foresee that. Then they might have modified their carbon-dioxide forecast, and concluded that emissions could easily be restored to below the 1902 level by the end of the century. But, again, that would only be because they could not possibly foresee the campaign against nuclear power, which would put a stop to its expansion (ironically, on environmental grounds) before it ever became a significant factor in reducing emissions. And so on. Time and again, the unpredictable factor of new human ideas, both good and bad, would make the scientific prediction useless. The same is bound to be true – even more so – of forecasts today for the coming century. Which brings me to my third observation about the current controversy.

It is not yet accurately known how sensitive the atmosphere's temperature is to the concentration of carbon dioxide – that is, how much a given increase in concentration increases the temperature. This number is important politically, because it affects how urgent the problem is: high sensitivity means high urgency; low sensitivity means the opposite. Unfortunately, this has led to the political debate being dominated by the side issue of how 'anthropogenic' (human-caused) the increase in temperature to date has been. It is as if people were arguing about how best to prepare for the next hurricane while all agreeing that the only hurricanes one should prepare for are human-induced ones. All sides seem to assume that if it turns out that a *random* fluctuation in the temperature is about to raise sea levels, disrupt agriculture, wipe out species and so on, our best plan would be simply to grin and bear it. Or if two-thirds of the increase is anthropogenic, we should not mitigate the effects of the other third.

Trying to predict what our net effect on the environment will be for the next century and then subordinating all policy decisions to optimizing that prediction cannot work. We cannot know how much to reduce emissions by, nor how much effect that will have, because we cannot know the future discoveries that will make some of our present actions seem wise, some counter-productive and some irrelevant, nor how much our efforts are going to be assisted or impeded by sheer luck. Tactics to delay the onset of foreseeable problems may help. But they cannot replace, and must be subordinate to, increasing our ability to intervene *after* events turn out as we did not foresee. If that does not happen in regard to carbon-dioxide-induced warming, it will happen with something else.

Indeed, we did not foresee the global-warming disaster. I call it a disaster because the prevailing theory is that our best option is to prevent carbon-dioxide emissions by spending vast sums and enforcing severe worldwide restrictions on behaviour, and that is already a disaster by any reasonable measure. I call it unforeseen because we now realize that it was already under way even in 1971, when I attended that lecture. Ehrlich did tell us that agriculture was soon going to be devastated by rapid climate change. But the change in question was going to be global *cooling*, caused by smog and the condensation trails of supersonic aircraft. The possibility of warming caused by gas emissions had already been mooted by some scientists, but Ehrlich did not consider it worth mentioning. He told us that the evidence was that a general cooling trend had already begun, and that it would continue with catastrophic effects, though it would be reversed in the very long term because of 'heat pollution' from industry (an effect that is currently at least a hundred times smaller than the global warming that preoccupies us).

There is a saying that an ounce of prevention equals a pound of cure. But that is only when one knows what to prevent. No precautions can avoid problems that we do not yet foresee. To prepare for those, there is nothing we can do but increase our ability to put things right if they go wrong. Trying to rely on the sheer good luck of avoiding bad outcomes indefinitely would simply guarantee that we would eventually fail without the means of recovering.

The world is currently buzzing with plans to force reductions in gas emissions at almost any cost. But it ought to be buzzing much more with plans to reduce the temperature, or for how to thrive at a higher temperature. And not at all costs, but efficiently and cheaply. Some such plans exist – for instance to remove carbon dioxide from the atmosphere by a variety of methods; and to generate clouds over the oceans to reflect sunlight; and to encourage aquatic organisms to absorb more carbon dioxide. But at the moment these are very minor research efforts. Neither supercomputers nor international treaties nor vast sums are devoted to them. They are not central to the human effort to face this problem, or problems like it.

This is dangerous. There is as yet no serious sign of retreat into a sustainable lifestyle (which would really mean achieving only the *semblance* of sustainability), but even the aspiration is dangerous. For what would we be aspiring to? To forcing the future world into our image, endlessly reproducing our lifestyle, our misconceptions and our mistakes. But if we choose instead to embark on an open-ended journey of creation and exploration whose every step is unsustainable until it is redeemed by the next – if this becomes the prevailing ethic and aspiration of our society – then the ascent of man, the beginning of infinity, will have become, if not secure, then at least sustainable.

TERMINOLOGY

The ascent of man   The beginning of infinity. Moreover, Jacob Bronowski's *The Ascent of Man* was one of the inspirations for this book.

Sustain   The term has two almost opposite, but often confused, meanings: to provide someone with what they need, and to prevent things from changing.

MEANINGS OF 'THE BEGINNING OF INFINITY' ENCOUNTERED IN THIS CHAPTER

– Rejecting (the semblance of) sustainability as an aspiration or a constraint on planning.

SUMMARY

Static societies eventually fail because their characteristic inability to create knowledge rapidly must eventually turn some problem into a catastrophe. Analogies between such societies and the technological civilization of the West today are therefore fallacies. Marx, Engels and Diamond's 'ultimate explanation' of the different histories of different societies is false: history is the history of ideas, not of the mechanical effects of biogeography. Strategies to prevent foreseeable disasters are bound to fail eventually, and cannot even address the unforeseeable. To prepare for those, we need rapid progress in science and technology and as much wealth as possible.

18

The Beginning

'This is Earth. Not the eternal and only home of mankind, but only a starting point of an infinite adventure. All you need do is make the decision [to end your static society]. It is yours to make.'

[With that decision] came the end, the final end of Eternity. – And the beginning of Infinity.

Isaac Asimov, *The End of Eternity* (1955)

The first person to measure the circumference of the Earth was the astronomer Eratosthenes of Cyrene, in the third century BCE. His result was fairly close to the actual value, which is about 40,000 kilometres. For most of history this was considered an enormous distance, but with the Enlightenment that conception gradually changed, and nowadays we think of the Earth as small. That was brought about mainly by two things: first, by the science of astronomy, which discovered titanic entities compared with which our planet is indeed unimaginably tiny; and, second, by technologies that have made worldwide travel and communication commonplace. So the Earth has become smaller both relative to the universe and relative to the scale of human action.

Thus, in regard to the *geography* of the universe and to our place in it, the prevailing world view has rid itself of some parochial misconceptions. We know that we have explored almost the whole surface of that formerly enormous sphere; but we also know that there are far more places left to explore in the universe (and beneath the surface of the Earth's land and oceans) than anyone imagined while we still had those misconceptions.

In regard to theoretical knowledge, however, the prevailing world view has not yet caught up with Enlightenment values. Thanks to the fallacy and bias of prophecy, a persistent assumption remains that our existing theories are at or fairly close to the limit of what it is knowable – that we are *nearly there*, or perhaps halfway there. As the economist David Friedman has remarked, most people believe that an income of about twice their own should be sufficient to satisfy any reasonable person, and that no genuine benefit can be derived from amounts above that. As with wealth, so with scientific knowledge: it is hard to imagine what it would be like to know twice as much as we do, and so if we try to prophesy it we find ourselves just picturing the next few decimal places of what we already know. Even Feynman made an uncharacteristic mistake in this regard when he wrote:

> I think there will certainly not be novelty, say for a thousand years. This thing cannot keep going on so that we are always going to discover more and more new laws. If we do, it will become boring that there are so many levels one underneath the other . . . We are very lucky to live in an age in which we are still making discoveries. It is like the discovery of America – you only discover it once.

> *The Character of Physical Law* (1965)

Among other things, Feynman forgot that the very concept of a 'law' of nature is not cast in stone. As I mentioned in Chapter 5, this concept was different before Newton and Galileo, and it may change again. The concept of levels of explanation dates from the twentieth century, and it too will change if I am right that, as I guessed in Chapter 5, there are fundamental laws that look emergent relative to microscopic physics. More generally, the most fundamental discoveries have always, and will always, not only consist of new explanations, but use new modes of explanation. As for being boring, that is merely a prophecy that criteria for judging problems will not evolve as fast as the problems themselves; but there is no argument for that other than a failure of imagination. Even Feynman cannot get round the fact that the future is not yet imaginable.

Shedding that kind of parochialism is something that will have to be done again and again in the future. A level of knowledge, wealth, computer power or physical scale that seems absurdly huge at any given instant will later be considered pathetically tiny. Yet we shall never reach anything like an unproblematic state. Like the guests at Infinity Hotel, we shall never be 'nearly there'.

There are two versions of 'nearly there'. In the dismal version, knowledge is bounded by laws of nature or supernatural decree, and progress has been a temporary phase. Though this is rank pessimism by my definition, it has gone under various names – including 'optimism' – and has been integral to most world views in the past. In the cheerful version, all remaining ignorance will soon be eliminated or confined to insignificant areas. This is optimistic in form, but the closer one looks, the more pessimistic it becomes in substance. In politics, for instance, utopians promise that a finite number of already-known changes can bring about a perfected human state, and that is a well-known recipe for dogmatism and tyranny.

In physics, imagine that Lagrange had been right that 'the system of the world can be discovered only once', or that Michelson had been right that all physics still undiscovered in 1894 was about 'the sixth place of decimals'. They were claiming to *know* that anyone who subsequently became curious about what underlay that 'system of the world' would be enquiring futilely into the incomprehensible. And that anyone who ever wondered at an anomaly, and suspected that some fundamental explanation contained a misconception, would be mistaken.

Michelson's future – our present – would have been lacking in explanatory knowledge to an extent that we can no longer easily imagine. A vast range of phenomena already known to him, such as gravity, the properties of the chemical elements, and the luminosity of the sun, remained to be explained. He was claiming that these phenomena would only ever appear as list of facts or rules of thumb, to be memorized but never understood or fruitfully questioned. Every such frontier of fundamental knowledge that existed in 1894 would have been a barrier beyond which nothing would ever be amenable to explanation. There would be no such thing as the internal structure of atoms, no dynamics of space and time, no such subject as cosmology, no explanation for the equations governing gravitation or electromagnetism, no connections between physics and the theory of computation . . . The deepest *structure* in the world would be an inexplicable, anthropocentric boundary, coinciding with the boundary of what the physicists of 1894 thought they understood. And nothing inside that boundary – like, say, the existence of a force of gravity – would ever turn out to be profoundly false.

Nothing very important would ever be discovered in the laboratory that Michelson was opening. Each generation of students who studied there, instead of striving to understand the world more deeply than their teachers, could aspire to nothing better than to emulate them – or, at best, to discover the seventh decimal place of some constant whose sixth was already known. (But how? The most sensitive scientific instruments today depend on fundamental discoveries made after 1894.) Their system of the world would for ever remain a tiny, frozen island of explanation in an ocean of incomprehensibility. Michelson's 'fundamental laws and facts of physical science', instead of being the beginning of an infinity of further understanding, as they were in reality, would have been the last gasp of reason in the field.

I doubt that either Lagrange or Michelson thought of himself as pessimistic. Yet their prophecies entailed the dismal decree that *no matter what you do, you will understand no further*. It so happens that both of them had made discoveries which could have led them to the very progress whose possibility they denied. They should have been seeking that progress, should they not? But almost no one is creative in fields in which they are pessimistic.

I remarked at the end of Chapter 13 that the desirable future is one where we progress from misconception to ever better (less mistaken) misconception. I have often thought that the nature of science would be better understood if we called theories 'misconceptions' from the outset, instead of only after we have discovered their successors. Thus we could say that Einstein's Misconception of Gravity was an improvement on Newton's Misconception, which was an improvement on Kepler's. The neo-Darwinian Misconception of Evolution is an improvement on Darwin's Misconception, and his on Lamarck's. If people thought of it like that, perhaps no one would need to be reminded that science claims neither infallibility nor finality.

Perhaps a more practical way of stressing the same truth would be to frame the growth of knowledge (all knowledge, not only scientific) as a continual transition from *problems* to *better problems*, rather than from problems to solutions or from theories to better theories. This is the positive conception of 'problems' that I stressed in Chapter 1. Thanks to Einstein's discoveries, our current problems in physics embody more knowledge than Einstein's own problems did. His problems were rooted in the discoveries of Newton and Euclid, while most problems that preoccupy physicists today are rooted in – and would be inaccessible mysteries without – the discoveries of twentieth-century physics.

The same is true in mathematics. Although mathematical theorems are rarely proved *false* once they have been around for a while, what does happen is that mathematicians' understanding of what is fundamental improves. Abstractions that were originally studied in their own right are understood as aspects of more general abstractions, or are related in unforeseen ways to other abstractions. And so progress in mathematics also goes from problems to better problems, as does progress in all other fields.

Optimism and reason are incompatible with the conceit that our knowledge is 'nearly there' in any sense, or that its foundations are. Yet comprehensive optimism has always been rare, and the lure of the prophetic fallacy strong. But there have always been exceptions. Socrates famously claimed to be deeply ignorant. And Popper wrote:

> I believe that it would be worth trying to learn something about the world even if in trying to do so we should merely learn that we do not know much ... It might be well for all of us to remember that, while differing widely in the various little bits we know, in our infinite ignorance we are all equal.
>
> *Conjectures and Refutations* (1963)

Infinite ignorance is a necessary condition for there to be infinite potential for knowledge. Rejecting the idea that we are 'nearly there' is a necessary condition for the avoidance of dogmatism, stagnation and tyranny.

In 1996 the journalist John Horgan caused something of a stir with his book *The End of Science: Facing the Limits of Knowledge in the Twilight of the Scientific Age*. In it, he argued that the final truth in all fundamental areas of science – or at least as much of it as human minds would ever be capable of grasping – had already been discovered during the twentieth century.

Horgan wrote that he had originally believed science to be 'open-ended, even infinite'. But he became convinced of the contrary by (what I would call) a series of misconceptions and bad arguments. His basic misconception was empiricism. He believed that what distinguishes science from unscientific fields such as literary criticism, philosophy or art is that science has the ability to 'resolve questions' objectively (by comparing theories with reality), while other fields can produce only multiple, mutually incompatible interpretations of any issue. He was mistaken in both respects. As I have explained throughout this book, there is objective truth to be found in all those fields, while finality or infallibility cannot be found anywhere.

Horgan accepts from the bad philosophy of 'postmodern' literary criticism its wilful confusion between two kinds of 'ambiguity' that can exist in philosophy and art. The first is the 'ambiguity' of multiple true meanings, either intended by the author or existing because of the reach of the ideas. The second is the ambiguity of deliberate vagueness, confusion, equivocation or self-contradiction. The first is an attribute of deep ideas, the second an attribute of deep silliness. By confusing them, one ascribes to the best art and philosophy the qualities of the worst. Since, in that view, readers, viewers and critics can attribute any meaning they choose to the second kind of ambiguity, bad philosophy declares the same to be true of all knowledge: all meanings are equal, and none of them is objectively true. One then has a choice between complete nihilism or regarding *all* 'ambiguity' as a good thing in those fields. Horgan chooses the latter option: he classifies art and philosophy as 'ironic' fields, irony being the presence of multiple conflicting meanings in a statement.

However, unlike the postmodernists, Horgan thinks that science and mathematics are the shining exceptions to all that. They alone are capable of non-ironic knowledge. But there is also, he concludes, such a thing as *ironic science* – the kind of science that cannot 'resolve questions' because, essentially, it is just philosophy or art. Ironic science *can* continue indefinitely, but that is precisely because it never resolves anything; it never discovers objective truth. Its only value is in the eye of the beholder. So the future, according to Horgan, belongs to ironic knowledge. Objective knowledge has already reached its ultimate bounds.

Horgan surveys some of the open questions of fundamental science, and judges them all either 'ironic' or non-fundamental, in support of his thesis. But that conclusion was made inevitable by his premises alone. For consider the prospect of *any* future discovery that would constitute fundamental progress. We cannot know what it is, but bad philosophy can already split it, on principle, into a new rule of thumb and a new 'interpretation' (or explanation). The new rule of thumb cannot possibly be fundamental: it will just be another equation. Only a trained expert could tell the difference between it and the old equation. The new 'interpretation' will by definition be pure philosophy, and hence must be 'ironic'. By this method, any potential progress can be pre-emptively reinterpreted as non-progress.

Horgan rightly points out that his prophecy cannot be proved false by placing it in the context of previous failed prophecies. The fact that Michelson was wrong about the achievements of the nineteenth century, and Lagrange about those of the seventeenth, does not imply that Horgan was wrong about those of the twentieth. However, it so happens that our current scientific *knowledge* includes a historically unusual number of deep, fundamental problems. Never before in the history of human thought has it been so obvious that our knowledge is tiny and our ignorance vast. And so, unusually, Horgan's pessimism contradicts existing knowledge as well as being a prophetic fallacy. For example, the problem-situation of fundamental physics today has a radically different structure from that of 1894. Although physicists then were aware of some phenomena and theoretical issues which we now recognize as harbingers of the revolutionary explanations to come, their importance was unclear at the time. It was hard to distinguish those harbingers from anomalies that would eventually be cleared up with existing explanations plus the tweaking of the 'sixth place of decimals' or minor terms in a formula. But today there is no such excuse for denying that some of our problems are fundamental. Our best theories are telling us of profound mismatches between themselves and the reality that they are supposed to explain.

One of the most blatant examples of that is that physics currently has *two* fundamental 'systems of the world' – quantum theory and the general theory of relativity – and they are radically inconsistent. There are many ways of characterizing this inconsistency – known as the problem of quantum gravity – corresponding to the many proposals for solving it that have been tried without success. One aspect is the ancient tension between the discrete and the continuous. The resolution that I described in Chapter 11, in terms of continuous clouds of fungible instances of a particle with diverse discrete attributes, works only if the spacetime in which this happens is itself continuous. But if spacetime is affected by the gravitation of the cloud, then it would acquire discrete attributes.

In cosmology, there has been revolutionary progress even in the few years since *The End of Science* was written – and also since I wrote *The Fabric of Reality* soon afterwards. At the time, all viable cosmological theories had the expansion of the universe gradually slowing down, due to gravity, ever since the initial explosion at the Big Bang and for ever in the future. Cosmologists were trying to determine whether, despite slowing down, its expansion rate was sufficient to make the universe expand for ever (like a projectile that has exceeded escape velocity) or whether it would eventually recollapse in a 'Big Crunch'. Those were believed to be the only two possibilities. I discussed them in *The Fabric of Reality* because they were relevant to the question: is there a bound on the number of computational steps that a computer can execute during the lifetime of the universe? If there is, then physics will also impose a bound on the amount of knowledge that can be created – knowledge-creation being a form of computation.

Everyone's first thought was that unbounded knowledge-creation is possible only in a universe that does not recollapse. However, on analysis it turned out that the reverse is true: in universes that expand for ever, the inhabitants would run out of energy. But the cosmologist Frank Tipler discovered that in certain types of recollapsing universes the Big Crunch singularity is suitable for performing the faster-and-faster trick that we used in Infinity Hotel: an infinite sequence of computational steps could be executed in a finite time before the singularity, powered by the ever-increasing tidal effects of the gravitational collapse itself. To the inhabitants – who would eventually have to upload their personalities into computers made of something like pure tides – the universe would last for ever because they would be thinking faster and faster, without limit, as it collapsed, and storing their memories in ever smaller volumes so that access times could also be reduced without limit. Tipler called such universes 'omega-point universes'. At the time, the observational evidence was consistent with the real universe being of that type.

A small part of the revolution that is currently overtaking cosmology is that the omega-point models have been ruled out by observation. Evidence – including a remarkable series of studies of supernovae in distant galaxies – has forced cosmologists to the unexpected conclusion that the universe not only will expand for ever but has been expanding *at an accelerating rate*. Something has been counteracting its gravity.

We do not know what. Pending the discovery of a good explanation, the unknown cause has been named 'dark energy'. There are several proposals for what it might be, including effects that merely give the appearance of acceleration. But the best working hypothesis at present is that in the equations for gravity there is an additional term, of a form first mooted by Einstein in 1915 and then dropped because he realized that his explanation for it was bad. It was proposed again in the 1980s as a possible effect of quantum field theory, but again there is no theory of the physical meaning of such a term that is good enough to predict, for instance, its magnitude. The problem of the nature and effects of dark energy is no minor detail, nor does anything about it suggest a perpetually unfathomable mystery. So much for cosmology being a fundamentally completed science.

Depending on what dark energy turns out to be, it may well be possible to harness it in the distant future, to provide energy for knowledge-creation to continue for ever. Because this energy would have to be collected over ever greater distances, the computation would have to become ever slower. In a mirror image of what would happen in omega-point cosmologies, the inhabitants of the universe would notice no slowdown, because, again, they would be instantiated as computer programs whose total number of steps would be unbounded. Thus dark energy, which has ruled out one scenario for the unlimited growth of knowledge, would provide the literal driving force of another.

The new cosmological models describe universes that are infinite in their spatial dimensions. Because the Big Bang happened a finite time ago, and because of the finiteness of the speed of light, we shall only ever see a finite portion of infinite space – but that portion will continue to grow for ever. Thus, eventually, ever more unlikely phenomena will come into view. When the total volume that we can see is a million times larger than it is now, we shall see things that have a probability of one in a million of existing in space as we see it today. Everything physically possible will eventually be revealed: watches that came into existence spontaneously; asteroids that happen to be good likenesses of William Paley; everything. According to the prevailing theory, all those things *exist today*, but many times too far away for light to have reached us from them – yet.

Light becomes fainter as it spreads out: there are fewer photons per unit area. That means that ever larger telescopes are needed to detect a given object at ever larger distances. So there may be a limit to how distant – and therefore how unlikely – a phenomenon we shall ever be able to see. Except, that is, for one type of phenomenon: a beginning of infinity. Specifically, any civilization that is colonizing the universe in an unbounded way will eventually reach our location.

Hence a single infinite space could play the role of the infinitely many universes postulated by anthropic explanations of the fine-tuning coincidences. In some ways it could play that role better: if the probability that such a civilization could form is not zero, there must be infinitely many such civilizations in space, and they will eventually encounter each other. If they could estimate that probability from theory, they could test the anthropic explanation.

Furthermore, anthropic arguments could not only dispense with all those parallel universes,* they could dispense with the variant laws of physics too. Recall from Chapter 6 that all the mathematical functions that occur in physics belong to a relatively narrow class, the *analytic functions*. They have a remarkable property: if an analytic function is non-zero at even one point, then over its entire range it can pass through zero only at isolated points. So this must be true of 'the probability that an astrophysicist exists' expressed as a function of the constants of physics. We know little about this function, but we do know that it is non-zero for at least one set of values of the constants, namely ours. Hence we also know that it is non-zero for almost any values. It is

*Let me remind the reader that these highly speculative parallel universes have nothing to do with the universes or histories in the quantum multiverse, for whose existence there is overwhelming evidence. Strictly speaking, the standard anthropic explanations postulate infinitely many quantum *multiverses*.

presumably unimaginably tiny for almost all sets of values – but, never-theless, non-zero. And hence, almost whatever the constants were, there would be infinitely many astrophysicists in our single universe.

Unfortunately, at this point the anthropic explanation of fine-tuning has cancelled itself out: astrophysicists exist whether there is fine-tuning or not. So, in the new cosmology even more than in the old one, the anthropic argument does not explain the fine-tuning. Nor, therefore, can it solve the Fermi problem, 'Where are they?' It may turn out to be a necessary part of the explanation, but it can never explain anything by itself. Also, as I explained in Chapter 8, any theory involving an anthropic argument must provide a measure for defining probabilities in an infinite set of things. It is unknown how to do that in the spatially infinite universe that cosmologists currently believe we live in.

That issue has a wider scope. For example, there is the so-called 'quantum suicide argument' in regard to the multiverse. Suppose you want to win the lottery. You buy a ticket and set up a machine that will automatically kill you in your sleep if you lose. Then, in all the histories in which you do wake up, you are a winner. If you do not have loved ones to mourn you, or other reasons to prefer that most histories not be affected by your premature death, you have arranged to get something for nothing with what proponents of this argument call 'subjective certainty'. However, that way of applying probabilities does not follow directly from quantum theory, as the usual one does. It requires an additional assumption, namely that when making de-cisions one should ignore the histories in which the decision-maker is absent. This is closely related to anthropic arguments. Again, the theory of probability for such cases is not well understood, but my guess is that the assumption is false.

A related assumption occurs in the so-called *simulation argument*, whose most cogent proponent is the philosopher Nick Bostrom. Its premise is that in the distant future the whole universe as we know it is going to be simulated in computers (perhaps for scientific or historical research) many times – perhaps infinitely many times. Therefore virtu-ally all instances of us are in those simulations and not the original world. And therefore we are almost certainly living in a simulation. So the argument goes. But is it really valid to equate 'most instances' with 'near certainty' like that?

For an inkling of why it might not be, consider a thought experiment. Imagine that physicists discover that space is actually many-layered like puff pastry; the number of layers varies from place to place; the layers split in some places, and their contents split with them. Every layer has identical contents, though. Hence, although we do not feel it, instances of us split and merge as we move around. Suppose that in London space has a million layers, while in Oxford it has only one. I travel frequently between the two cities, and one day I wake up having forgotten which one I am in. It is dark. Should I bet that I am much more likely to be in London, just because a million times as many instances of me ever wake up in London as in Oxford? I think not. In that situation it is clear that counting the number of instances of oneself is no guide to the probability one ought to use in decision-making. We should be counting histories not instances. In quantum theory, the laws of physics tell us how to count histories by measure. In the case of multiple simulations, I know of no good argument for *any* way of counting them: it is an open question. But I do not see why repeating the same simulation of me a million times should in any sense make it 'more likely' that I am a simulation rather than the original. What if one computer uses a million times as many electrons as another to represent each bit of information in its memory? Am I more likely to be 'in' the former computer than the latter?

A different issue raised by the simulation argument is this: will the universe as we know it really be simulated often in the future? Would that not be immoral? The world as it exists today contains an enormous amount of suffering, and whoever ran such a simulation would be responsible for recreating it. Or would they? Are two identical instances of a quale the same thing as one? If so, then creating the simulation would not be immoral – no more so than reading a book about past suffering is immoral. But in that case how different do two simulations of people have to be before they count as two people for moral purposes? Again, I know of no good answer to those questions. I suspect that they will be answered only by the explanatory theory from which AI will also follow.

Here is a related but starker moral question. Take a powerful computer and set each bit randomly to 0 or 1 using a quantum randomizer. (That means that 0 and 1 occur in histories of equal measure.) At that point *all possible contents* of the computer's memory exist in the multiverse. So there are necessarily histories present in which the computer contains an AI program – indeed, all possible AI programs in all possible states, up to the size that the computer's memory can hold. Some of them are fairly accurate representations of you, living in a virtual-reality environment crudely resembling your actual environment. (Present-day computers do not have enough memory to simulate a realistic environment accurately, but, as I said in Chapter 7, I am sure that they have more than enough to simulate a person.) There are also people in every possible state of suffering. So my question is: is it wrong to switch the computer on, setting it executing all those programs simultaneously in different histories? Is it, in fact, the worst crime ever committed? Or is it merely inadvisable, because the combined measure of all the histories containing suffering is very tiny? Or is it innocent and trivial?

An even more dubious example of anthropic-type reasoning is the *doomsday argument*. It attempts to estimate the life expectancy of our species by assuming that the typical human is roughly halfway through the sequence of all humans. Hence we should expect the total number who will ever live to be about twice the number who have lived so far. Of course this is prophecy, and for that reason alone cannot possibly be a valid argument, but let me briefly pursue it in its own terms. First, it does not apply at all if the total number of humans is going to be infinite – for in that case every human who ever lives will live unusually early in the sequence. So, if anything, it suggests that we are at the beginning of infinity.

Also, how long is a human lifetime? Illness and old age are going to be cured soon – certainly within the next few lifetimes – and technology will also be able to prevent deaths through homicide or accidents by creating backups of the states of brains, which could be uploaded into new, blank brains in identical bodies if a person should die. Once that technology exists, people will consider it considerably more foolish not to make frequent backups *of themselves* than they do today in regard to their computers. If nothing else, evolution alone will ensure that, because those who do not back themselves up will gradually die out. So there can be only one outcome: effective immortality for the whole human population, with the present generation being one of the last that will have short lives. That being so, if our species will nevertheless have a finite lifetime, then knowing the total number of humans who will ever live provides no upper bound on that lifetime, because it cannot tell us how long the potentially immortal humans of the future will live before the prophesied catastrophe strikes.

In 1993 the mathematician Vernor Vinge wrote an influential essay entitled 'The Coming Technological Singularity', in which he estimated that, within about thirty years, predicting the future of technology would become impossible – an event that is now known simply as 'the Singularity'. Vinge associated the approaching Singularity with the achievement of AI, and subsequent discussions have centred on that. I certainly *hope* that AI is achieved by then, but I see no sign yet of the theoretical progress that I have argued must come first. On the other hand, I see no reason to single out AI as a mould-breaking technology: we already have billions of humans.

Most advocates of the Singularity believe that, soon after the AI breakthrough, *superhuman* minds will be constructed and that then, as Vinge put it, 'the human era will be over.' But my discussion of the universality of human minds rules out that possibility. Since humans are already universal explainers and constructors, they can already transcend their parochial origins, so there can be no such thing as a superhuman mind as such. There can only be further automation, allowing the existing kind of human thinking to be carried out faster, and with more working memory, and delegating 'perspiration' phases to (non-AI) automata. A great deal of this has already happened with computers and other machinery, as well as with the general increase in wealth which has multiplied the number of humans who are able to spend their time thinking. This can indeed be expected to continue. For instance, there will be ever-more-efficient human–computer interfaces, no doubt culminating in add-ons for the brain. But tasks like internet searching will never be carried out by super-fast AIs scanning billions of documents creatively for meaning, because they will not want to perform such tasks any more than humans do. Nor will artificial scientists, mathematicians and philosophers ever wield concepts or arguments that humans are inherently incapable of understanding. Universality implies that, in every important sense, humans and AIs will never be other than equal.

Similarly, the Singularity is often assumed to be a moment of unprecedented upheaval and danger, as the rate of innovation becomes too rapid for humans to cope with. But this is a parochial misconception. During the first few centuries of the Enlightenment, there has been a constant feeling that rapid and accelerating innovation is getting out of hand. But our capacity to cope with, and enjoy, changes in our technology, lifestyle, ethical norms and so on has been increasing too, with the weakening and extinction of some of the anti-rational memes that used to sabotage it. In future, when the rate of innovation will also increase due to the sheer increasing clock rate and throughput of brain add-ons and AI computers, then our capacity to cope with that will increase at the same rate or faster: if everyone were suddenly able to think a million times as fast, no one would feel hurried as a result. Hence I think that the concept of the Singularity as a sort of discontinuity is a mistake. Knowledge will continue to grow exponentially or even faster, and that is astounding enough.

The economist Robin Hanson has suggested that there have been several singularities in the history of our species, such as the agricultural revolution and the industrial revolution. Arguably, even the early Enlightenment was a 'singularity' by that definition. Who could have predicted that someone who lived through the English Civil War – a bloody struggle of religious fanatics versus an absolute monarch – and through the victory of the religious fanatics in 1651, might also live through the peaceful birth of a society that saw liberty and reason as its principal characteristics? The Royal Society, for instance, was founded in 1660 – a development that would hardly have been conceivable a generation earlier. Roy Porter marks 1688 as the beginning of the English Enlightenment. That is the date of the 'Glorious Revolution', the beginning of predominantly constitutional government along with many other rational reforms which were part of that deeper and astonishingly rapid shift in the prevailing world view.

Also, the time beyond which scientific prediction has no access is different for different phenomena. For each phenomenon it is the moment at which the creation of new knowledge may begin to make a significant difference to what one is trying to predict. Since our estimates of that, too, are subject to the same kind of horizon, we should really understand *all* our predictions as implicitly including the proviso 'unless the creation of new knowledge intervenes'.

Some explanations do have reach into the distant future, far beyond the horizons that make most other things unpredictable. One of them is that fact itself. Another is the infinite potential of explanatory knowledge – the subject of this book.

To attempt to predict anything beyond the relevant horizon is futile – it is prophecy – but *wondering* what is beyond it is not. When wondering leads to conjecture, that constitutes *speculation*, which is not irrational either. In fact it is vital. Every one of those deeply unforeseeable new ideas that make the future unpredictable will begin as a speculation. And every speculation begins with a problem: *problems* in regard to the future can reach beyond the horizon of prediction too – and problems have solutions.

In regard to understanding the physical world, we are in much the same position as Eratosthenes was in regard to the Earth: he could measure it remarkably accurately, and he knew a great deal about certain aspects of it – immensely more than his ancestors had known only a few centuries before. He must have known about such things as seasons in regions of the Earth about which he had no evidence. But he also *knew* that most of what was out there was far beyond his theoretical knowledge as well as his physical reach.

We cannot yet measure the universe as accurately as Eratosthenes measured the Earth. And we, too, *know* how ignorant we are. For instance, we know from universality that AI is attainable by writing computer programs, but we have no idea how to write (or evolve) the right one. We do not know what qualia are or how creativity works, despite having working examples of qualia and creativity inside all of us. We learned the genetic code decades ago, but have no idea why it has the reach that it has. We know that both of the deepest prevailing theories in physics must be false. We know that *people* are of fundamental significance, but we do not know whether we are among those people: we may fail, or give up, and intelligences originating elsewhere in the universe may be the beginning of infinity. And so on for all the problems I have mentioned and many more.

Wheeler once imagined writing out all the equations that might be the ultimate laws of physics on sheets of paper all over the floor. And then:

Stand up, look back on all those equations, some perhaps more hopeful than others, raise one's finger commandingly, and give the order 'Fly!' Not one of those equations will put on wings, take off, or fly. Yet the universe 'flies'.

C. W. Misner, K. S. Thorne and J. A. Wheeler, *Gravitation* (1973)

We do not know why it 'flies'. What is the difference between laws that are instantiated in physical reality and those that are not? What is the difference between a computer simulation of a person (which must *be* a person, because of universality) and a recording of that simulation (which cannot be a person)? When there are two identical simulations under way, are there two sets of qualia or one? Double the moral value or not?

Our world, which is so much larger, more unified, more intricate and more beautiful than that of Eratosthenes, and which we understand and control to an extent that would have seemed godlike to him, is nevertheless just as mysterious, yet open, to us now as his was to him then. We have lit only a few candles here and there. We can cower in their parochial light until something beyond our ken snuffs us out, or we can resist. We already see that we do not live in a senseless world. The laws of physics make sense: the world is explicable. There are higher levels of emergence and higher levels of explanation. Profound abstractions in mathematics, morality and aesthetics are accessible to us. Ideas of tremendous reach are possible. But there is also plenty in the world that does not and will not make sense until we ourselves work out how to rectify it. Death does not make sense. Stagnation does not make sense. A bubble of sense within endless senselessness does not make sense. Whether the world ultimately does make sense will depend on how *people* – the likes of us – chose to think and to act.

Many people have an aversion to infinity of various kinds. But there are some things that we do not have a choice about. There is only one way of thinking that is capable of making progress, or of surviving in the long run, and that is the way of seeking good explanations through creativity and criticism. What lies ahead of us is in any case infinity. All we can choose is whether it is an infinity of ignorance or of knowledge, wrong or right, death or life.

Bibliography

Everyone should read these

Jacob Bronowski, *The Ascent of Man* (BBC Publications, 1973)

Jacob Bronowski, *Science and Human Values* (Harper & Row, 1956)

Richard Byrne, 'Imitation as Behaviour Parsing', *Philosophical Transactions of the Royal Society* B358 (2003)

Richard Dawkins, *The Selfish Gene* (Oxford University Press, 1976)

David Deutsch, 'Comment on Michael Lockwood, "'Many Minds' Interpretations of Quantum Mechanics"', *British Journal for the Philosophy of Science* 47, 2 (1996)

David Deutsch, *The Fabric of Reality* (Allen Lane, 1997)

Karl Popper, *Conjectures and Refutations* (Routledge, 1963)

Karl Popper, *The Open Society and Its Enemies* (Routledge, 1945)

Further reading

John Barrow and Frank Tipler, *The Anthropic Cosmological Principle* (Clarendon Press, 1986)

Susan Blackmore, *The Meme Machine* (Oxford University Press, 1999)

Nick Bostrom, 'Are You Living in a Computer Simulation?', *Philosophical Quarterly* 53 (2003)

David Deutsch, 'Apart from Universes', in S. Saunders, J. Barrett, A. Kent and D. Wallace, eds., *Many Worlds?: Everett, Quantum Theory, and Reality* (Oxford University Press, 2010)

David Deutsch, 'It from Qubit', in John Barrow, Paul Davies and Charles Harper, eds., *Science and Ultimate Reality* (Cambridge University Press, 2003)

David Deutsch, 'Quantum Theory of Probability and Decisions', *Proceedings of the Royal Society* A455 (1999)

David Deutsch, 'The Structure of the Multiverse', *Proceedings of the Royal Society* A458 (2002)

Richard Feynman, *The Character of Physical Law* (BBC Publications, 1965)

Richard Feynman, *The Meaning of It All* (Allen Lane, 1998)

Ernest Gellner, *Words and Things* (Routledge & Kegan Paul, 1979)

William Godwin, *Enquiry Concerning Political Justice* (1793)

Douglas Hofstadter, *Gödel, Escher, Bach: An Eternal Golden Braid* (Basic Books, 1979)

Douglas Hofstadter, *I am a Strange Loop* (Basic Books, 2007)

Bryan Magee, *Popper* (Fontana, 1973)

Pericles, 'Funeral Oration'

Plato, *Euthyphro*

Karl Popper, *In Search of a Better World* (Routledge, 1995)

Karl Popper, *The World of Parmenides* (Routledge, 1998)

Roy Porter, *Enlightenment: Britain and the Creation of the Modern World* (Allen Lane, 2000)

Martin Rees, *Just Six Numbers* (Basic Books, 2001)

Alan Turing, 'Computing Machinery and Intelligence', *Mind*, 59, 236 (October 1950)

Jenny Uglow, *The Lunar Men* (Faber, 2002)

Vernor Vinge, 'The Coming Technological Singularity', *Whole Earth Review*, winter 1993

Index

and measurement 40, 229
visions, perception and 229, 241–2
consciousness 153–4, 157, 162–3, 318, 415
test for judging claims to have understood 154
consent 155, 266, 343
conspicuous consumption 433
constants of nature *see* physics, constants of
Constitution, US 335
and apportionment paradoxes 326–33
constructors, universal 58–60, 62, 76, 429
Continental Enlightenment 65–6, 313
continuum 43, 140–2, 159, 164, 181, 274, 298, 450
infinity of the 164, 170, 182, 195
control 45, 55–6, 62, 69–71, 88, 111, 130, 134–9, 159, 241–2, 384, 391–2, 415, 459
convergence
between Spaceship Earth and the Principle of Mediocrity 45, 53
convergent evolution 95
and error correction 350
upon the truth 231, 257, 350, 368
Conway, John 166
Copenhagen interpretation of quantum theory 308–10, 312, 315, 322, 324, 325
copying
memes not replicated by imitation 402–10
replicators *see* memes; replicators
cork 72–4
correspondence 39, 241
of theories with objective truth 353
one-to-one 167, 170, 171–2, 181, 193; tallying 128, 129, 130–31, 134, 140–41, 193, 356
cosmic rays 68, 293–4
cosmic significance *see* significance
cosmology 68, 81, 113, 445, 450–53
evolutionary cosmologies 178–9
see also astrophysics
cow, size of 35
creation of knowledge 78–105
and the argument from design 83–7
impeded by bad philosophy 305–25, 448

causing convergence 350
creationism and 79–81, 104
about environments 74–5
and error correction 140–2, 147, 271, 302
ex nihilo 104, 345
fine-tuning of the universe and 96–103, 104, 106
as a transition process from problems to better problems 446–7
Lamarckism and 87–9
and the limitation on predictability 104, 193, 197–8, 212
neo-Darwinism and 89–96
not affected by distance 275, 427
open-ended/unbounded vii, 60–65, 66, 67, 69, 81, 146, 165, 175, 450–52
optimism and 196–222, 431
requirements for 61
spark for 75
spontaneous generation and 81–3
creationism 79–81, 86, 104, 193
fine-tuning as supposed evidence for 97
and spontaneous generation 82
see also Paley, William
creativity 30
artificial 148–63; *see also* artificial intelligence (AI)
artistic 355–7; *see also* aesthetics
creative conjecture 412; *see also* conjecture
creatively changing the options 351
in discovering new explanations 7–8
as an evolutionary process in the brain 373; evolution of 398–400, 402–15, 416
future of 415–16
mutual enhancement of meme transmission and 400
needed to improve explanations 342
puzzle of what use it was in non-innovating cultures 398–402, 410–15
and scientific toil 41, 355–6
as a hideous joke played on humans 416
criterion of demarcation (Popper) 14 *see also* testability
criticism 114, 119, 233
conjecture and 58, 192, 203, 239–40, 352, 412

AVAILABLE FROM PENGUIN

The Fabric of Reality
The Science of Parallel Universes—and Its Implications

ISBN 978-0-14-027541-4

For David Deutsch, an award-winning physicist of unusual originality, quantum theory contains our most fundamental knowledge of the physical world. Taken literally, it implies that there are many universes "parallel" to the one we see around us. This multiplicity of universes, according to Deutsch, turns out to be the key to achieving a new worldview—one which synthesizes the theories of evolution, computation, and knowledge with quantum physics.

PENGUIN
BOOKS